Forests: Ecology, Diversity and Conservation

Forests: Ecology, Diversity and Conservation

Edited by Pierre Simone

SYRAWOOD
PUBLISHING HOUSE
New York

Published by Syrawood Publishing House,
750 Third Avenue, 9th Floor,
New York, NY 10017, USA
www.syrawoodpublishinghouse.com

Forests: Ecology, Diversity and Conservation
Edited by Pierre Simone

International Standard Book Number: 978-1-68286-655-9 (Hardback)

Cataloging-in-Publication Data

Forests : ecology, diversity and conservation / edited by Pierre Simone.
p. cm.
Includes bibliographical references and index.
ISBN 978-1-68286-655-9
1. Forests and forestry. 2. Forest ecology. 3. Forest biodiversity. 4. Forest conservation.
I. Simone, Pierre.
SD131 .F673 2019
634.9--dc23

Printed in China.

TABLE OF CONTENTS

PREFACE

Every book is initially just a concept; it takes months of research and hard work to give it the final shape in which the readers receive it. In its early stages, this book also went through rigorous reviewing. The notable contributions made by experts from across the globe were first molded into patterned chapters and then arranged in a sensibly sequential manner to bring out the best results.

Forests form a primary ecosystem of our planet. They are responsible for almost 75% of the gross primary productivity of the Earth's biosphere. Temperate, boreal, tropical forests are some of the most commonly known types of forests. Apart from being sources of minerals, herbs and other natural resources, forests also help in maintaining the ecological balance of our planet. They provide habitat to numerous species of plants and animals. This book covers in detail some existent theories and innovative concepts revolving around forest, ecology, biodiversity and conservation. It is a compilation of chapters that discuss the most vital approaches and emerging trends in this field. From theories to research to practical applications, studies related to all contemporary topics of relevance to this field have been included in this book. The extensive content herein provides the readers with a thorough understanding of the subject. This book will prove to be immensely beneficial to students and researchers of this field.

It has been my immense pleasure to be a part of this project and to contribute my years of learning in such a meaningful form. I would like to take this opportunity to thank all the people who have been associated with the completion of this book at any step.

Editor

Direct Measurement of Tree Height Provides Different Results on the Assessment of LiDAR Accuracy

Emanuele Sibona [1], Alessandro Vitali [2], Fabio Meloni [1], Lucia Caffo [3], Alberto Dotta [3], Emanuele Lingua [4], Renzo Motta [1] and Matteo Garbarino [1,2,*]

1 DISAFA: Dipartimento di Scienze Agrarie, Forestali e Alimentari, University of Torino, Largo Paolo Braccini 2, 10095 Grugliasco (TO), Italy; emanuelemarco.sibona@unito.it (E.S.); fabio.meloni@unito.it (F.M.); renzo.motta@unito.it (R.M.)

2 D3A: Department of Agricultural, Food and Environmental Sciences, Marche Polytechnic University, Via Brecce Bianche 10, 60131 Ancona (AN), Italy; a.vitali@pm.univpm.it

3 CFAVS: Consorzio Forestale Alta Val di Susa, Via Pellousiere 6, 10056 Oulx (TO), Italy; foreste@cfavs.it (L.C.); direzione@cfavs.it (A.D.)

4 TESAF: Department of Land, Environment, Agriculture and Forestry, University of Padova, Viale dell'Università 16, 35020 Legnaro (PD), Italy; emanuele.lingua@unipd.it

* Correspondence: m.garbarino@univpm.it

Abstract: In this study, airborne laser scanning-based and traditional field-based survey methods for tree heights estimation are assessed by using one hundred felled trees as a reference dataset. Comparisons between remote sensing and field-based methods were applied to four circular permanent plots located in the western Italian Alps and established within the Alpine Space project NewFor. Remote sensing (Airborne Laser Scanning, ALS), traditional field-based (indirect measurement, IND), and direct measurement of felled trees (DIR) methods were compared by using summary statistics, linear regression models, and variation partitioning. Our results show that tree height estimates by Airborne Laser Scanning (ALS) approximated to real heights (DIR) of felled trees. Considering the species separately, *Larix decidua* was the species that showed the smaller mean absolute difference (0.95 m) between remote sensing (ALS) and direct field (DIR) data, followed by *Picea abies* and *Pinus sylvestris* (1.13 m and 1.04 m, respectively). Our results cannot be generalized to ALS surveys with low pulses density (<5/m^2) and with view angles far from zero (nadir). We observed that the tree heights estimation by laser scanner is closer to actual tree heights (DIR) than traditional field-based survey, and this was particularly valid for tall trees with conical shape crowns.

Keywords: ALS; conifer forest; forest inventory; tree height; felled trees; ground control point

---

1. Introduction

Sustainable forest management needs a huge amount of tree parameters such as species distribution, timber volume, and average tree height as the basis of broad scale forest inventories [1]. Among these, tree height is one of the most important variables in forest inventory, as it is often used in the estimation of forest growth, biomass, carbon stock, and site productivity [2]. Traditional forest inventory field methods for forest height estimation are expensive, time consuming, and almost impossible to perform over large areas [3]. When dealing with large areas, a sampling plot approach is commonly arranged in order to reduce costs, and then diameter-height relations are constructed to predict individual tree heights for those areas that were not surveyed. Using remote sensing (RS) techniques allows for the reduction of those constraints, thereby obtaining spatially continuous data over large study areas on a cost efficient basis [4]. Tree height estimation from the ground is difficult due to treetop placement, which is often hidden by the branches of other trees. While with RS survey from above, the problem of hidden treetops is reduced, at least for canopy trees. Among RS techniques, the potential of Airborne Laser Scanning (ALS) in tree heights estimation has been widely demonstrated [5–7]; it has been shown to be useful for the estimation of other forest stand attributes such as stand volume [6] and basal area [8] as well. The impact of errors in tree height measurements has been shown to have a stronger effect on individual tree volume estimates than errors in species composition [9], but this depends on the surveyed species and the obtained height errors. For this reason, the accuracy assessment of tree height measurement derived from ALS is a very important issue in the forestry field.

The accuracy of forest attributes derived from RS data is usually assessed using traditional field surveys that can be area-based [4] or tree-based [10]. Field-based surveys always take advantage of trigonometric approaches using rangefinder devices. These field-based data are commonly assumed as being the ground truth for RS data, but in actuality they are measured with errors. In spite of the huge amount of recent papers that use field-based tree height estimations as ground truth for RS data assessment [2,4], the accuracy of this indirect measurement used as reference is often not declared.

In this paper, we considered as ground truth (Ground Control Points, GCPs) one hundred felled trees to test tree heights measured with two estimation approaches: Airborne Laser Scanning (ALS) and classical field surveys. Specific goals of the paper are: (1) to evaluate the use of ALS data for tree height estimation in a comparison with traditional field-based survey method; (2) to use tree height of felled trees to assess the accuracy and bias of field-based survey method and its influence on ALS testing; (3) to study the relative influence of tree size and species on the divergence between ALS remote sensing and direct field-based estimations.

2. Methods

2.1. Study Area

The Cotolivier forest is about 4000 ha and is located on the right bank of the Upper Susa Valley (Piedmont, Italy), between Oulx and Bardonecchia (45°2′ N, 6°46′ E). It is a north facing slope ranging from the montane belt (about 1000 m a.s.l.) up to the treeline (about 2200 m a.s.l.). The mean annual temperature is 6.7 °C and the total precipitation is about 614.9 mm (Bardonecchia Pian del Sole, 1585 m a.s.l., 2004–2015 period). The topography of the area is complex with steep slopes and high roughness. Both productive and protective forests are present, with intense forest exploitation on more accessible sites, but topography limits the use of a massive mechanization. *Pinus sylvestris* L. stands dominate at lower elevation, with *Picea abies* (L.) H.Karst. and *Larix decidua* Mill. Prevalent at higher elevations. This mixed conifers forest is characterized by a high structural diversity: *Larix* stands range from wood pasture with sparse big trees on more accessible sites [11] to dense and young stands, while *Pinus* and *Picea* stands show less structural diversity.

2.2. Sampling Design and Data Collection

Four permanent circular plots with 20 m radiuses were established during a field campaign in 2013 within the "NewFor" Alpine Space project [12]. Within these plots, we estimated the total height of 100 trees (28% of the surveyed trees within the plots) by using three different survey methods: (a) traditional field-based survey here defined as "indirect" measurement (IND); (b) remote sensing measurement through Airborne Laser Scanning (ALS); (c) direct measurement on felled trees, referred as "direct" measurement (DIR).

2.2.1. Indirect Field-Based Survey (IND)

Field-Map™ integrated system [13] was used to accomplish the traditional field-based survey that was done during June–July 2012. It includes a GPS receiver (*GeoXM 2005* [14]) to obtain georeferenced data of the plot center (differential correction in laboratory with post-processing method), and a laser rangefinder (*TruPulse*™ *360b* [15]) to measure tree heights and relative positions (distances and angles) to the plot center. A field computer (*ARMOR* [16]) equipped with mobile GIS software was used for real time data mapping and 3-D visualization of trees. Diameter (DBH), crown projections measure, and species have been collected for all 100 trees (DBH > 4 cm).

2.2.2. LiDAR Survey (ALS)

ALS data were acquired in June 2012 by a helicopter flight, which covered the whole study area. The LiDAR sensor used was an *Optech*™ ALTM 3100EA, that allowed a planimetric position accuracy of less than 0.30 m and an elevation accuracy (Z) of less than 0.15 m. Four echoes of each laser pulse were recorded flying at an average altitude of 400 m, resulting in an average point sampling density of 10 points/m^2. The vendor provided the ALS data classified as ground and non-ground points using *Terrascan* software [17]. The original data (ALS10) were thinned with the module *lasthin* of the *LAStools* software [18], obtaining two new point densities: 5 (ALS5) and 1 (ALS1) points/m^2. This data thinning methodology is commonly adopted to test the effects of point density on forest attributes [19–21]. Digital Terrain Model (DTM) and DSM (Digital Surface Model) of ALS10 were generated by the vendor with a grid size of 1 m using *TerraScan* software. The thinned ALS point clouds were used to generate raster data of the DTMs and DSMs of the area with a grid size of 1 m using the "las2dem" tool. These layers were subtracted to produce the CHM (Canopy Height Model) that was employed as a base raster for trees positions and heights detection [1,22–24]. The CHMs have undergone a surface smoothing using a Gaussian kernel filter [22]. Then focal statistics with a local maximum detection algorithm ("Pouring" algorithm implemented in *HALCON*, similar to an inverted, classical watershed-algorithm [1]) were used to extract potential treetops from the smoothed CHMs. The detected trees were then analyzed through a conditional script that considers a minimum distance and height difference from the nearest trees in order to identify and delete potential false positives [24]. The method is fully automated and has been implemented as a workflow of geoprocessing tools (Model Builder) within the software *ESRI ArcGIS*™ [25]. The output is a shapefile containing tree position and height [20]. We manually co-registered the remote sensing data to the field dataset by using a very high spatial resolution (20 cm) aerial image acquired a month before the ALS data in order to correctly match the trees detected remotely to each GPS-positioned tree [24,26]. False negatives have been consequently detected by experienced human interpreter [22]. Only correctly assigned trees (true positives) were used in this work.

2.2.3. Direct Field-Based Survey (DIR)

The 100 felled trees were the result of two different harvesting methods: A clearcut on an almost pure *L. decidua* stand (plot 1) and thinning in multilayered mixed conifer stands (plots 2, 3, 4) (Table 1). Tree length plus the stump height was measured for each felled tree to get "direct" data, which have been considered in the analysis process as ground truth.

Table 1. Site characteristics, stand structure descriptors, and species composition of the four permanent plots as a whole and considering only the felled trees. Tree heights were obtained from DIR measurement, instead diameter at breast height (DBH), basal area (BA), bole height (Crown H), and crown area (Crown A) were obtained from IND survey method.

Whole Plot	Plot 1	Plot 2	Plot 3	Plot 4
Slope (°)	11	21	28	28
Elevation (m a.s.l.)	1588	1365	1364	1407
Tree Density (tree/ha)	477	772	716	883
BA (m^2/ha)	40.40	37.77	42.79	37.58
Vertical Structure	single layered	multilayered	multilayered	multilayered
Larix decidua (BA %)	99	20	11	44
Picea abies (BA %)	1	65	42	26
Pinus sylvestris (BA %)	0	9	9	30
Other Species (BA %)	0	6	38	0

Table 1. *Cont.*

Whole Plot	Plot 1	Plot 2	Plot 3	Plot 4
Felled trees				
N	47	18	16	19
Tree Height Mean (m)	28.61	19.64	18.69	17.17
Tree Height Min (m)	19.80	14.00	13.70	12.00
Tree Height Max (m)	34.90	25.50	25.50	23.00
Tree Height SD (m)	3.18	2.93	3.39	3.10
DBH (cm)	34.04	28.56	30.71	28.68
BA (m^2/ha)	4.84	1.23	1.40	1.30
Crown H (m)	18.40	9.01	11.32	6.94
Crown A (m^2)	8.87	11.84	13.06	10.13

2.3. Data Analysis

We obtained five datasets that were used to compare remote sensing ALS data with three different point densities (ALS10, ALS5, and ALS1) and field-based estimation methods (IND) by using direct measurements on felled trees (DIR) as ground control measure. We tested the statistical difference between the five measurements by using a Kruskal-Wallis chi-squared non-parametric test. We calculated standard summary statistics of tree height measures for both the whole sample and the three main species (*Larix, Picea, Pinus*) separately. The accuracy of tree heights as measured by ALS data was assessed by mean differences (MD) and root-mean-square errors (RMSE) between IND-ALS and DIR-ALS for the whole dataset and divided by species. We performed a regression analysis between ALS and IND data against our ground truth (DIR), again both for the entire sample and for each species separately to further investigate the observed differences. Then we regressed heights from the two field datasets (IND and DIR) against each other. We applied the hierarchical partitioning approach [27] through the statistical package *hier.part* [28] to calculate the independent contribution of tree attributes (real height, DBH, crown surface area and species) in determining the difference in absolute value between DIR and ALS estimates. Because of a non-uniform distribution of species along the range of tree heights, the hierarchical partitioning was applied to the whole dataset and to the species separately. Summary statistics, regression analyses, and hierarchical partitioning were performed in *R* software [29].

3. Results

L. decidua was the tallest species in our dataset followed by *P. abies* and *P. silvestris* (Figure 1), and some differences between estimation methods existed. However, the five estimation methods were not statistically different from each other (Kruskal-Wallis test, $p = 0.92$). Data thinning on the original ALS10 data had a strong effect on the number of the detected trees. Almost half (56) of the 100 felled trees were detected with ALS1 and 73 with ALS5 (Table 2). The more affected species was *P. sylvestris* (38% with ALS1) and the less influenced was *P. abies* (68% with ALS1). From a comparison between mean tree heights measured by ALS data and IND (Table 3), emerged that ALS10 slightly overestimated tree heights (23.51 m vs. 23.44 m) and the opposite trend was observed for *Picea* (19.64 m vs. 20.26 m). Considering the mean tree heights of the whole dataset (All) an overestimation was obtained by using the ALS1 point cloud, this trend was particularly evident for *Picea* and *Pinus* species (Table 3). A comparison between ALS10 and real tree heights (DIR) revealed that the mean value was identical (23.51 m), but some differences emerged for *Picea* and *Pinus* species. The biggest difference between ALS10 and DIR height average measurements was observed for *Pinus* (0.58 m) and *Picea* (−0.33 m), but considering the difference between ALS1 and DIR those values increased (*Pinus* 1.18 m and *Picea* 0.55 m).

Table 2. Total number (N) and percent value of detected trees by ALS from the original dataset of felled trees. ALS data are presented divided by pulse densities (10, 5, 1 points/m^2) obtained through point cloud thinning.

Species	ALS10		ALS5		ALS1	
	N	%	*N*	%	*N*	%
All	100	100	73	73	56	56
L. decidua	48	100	35	73	27	56
P. abies	31	100	27	87	21	68
P. sylvestris	21	100	11	52	8	38

Table 3. Average (Mean) and standard deviation (SD) values of tree heights measured by three estimation methods: field direct (DIR), indirect (IND), and Airborne Laser Scanning (ALS) for the whole dataset (All) and divided by species. ALS data are presented divided by pulse densities (10, 5, 1 points/m^2) obtained through point cloud thinning. *N* indicates the total of detected trees by ALS from the original dataset of felled trees.

Species	Metric	DIR	IND	ALS10	ALS5	ALS1
All	*N*	100	100	100	73	56
	Mean (m)	23.51	23.44	23.51	23.77	24.14
	SD (m)	6.10	5.98	5.96	5.81	5.23
L. decidua	*N*	48	48	48	35	27
	Mean (m)	28.99	28.56	28.96	28.98	28.94
	SD (m)	2.82	3.44	2.77	2.80	1.88
P. abies	*N*	31	31	31	27	21
	Mean (m)	19.97	20.26	19.64	19.84	20.52
	SD (m)	2.97	3.05	2.69	2.92	2.28
P. sylvestris	*N*	21	21	21	11	8
	Mean (m)	16.21	16.41	16.79	16.86	17.40
	SD (m)	2.22	2.15	2.29	2.34	2.55

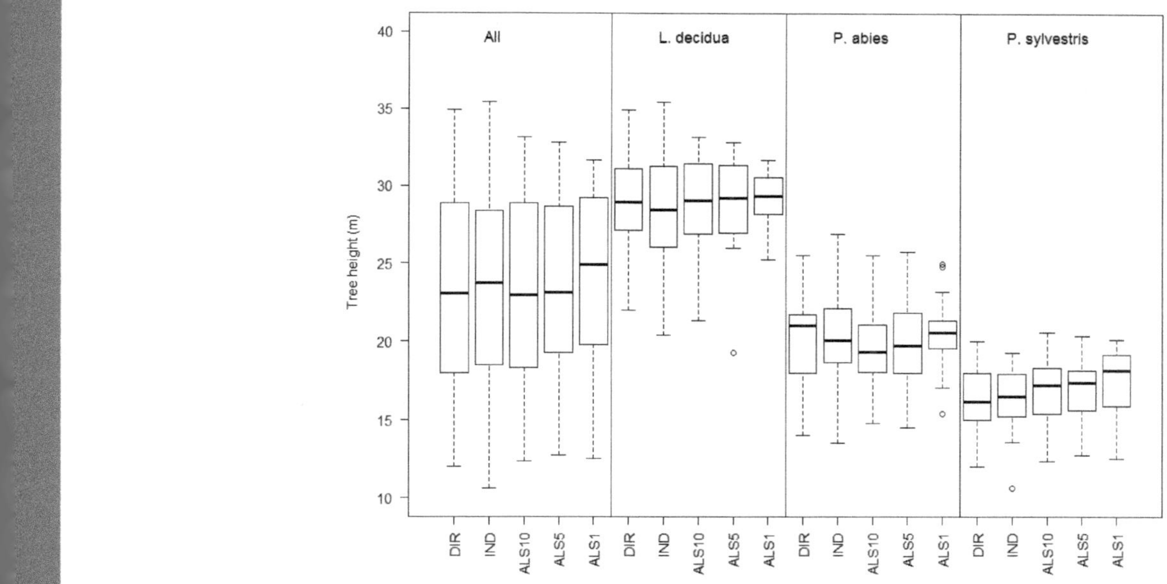

Figure 1. Box plots of tree heights distribution as measured by three survey methods (DIR = field direct measurement of felled trees; IND = traditional field-based; ALS = Airborne Laser Scanning) of the whole dataset and considering the species. ALS data are presented divided by pulse densities (10, 5, 1 points/m^2) obtained through point cloud thinning.

Every model obtained through regression analysis showed an overestimation of small trees and an underestimation of the taller ones. The differences between the models lie in the magnitude of these over/underestimates. All the regressions were significant ($p < 0.001$) and every model gave very good fit if we consider the population as a whole ($R^2 > 0.9$). In the ALS10 vs. IND models (Figure 2) *P. sylvestris* achieved a poorer fit ($R^2 = 0.67$), while *L. decidua* achieved a better fit ($R^2 = 0.73$) and *P. abies* emerged as being the best fitting species ($R^2 = 0.95$). All *P. abies* trees were underestimated by ALS as well as *L. decidua* taller trees (Figure 2). A similar pattern was observed in the ALS10 vs. DIR model, but with less underestimation (Figure 3a). In the ALS10 vs. DIR model (Figure 3a) among the studied species *P. sylvestris* achieved a poorer fit ($R^2 = 0.69$), while *P. abies* and *L. decidua* performed better ($R^2 = 0.79$ and 0.81, respectively). In the regression between DIR and IND measurements (Figure 3b) *P. sylvestris* and *P. abies* were the best fitting species ($R^2 = 0.80$) followed by *L. decidua* ($R^2 = 0.71$). In this model *L. decidua* was underestimated by IND, instead *P. sylvestris* and *P. abies* showed a high correspondence between the two measurements (Figure 3b).

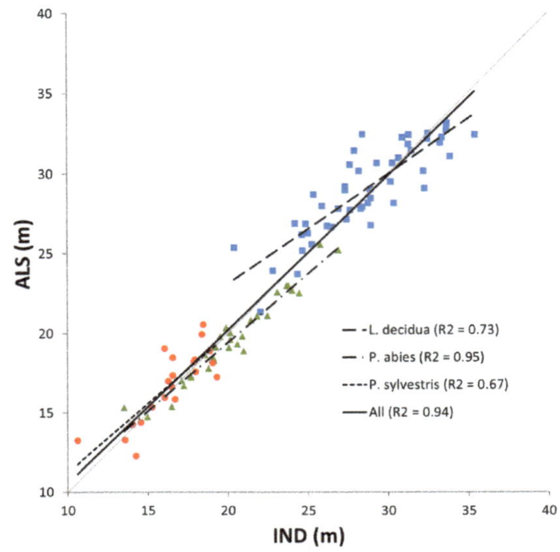

Figure 2. Regression analysis between tree height derived by airborne laser scanner data (ALS = ALS10) and tree height derived from indirect field-based measurement (IND), considering the whole dataset (All) and the three species separately. Each dot corresponds to a single tree (blue square = *Larix*; green triangle = *Picea*; red circle = *Pinus*).

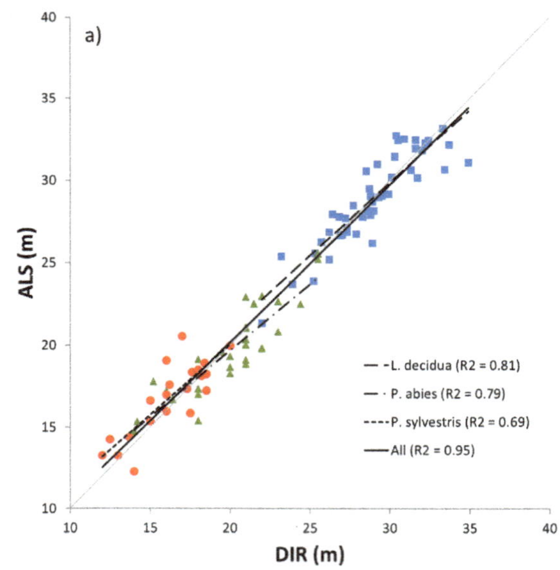

Figure 3. *Cont.*

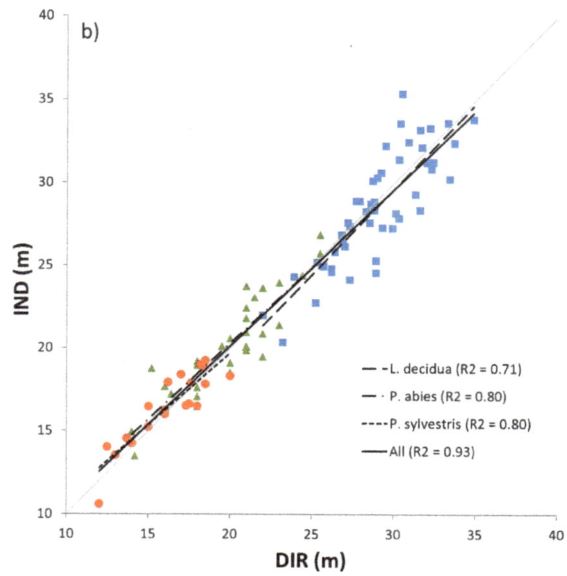

Figure 3. Regression analysis between (**a**) tree heights derived by airborne laser scanner data (ALS = ALS10) and tree heights derived from direct field-based measurement (DIR); (**b**) tree heights derived from indirect field-based measurement (IND) and DIR, considering the whole dataset (All) and the three species separately. Each dot corresponds to a single tree (blue square = *Larix*; green triangle = *Picea*; red circle = *Pinus*).

The error in the individual tree height estimation was higher (*RMSE* = 1 m) when ALS was tested against IND (Table 4) than in the case of DIR (*RMSE* = 0.76 m, Table 5). Overall, *RMSE* increased with thinned ALS1 data both by using IND and DIR data as ground truth, and this tendency was particularly evident for *Picea* and *Pinus*. The accuracy assessment between IND and DIR measurements gave an RMSE of 1.08 for All trees, 1.84 for *Larix*, 1.34 for *Picea*, and 0.94 for *Pinus*.

Table 4. Mean difference (MD) and Root mean square error (*RMSE*) between field indirect (IND) and ALS data for the whole dataset (All) and divided by species. ALS data are presented divided by pulse densities (10, 5, 1 points/m^2) obtained through point cloud thinning.

Species	IND vs. ALS10		IND vs. ALS5		IND vs. ALS	
	MD	*RMSE*	*MD*	*RMSE*	*MD*	*RMSE*
All	0.08	1.00	−0.31	1.16	−0.10	1.41
L. decidua	0.39	1.42	−0.13	1.51	−0.22	1.18
P. abies	−0.62	0.62	−0.83	1.42	−0.43	1.00
P. sylvestris	0.38	1.28	0.39	1.15	1.20	0.90

Table 5. Mean difference (MD) and Root mean square error (*RMSE*) between field direct (DIR) and ALS data for the whole dataset (All) and divided by species. ALS data are presented divided by pulse densities (10, 5, 1 points/m^2) obtained through point cloud thinning.

Species	DIR vs. ALS10		DIR vs. ALS5		DIR vs. ALS1	
	MD	*RMSE*	*MD*	*RMSE*	*MD*	*RMSE*
All	0.00	0.76	−0.48	0.97	−0.43	1.11
L. decidua	−0.03	1.21	−0.63	1.45	−1.06	1.18
P. abies	−0.33	1.21	−0.74	1.83	−0.17	1.28
P. sylvestris	0.58	1.25	0.60	1.24	1.05	1.26

From the variation partitioning approach on the whole dataset, it was revealed that species had the highest independent effect (54.8%) on the absolute value of the difference between DIR and ALS, followed by real tree height (DIR, 41.4%), DBH (2.8%), and crown surface area (1%). Applying the variation partitioning to the species separately we observed that the independent effect of crown surface area was much higher for *P. sylvestris* (22.4%) than for the other species (*L. decidua* 2.2%; *P. abies* 8%). Real tree height (DIR) showed a stronger independent effect on delta estimations in *L. decidua* (95.6%) than in other species (*P. abies* 73.3%; *P. sylvestris* 57.5%).

4. Discussion

Field-based tree height measurements are a fundamental data source for forest inventories and are generally very time consuming. Airborne Laser Scanning (ALS) is an interesting remote sensing technique that can be used to make fieldworks more efficient. When using field measurements as reference data for the evaluation of remote sensing tree height data the question of "what is the truth" is still open. We used one hundred felled trees as ground truth to investigate the impact of species composition on the accuracy of individual tree height estimates. Our results show that tree height measurements performed with Airborne Laser Scanning (ALS) approximated to real heights (DIR) of felled trees. Interestingly, ALS estimates were closer to DIR ones than those obtained through traditional field-based surveys (IND). This was also highlighted by the lower error (*RMSE*) observed when the DIR measurement was used as ground truth. Considering the species separately, *L. decidua* showed the smaller mean difference in absolute values (0.95 m) between remote sensing (ALS) and direct field (DIR) data, followed by *P. abies* and *P. sylvestris* (1.13 and 1.04, respectively). *P. abies* was the only species showing a smaller difference in tree height between ALS and IND (0.80 m) than between ALS and DIR (1.13 m). This was probably because *P. abies* has a simple and dense canopy that is easier to collimate from the ground.

Tree height obtained through direct measurements on felled trees (DIR) emerged as being an important factor influencing the accuracy of remote and field surveys. Our regression linear models showed that ALS-derived height estimations of tall trees were underestimated while small ones were overestimated by ALS technique. The underestimation of tall trees by ALS has been commonly found in several studies and relates to the missing of points corresponding to the highest point of the crown [30]. Underestimation of tree height using small-footprint laser scanner systems (ALS) was documented early in the 1980s [31]. The main factors affecting underestimation of tree heights are as follows: density and coverage of laser pulses; the algorithm used to obtain DTM and CHM; pulse penetration into the canopy; amount and height of understory vegetation; tree shape and species [32]. The overestimation by ALS of small trees is an artefact of receiving pulses from nearby taller trees [33], but other causes can be related to DTM errors related to filtering and interpolation methods. The architecture of the crown can also determine an overestimation, especially for those species without a strong monopodial shape, such as broadleaves and some pine species (e.g., *P. sylvestris*). This fact had an important influence on the observed differences between species. *Pinus* trees were smaller than the other two species and this is likely an explanation for the height overestimation by ALS.

From a variation partitioning analysis, it was revealed that species attribute was more important than tree height and DBH for influencing the difference between ALS and DIR measurements; this was probably because our dataset was composed of small *Pinus*, intermediate *Picea*, and tall *Larix* trees. However, from the variation partitioning analysis applied to each species we observed that, for *Pinus*, crown surface area was an important variable for influencing the difference between ALS and DIR measurements; this was probably due to the complex crown shape of this species. Instead, *Larix* and *Picea* have a simpler pyramid-shape crown architecture [34]; this can explain why real height (DIR) become the most important factor for influencing the difference between ALS and DIR measurements. Similar differences between crown architecture of pines and fir-spruce species were found by other authors [2,10,35] and recently the importance of tree crown architecture for wood volume estimation has been outlined [36].

In spite of the small size of the sample, we observed that tree height estimation by laser scanner is closer to real tree heights (DIR) than traditional field-based survey. This was particularly valid for tall trees having regular crowns, and this has important implications for the use of ALS techniques in forest inventories [9,37]. We are aware that our results cannot be generalized to ALS surveys with low pulses density ($<5/m^2$) and with view angles far from zero (nadir). Pulse density is a limit for singletree height estimation [32], but is not a constraint at the plot scale until pulse densities drop below 1 pulse/m^2 [38]. From our experiment on thinned data we observed that pulse density had a strong influence on the number of detected trees. With a density of 1 pulse/m^2 only 56% of trees were detected and this percentage was even lower (38%) for *Pinus* trees. This is probably because the point cloud thinning affected shorter trees with complex crown shape. Regarding the influence of point density on the accuracy of tree height estimation we obtained higher RMSE (1.41 m) with low densities (1 pulse/m^2), but this value is not too far from those indicated in the literature for the same species with 10 pulses/m^2 (RMSE = 0.98 m) and 4 pulses/m^2 (RMSE = 0.63 m) [10,39].

The source of error on tree height field measurement can be related to the instrument used, to the skills and experience of surveyor, and to the tree and stand characteristics [40]. There are only a few papers that compare tree height measurements obtained by handled instruments with actual height obtained by direct measurement [41], but as far as we know there are none comparing LiDAR derived height with direct measurements. Field-based surveying methods can influence the accuracy of tree height estimation and usually are affected by lack of precision [42,43]. The model of handheld laser rangefinder (e.g., Impulse or Truepulse) affects the quality of compass measurements [44,45]. In our study, we used a Truepulse 360B device that is affected by a length measurement error of 30 cm and azimuth error of 3 degrees [45]. When using *Suunto* hypsometers the standard error in tree height estimation has proven to be between 0.4 and 0.8 m [46,47]. Moreover, field crew judgement and precision is another important factor affecting the accuracy of field-based surveys with handheld laser rangefinders [48]. The use of tacheometer, total station, and accurate GPS receiver can significantly improve field measurements, and thus the accuracy assessment of ALS data [47,49,50]. Other authors have found that field methods were more accurate than LiDAR data [2,41], but point density was almost half of the present study. In any case, the authors agreed in considering that the discrepancy in height accuracy was greatly compensated by the higher cost efficiencies and wider coverage guaranteed by LiDAR. Further case study, species, instruments, and field crew should be tested in order to have a clearer idea of the issue.

5. Conclusions

Our study demonstrates that direct field measurements on felled trees improve the assessment of small-footprint airborne laser scanner (ALS) estimation of tree heights. Adopting direct measurements as ground truth for remote sensing tree height estimation we obtained a reduction in the magnitude of errors (RMSE). The proposed method of using felled trees as ground control points is time consuming and expensive, but highlights the fact that traditional field-based surveys methods are estimations in turn. ALS can be considered reliable for height estimation of dominant trees, but its accuracy with shorter trees having irregular canopy is lower. A solution could be to improve the number of GCPs and the number of repeated field-based measurements on dominated trees to obtain a more reliable assessment of remote sensing data. The use of felled trees as direct field measurements in forest inventory obviously cannot be adopted as a routine approach. Traditional field-based measurements should still be considered an essential survey methodology, but an increased awareness of their accuracy, especially when used for the validation of other indirect measurements such as RS methods, should be promoted.

Acknowledgments: This work was funded by the European Commission (project Alpine Space 2-3-2-FR NEWFOR) within the European Territorial Cooperation program "Alpine Space" and partially funded by the Italian Ministry of University and Research (project PRIN 2012, grant no. 2012EWEY 2S). The authors would like to thank the NEWFOR partners, the project observer Consorzio Forestale Alta Valle di Susa (CFAVS) for logistic support and providing the forest inventory data, and three anonymous reviewers that contributed substantially to enhance the quality of the manuscript.

Author Contributions: E.S.: paper writing; data quality check; preparation of results. F.M., A.D., and L.C.: data collection; logistics; paper revision. A.V. and E.L.: data analysis; paper revision. R.M.: supervision. M.G.: paper writing; supervision.

Conflicts of Interest: The authors declare no conflict of interest.

References

1. Koch, B.; Heyder, U.; Weinacker, H. Detection of individual tree crowns in airborne LiDAR data. *Photogramm. Eng. Remote Sens.* **2006**, *72*, 357–363. [CrossRef]
2. Andersen, H.E.; Reutebuch, S.E.; McGaughey, R.J. A rigorous assessment of tree height measurements obtained using airborne LiDAR and conventional field methods. *Can. J. Remote Sens.* **2006**, *32*, 355–366. [CrossRef]
3. Fieber, K.D.; Davenport, I.J.; Tanase, M.A.; Ferryman, J.M.; Gurney, R.J.; Becerra, V.M.; Walker, J.P.; Hacker, J.M. Validation of Canopy Height Profile methodology for small-footprint full-waveform airborne LiDAR data in a discontinuous canopy environment. *ISPRS J. Photogramm. Remote Sens.* **2015**, *104*, 144–157. [CrossRef]
4. Næsset, E. Predicting forest stand characteristics with airborne scanning laser using a practical two-stage procedure and field data. *Remote Sen. Environ.* **2002**, *80*, 88–99. [CrossRef]
5. Nilsson, M. Estimation of tree heights and stand volume using an airborne LiDAR system. *Remote Sen. Environ.* **1996**, *56*, 1–7. [CrossRef]
6. Næsset, E. Estimating timber volume of forest stands using airborne laser scanner data. *Remote Sens. Environ.* **1997**, *61*, 246–253. [CrossRef]
7. Magnussen, S.; Boudewyn, P. Derivations of stand heights from airborne laser scanner data with canopy-based quantile estimators. *Can. J. For. Res.* **1998**, *28*, 1016–1031. [CrossRef]
8. Lefsky, M.A.; Cohen, W.B.; Acker, S.A.; Parker, G.G.; Spies, T.A.; Harding, D. LiDAR remote sensing of the canopy structure and biophysical properties of Douglas-fir western hemlock forests. *Remote Sens. Environ.* **1999**, *70*, 339–361. [CrossRef]
9. Tompalski, P.; Coops, N.C.; White, J.C.; Wulder, M.A. Simulating the impacts of error in species and height upon tree volume derived from airborne laser scanning data. *For. Ecol. Manag.* **2014**, *327*, 167–177. [CrossRef]
10. Persson, A.; Holmgren, J.; Söderman, U. Detecting and measuring individual trees using an airborne laser scanner. *Photogramm. Eng. Remote Sens.* **2002**, *68*, 925–932.
11. Garbarino, M.; Lingua, E.; Weisberg, P.J.; Bottero, A.; Meloni, F.; Motta, R. Land-use history and topographic gradients as driving factors of subalpine *Larix decidua* forests. *Landsc. Ecol.* **2013**, *28*, 805–817. [CrossRef]
12. Berger, F.; Dupire, S.; Monnet, J.-M.; Hainzer, E.; Stoehr, D.; Nemestóthy, N.; Sotier, B.; Lerch, T.; Maier, B.; Malin, H.; et al. *Forest Logistic Planning Strategies. Good Practices for the Alpine Forests*; TESAF e-Book: Padova, Italy, 2014; p. 111.
13. Field-Map. IFER: Jilove, Czech Republic. Available online: http://www.fieldmap.cz (accessed on 25 May 2015).
14. GeoXM 2005. Trimble Inc.: Sunnyvale, CA, USA. Available online: http://investor.trimble.com/releasedetail.cfm?releaseid=190963 (accessed on 15 July 2014).
15. TruPulse. Laser Technology Inc.: Centennial, CO, USA. Available online: http://www.lasertech.com/TruPulse-Laser-Rangefinder.aspx (accessed on 25 May 2015).
16. ARMOR. Technologies Inc.: Arlington, VA, USA. Available online: http://www.drs.com/products-and-services/armor-customer-support/DRS (accessed on 3 July 2015).
17. Terrascan. Terrasolid LTD.: Helsinki, Finland. Available online: http://www.terrasolid.com/products/terrascanpage.php (accessed on 14 March 2014).
18. LasTools. GmbH: Gilching, Germany. Available online: http://lastools.org/download/LAStools.ziprapidlasso (accessed on 10 February 2014).
19. Vauhkonen, J.; Tokola, T.; Maltamo, M.; Packalén, P. Effects of pulse density on predicting characteristics of individual trees of Scandinavian commercial species using alpha shape metrics based on airborne laser scanning data. *Can. J. Remote Sens.* **2008**, *34*, S441–S459. [CrossRef]

20. Hansen, E.H.; Gobakken, T.; Næsset, E. Effects of pulse density on digital terrain models and canopy metrics using airborne laser scanning in a tropical forest. *Remote Sens.* **2015**, *7*, 8453–8468. [CrossRef]
21. Kandare, K.; Orka, H.O.; Chan, J.C.-W.; Dalponte, M. Effects of forest structure and airborne laser scanning point cloud density on 3D delineation of individual tree crowns. *Eur. J. Remote Sens.* **2016**, *49*, 337–359. [CrossRef]
22. Hyyppä, H.J.; Hyyppä, J.M. Effects of stand size on the accuracy of remote sensing based forest inventory. *IEEE Trans. Geosci. Remote Sens.* **2001**, *39*, 2613–2621. [CrossRef]
23. Sambugaro, M.; Colpi, C.; Marzano, R.; Pellegrini, M.; Pirotti, F.; Lingua, E. Utilizzo Del Telerilevamento Per l'Analisi Della Biodiversità Strutturale: Il Caso Studio Della Riserva Forestale di Clöise (Asiago, VI). In Proceedings of the 17th Conferenza Nazionale ASITA, Riva del Garda, Italy, 5–7 November 2013; pp. 1171–1178.
24. St-Onge, B. Estimating individual tree heights of the boreal forest using airborne laser altimetry and digital videography. *Int. Arch. Photogramm. Remote Sens.* **1999**, *32*, 179–184.
25. ArcGIS. ESRI: Redlands, CA, USA. Available online: http://www.esri.com/arcgis/products/arcgis-pro/Overview (accessed on 25 June 2015).
26. Eysn, L.; Hollaus, M.; Lindberg, E.; Berger, F.; Monnet, J.-M.; Dalponte, M.; Kobal, M.; Pellegrini, M.; Lingua, E.; Mongus, D.; Pfeifer, N. A benchmark of LiDAR-based single tree detection methods using heterogeneous forest data from the Alpine space. *Forests* **2015**, *6*, 1721–1747. [CrossRef]
27. Chevan, A.; Sutherland, M. Hierarchical partitioning. *Am. Stat.* **1991**, *45*, 90–96. [CrossRef]
28. Walsh, C.; Mac Nally, R. Hier.Part: Hierarchical Partitioning. R Package Version 1.0–4. Available online: https://www.CRAN.R-project.org/package=hier.part (accessed on 20 May 2013).
29. R Core Team. *R: A Language and Environment for Statistical Computing*; R Found. Stat. Comput.: Vienna, Austria, 2016.
30. Wing, M.G.; Eklund, A.; Sessions, J. Applying LiDAR technology for tree measurements in burned landscapes. *Int. J. Wildland Fire* **2010**, *19*, 104–114. [CrossRef]
31. Nelson, R.; Krabill, W.; Maclean, G. Estimating forest biomass and volume using airborne laser data. *Remote Sens. Environ.* **1988**, *24*, 247–267. [CrossRef]
32. Hyyppä, J.M.; Hyyppä, H.J.; Leckie, D.; Gougeon, F.; Yu, X.; Maltamo, M. Review of methods of small-footprint airborne laser scanning for extracting forest inventory data in boreal forests. *Int. J. Rem. Sens.* **2008**, *29*, 1339–1366.
33. Farid, A.; Goodrich, D.C.; Sorooshian, S. Using airborne LiDAR to discern age classes of cottonwood trees in a riparian area. *West. J. Appl. For.* **2006**, *21*, 149–158.
34. Korpela, I.; Ørka, H.O.; Maltamo, M.; Tokola, T.; Hyyppä, J. Tree species classification using airborne LiDAR—Effects of stand and tree parameters, downsizing of training set, intensity normalization, and sensor type. *Silva Fenn.* **2010**, *44*, 319–339. [CrossRef]
35. Yu, X.; Hyyppä, J.; Hyyppä, H.; Maltamo, M. Effects of flight altitude on tree height estimation using airborne laser scanning. In Proceedings of the International Conference NATSCAN Laser-Scanners for Forest and Landscape Assessment—Instruments, Processing Methods and Applications, Freiburg, Germany, 3–6 October 2004; pp. 96–101.
36. Hess, C.; Bienert, A.; Härdtle, W.; Von Oheimb, G. Does Tree Architectural Complexity Influence the Accuracy of Wood Volume Estimates of Single Young Trees by Terrestrial Laser Scanning? *Forests* **2015**, *6*, 3847–3867. [CrossRef]
37. Chave, J.; Andalo, C.; Brown, S.; Cairns, M.A.; Chambers, J.Q.; Eamus, D.; Fölster, H.; Fromard, F.; Higuchi, N.; Kira, T.; et al. Tree allometry and improved estimation of carbon stocks and balance in tropical forests. *Oecologia* **2005**, *145*, 87–99. [CrossRef] [PubMed]
38. Jakubowski, M.K.; Guo, Q.; Kelly, M. Tradeoffs between LiDAR pulse density and forest measurement accuracy. *Remote Sens. Environ.* **2013**, *130*, 245–253. [CrossRef]
39. Hyyppä, J.M.; Inkinen, M. Detecting and estimating attributes for single trees using laser scanner. *Photogramm. J. Finl.* **1999**, *16*, 27–42.
40. Gaudin, S.; Richard, J.B. Comparaison des dendrometres VertexIII et TruPulse 200B pour la mesure de la hauteur totale des arbres. *Rev. For. Fr.* **2014**, *66*, 163–181.
41. Larjavaara, M.; Muller-Landau, H.C. Measuring tree height: A quantitative comparison of two common field methods in a moist tropical forest. *Methods Ecol. Evol.* **2013**, *4*, 793–801. [CrossRef]

42. Nelson, R.; Parker, G.; Hom, M. A portable airborne laser system for forest inventory. *Photogramm. Eng. Remote Sens.* **2003**, *69*, 267–273. [CrossRef]
43. West, P.W. *Tree and Forest Measurement*, 2nd ed.; Springer: Berlin, Germany, 2009; p. 191.
44. Brach, M.; Bielak, K.; Drozdowski, S. Measurements accuracy of selected laser rangefinders in the forest environment. *Sylwan* **2013**, *157*, 671–677.
45. Tomaštik, J.; Tunák, D. Compass measurement—Still a suitable surveying method in specific conditions. *Geod. Cartogr.* **2015**, *41*, 31–40. [CrossRef]
46. Lindgren, O. *A Study of Circular Plot Sampling of Swedish Forest Compartments*; Swedish University of Agriculture Science: Umea, Sweden, 1984.
47. Holmgren, J.; Nilsson, M.; Olsson, H. Estimation of tree height and stem volume on plots using airborne laser scanning. *For. Sci.* **2003**, *49*, 419–428.
48. Ferrarese, J.A. *Characterizing Crown Structure of Three Interior Northwest Conifer Species Using Terrestrial Laser Scanning*; University of Montana: Missoula, MT, USA, 2013.
49. Maltamo, M.; Mustonen, K.; Hyyppä, J.; Pitkänen, J.; Yu, X. The accuracy of estimating individual tree variables with airborne laser scanning in a boreal nature reserve. *Can. J. For. Res.* **2004**, *34*, 1791–1801. [CrossRef]
50. Salas, C.; Ene, L.; Gregoire, T.G.; Næsset, E.; Gobakken, T. Modelling tree diameter from airborne laser scanning derived variables: A comparison of spatial statistical models. *Remote Sens. Environ.* **2010**, *114*, 1277–1285. [CrossRef]

2

Phenology-Based Method for Mapping Tropical Evergreen Forests by Integrating of MODIS and Landsat Imagery

Weili Kou [1,*], Changxian Liang [1], Lili Wei [1], Alexander J. Hernandez [2] and Xuejing Yang [1]

[1] School of Computer and Information, Southwest Forestry University, Kunming, Yunnan 650224 China; xnllcx@163.com (C.L.); lilywlwei@163.com (L.W.); yangxuejing0822@163.com (X.Y.)

[2] Remote Sensing and GIS Laboratories, Quinney College of Natural Resources, Utah State University, Logan, UT 84322, USA; alex.hernandez@usu.edu

* Correspondence: kwl@swfu.edu.cn

Abstract: Updated extent, area, and spatial distribution of tropical evergreen forests from inventory data provides valuable knowledge for research of the carbon cycle, biodiversity, and ecosystem services in tropical regions. However, acquiring these data in mountainous regions requires labor-intensive, often cost-prohibitive field protocols. Here, we report about validated methods to rapidly identify the spatial distribution of tropical forests, and obtain accurate extent estimates using phenology-based procedures that integrate the Moderate Resolution Imaging Spectroradiometer (MODIS) and Landsat imagery. Firstly, an analysis of temporal profiles of annual time-series MODIS Normalized Difference Vegetation Index (NDVI), Enhanced Vegetation Index (EVI), and Land Surface Water Index (LSWI) was developed to identify the key phenology phase for extraction of tropical evergreen forests in five typical lands cover types. Secondly, identification signatures of tropical evergreen forests were selected and their related thresholds were calculated based on Landsat NDVI, EVI, and LSWI extracted from ground true samples of different land cover types during the key phenology phase. Finally, a map of tropical evergreen forests was created by a pixel-based thresholding. The developed methods were tested in Xishuangbanna, China, and the results show: (1) Integration of Landsat and MODIS images performs well in extracting evergreen forests in tropical complex mountainous regions. The overall accuracy of the resulting map of the case study was 92%; (2) Annual time series of high-temporal-resolution remote sensing images (MODIS) can effectively be used for identification of the key phenology phase (between Julian Date 20 and 120) to extract tropical evergreen forested areas through analysis of NDVI, EVI, and LSWI of different land cover types; (3) NDVI and LSWI are two effective metrics ($NDVI \geq 0.670$ and $0.447 \geq LSWI \geq 0.222$) to depict evergreen forests from other land cover types during the key phenology phase in tropical complex mountainous regions. This method can make full use of the Landsat and MODIS archives as well as their advantages for tropical evergreen forests geospatial inventories, and is simple and easy to use. This method is suggested for use with other similar regions.

Keywords: tropical evergreen forests; phenology features; remote sensing; MODIS; Landsat

---

1. Introduction

Global forests occupy 31% (4.0×10^7 km^2) of the total land area [1]. Tropical forests play a very important role in global carbon cycles, water and heat fluxes [2,3], biodiversity, and water and soil conservation, and produce half of the total Net Primary Production (NPP) of the world. However, across the world many tropical forests are being converted to farmlands, pastures, or other land cover types driven by the increasing demand for food, energy, and economic activity, among others [4]. Moreover, in the past few decades over 80% of agricultural expansions occurred in tropical forest regions [4,5]. Tropical forest loss and degradation also affect different ecosystem services, which oftentimes lead to natural disasters. Accurate and up-to-date tropical forest maps are very important for tropical forest management and proper monitoring [6]. In some tropical areas, the increasing expansion of rubber plantations, to mention but one example, continually encroach tropical evergreen forests [7] in such a way that their areas are gradually decreasing. This situation causes a series of environmental problems, such as loss of biodiversity, and drought with such a magnitude that it becomes a hotspot issue to monitor the extent and change of forests in these areas that are prone to transition from primary forests to rubber plantations.

Producing timely and accurate tropical evergreen forest inventory data is very important for decision making and forest management. In this context, field measurements are also needed to validate biophysical attributes of the vegetation such as volume, carbon stocks, tree height, etc. However, traditional field sampling methods are limited in their application to tropical mountainous regions, and the accuracy of inventory data is constrained by the quality and quantity of the field samples [8]. Fortunately, remote sensing methods have evolved sufficiently to support forest inventories in large landscapes [9], which allows rapid, automatic, and periodical estimates of many forest inventory attributes [8] and can also help to improve the precision of forest area estimates [10].

Phenology is a vegetation feature that may be used to identify different characteristics such as plant growth, external shape forms, and colors. Phenology patterns are results of how vegetation adapts to the physical environment and climate over a long period of time, and is a very stable feature. Phenological differences among land cover types can be used for classification by means of remote sensing technology. Some terrestrial objects are difficult to be discerned by optical images since their spectral properties are very similar. Nevertheless, these objects can be easily distinguished in a specific period during the year given their particular phenological features during that same period of the year. For instance, evergreen forests and rubber plantations have very similar characteristics across environmental gradients in Southeast Asia. However, they can be easily distinguished by assessing their leaf-on and leaf-off phases, since rubber plantations are deciduous in winter but evergreen forests are not. In recent years, phenology-based methods have been used extensively in the field of vegetation monitoring by remote sensing technology, particularly for crops and plantations [11,12]. Combining satellite observations and ground truth data is an efficient way for land cover classification. Further, extraction of vegetation spatial distribution patterns based on long-term time series of satellite data is an important application of remote sensing technology.

The Landsat sensor has acquired global observation records for more than four decades, and can be freely accessed. This is an invaluable data source for local, regional, or global forest monitoring. Several Landsat-based global forest products have been generated, including a global map of mangroves [13], a 30-m global land cover map between 1999 to 2011 [14], and a global forest cover change map [15]. Landsat missions (Thematic Mapper TM, Enhanced Thematic Mapper /ETM+, and the Operational Land Imager /OLI) have been extensively used to map forests [1,15,16], but due to frequent cloud cover, high-quality images are hard to capture in tropical regions. Moderate Resolution Imaging Spectroradiometer (MODIS) imagery with a temporal resolution of eight days increases the chance to obtain quality data for target objects, and helps alleviate the shortage of other optical remote sensing sensors. MODIS is extensively used in Land Use and Land Cover Change (LULCC) studies including crop and plantation monitoring [5,17,18]. For instance, inter-annual time-series MODIS Enhanced Vegetation Index (EVI) and Short Wave Infrared (SWIR) have been efficiently used to identify rubber plantations in tropical mountainous regions [19].

In recent years, there has been an increasing interest by researchers worldwide to integrate imagery from multiple sensors. Landsat and Radar have been integrated to map rubber plantations in tropical regions [11], and MODIS and Landsat have been jointly used to monitor crops [20]. Hence, integration of multi-source remote sensing images is a growing research trend in many fields.

This study aims to integrate 500-m MODIS and 30-m Landsat to identify tropical evergreen forests based on vegetation phenology characteristics. MODIS with a high-revisit period (eight-days) enables capturing abundant and high-quality data for phenology-based time series analysis. Landsat data with fine spatial resolution (30m) can provide detailed spatial information and reduce mixing pixels to improve classification accuracy. Annual time-series MODIS Normalized Difference Vegetation Index (NDVI), EVI, and Land Surface Water Index (LSWI) are used to identify the key phenology phase that best fits tropical evergreen forest through analysis of differences in NDVI, EVI, and LSWI between evergreen forests and other typical land cover types. During the key phenology phase, indices and relevant thresholds were also identified through analysis and statistics of the values of NDVI, EVI, and LSWI for different land cover types based on 30-m Landsat images.

The objective of this study was to develop a phenology-based method for quickly extracting tropical evergreen forests for inventories and exploring their spatial distribution patterns by integrating MODIS and Landsat images across large scales. Mengyang, Jinuo, and Yunjinghong in Xishuangbanna, China were selected as the study area. The developed method is able to get accurate extent, areas, and spatial distribution for large-scale tropical evergreen forest inventory at 30-m spatial resolutions. These datasets can support basic inventory data for forestry planning and managing, ecosystem monitoring, regional land management, and other relevant research and applications.

2. Materials and Methods

2.1. The Study Area

This study was conducted in the Xishuangbanna Dai Nationality Autonomous Prefecture (Xishuangbanna), Yunnan, China. The total land area is 1.97×10^4 km^2, and the majority of the land (95%) is considered mountainous. The elevation ranges 344 m to 2,605 m above sea level. The total population is 1.14 million, with 69% dedicated to agricultural activities according to the Year Statistical Book of Xishuangbanna in 2010 [21]. The average annual precipitation is 1,550 mm, and the mean annual temperature is 21.7 °C [22]. Two major seasons are recognized in the study area: dry (from November to April), and rainy (from May to October). Rubber plantations cover more than 50% of the Yunnan Province, and are the second largest in China. Three major towns are included in the study area (Figure 1): Mengyang, Jinuo, and Yunjinghong. Due to the economic importance of natural rubber and agriculture, land use conflicts are common in the study area between these two classes and tropical evergreen forests.

2.2. Materials and Processing

NDVI, EVI, and LSWI are frequently used to stratify different land cover types. The key phenology phase—defined as an optimal time window for distinguishing tropical evergreen forests from other land cover types—was identified through analysis of annual time series created from multi-temporal eight-day-composites 500-m MODIS MOD09A1. Landsat imagery was acquired from the United States Geological Survey (USGS) server, and converted to surface reflectance using Landsat Ecosystem Disturbance Adaptive Processing System (LEDAPS) [23]. The rectified imagery was further processed using Fmask [24] to deal with clouds and shadows. During the key phenology phase, tropical evergreen forest was distinguished from other land cover types by using the improved Landsat imagery. A collection of photo-interpreted samples for algorithm training, time-series analysis, and results verification were created from geo-referenced field photos and Google Earth high quality imagery. This method is very effective and extensively used in current studies [11,25,26]. Google Earth high quality imagery can provide a reliable data source for interpreting and extracting samples (Figure 1B–F). Five typical land cover types (evergreen forests, rubber plantations, built-land, farmlands, and water bodies) were classified in the study area. NDVI, EVI, and LSWI are calculated using formulas (1)–(3). Land cover data of 500-m MODIS MCD12Q1 was used to compare with the resultant map of tropical evergreen forests [27].

$$NDVI = \frac{\rho_{NIR} - \rho_{red}}{\rho_{NIR} + \rho_{red}} \tag{1}$$

$$EVI = 2.5 \times \frac{\rho_{NIR} - \rho_{red}}{\rho_{NIR} + 6 \times \rho_{red} - 7.5 \times \rho_{blue} + 1} \tag{2}$$

$$LSWI = \frac{\rho_{NIR} - \rho_{SWIR}}{\rho_{NIR} + \rho_{SWIR}} \tag{3}$$

where ϱ_{NIR}, ϱ_{red}, ϱ_{MIR}, and ϱ_{blue} are the surface reflectance values of near infrared, red, shortwave-infrared, and blue bands in MODIS or Landsat images.

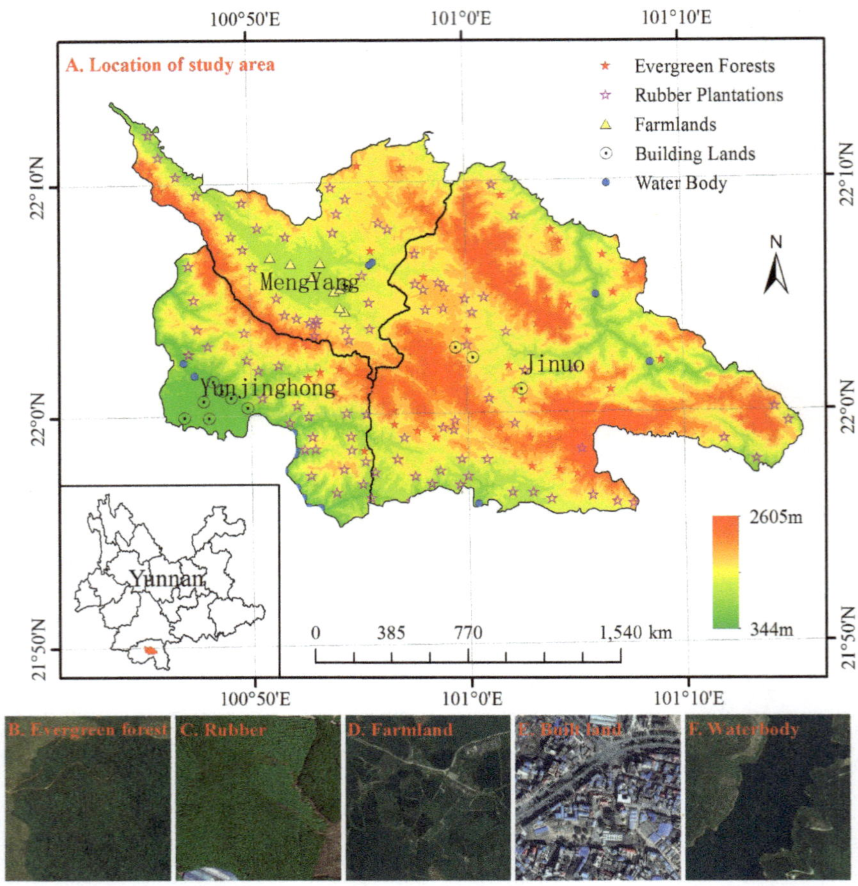

Figure 1. The study area location and samples for algorithm training and result verification. (**A**) is the location of three towns (Mengyang, Jinuo, and Yunjinghong) in Jinghong County Xishuangbanna, Yunnan, China, and spatial distributions for samples of five typical land cover types including evergreen forests, rubber plantations, farmlands, built-lands, and water bodies. They are created from geo-referenced field photos and Google Earth High quality imagery; (**B**–**F**) are high-resolution images of five typical land cover types from Google Earth.

2.3. Identifying the Key Phenology Phase

The key phenology phase is a relatively-fixed time window during which a particular vegetation type exhibits stable unique characteristics that are different from other land cover types. It is the optimal timing for distinguishing a kind of vegetation from others, and it was extensively used in previous studies [11,16,17]. Evergreen forests are stable ecosystems and have a strong resistance against influences from the physical environment (such as air temperature), whereas other land cover types (such as plantations and farmlands) are more susceptible to outside factors or mainly controlled by human activities. This phenomenon is generally reflected on the stability of annual

NDVI, EVI, and LSWI time-series curves of different land cover types. For instance, the NDVI curves of evergreen-forests are more stable than others.

Eight-day MODIS time-series have been extensively used to map paddy rice fields, forests, and plantations [17,28,29]. The key phenology phase can be identified through analysis of differences on annual time-series of MODIS NDVI, EVI, and LSWI of different land cover types. The fine temporal resolution is capable of capturing abundant, reliable, and continuous observation data in tropical regions, which supports annual time-series analysis for identification of the key phenology phase. Annual time-series profiles on NDVI, EVI, and LSWI of five typical land cover types were analyzed from temporal MODIS MOD09A1 between 2006 and 2014 based on samples created from geo-referenced field photos and Google Earth high resolution imagery. The method of creating these samples is very effective and extensively used in recent studies [11,16,26,30]. The phenological characteristics of different land cover types were used to discriminate evergreen forests from other land cover types during the key phenology phase. Considering the effects of mixing pixels caused by outside factors such as clouds, shadows, fragments, the minimum and maximum 3% were selected from 30-m Landsat NDVI, EVI, and LSWI time-series data of different land cover types [3]. This strategy was also used for analysis of the key phenology phase for extracting tropical evergreen forests.

2.4. Extracting Index Metrics and Relevant Thresholds

To extract tropical evergreen forests, high-quality Landsat 5 TM images captured on Julian Date 45 in 2010 were selected during the key phenology phase identified through analysis of MODIS annual time series. NDVI, EVI, and LSWI of evergreen forests, rubber plantations, farmlands, built-lands, and water bodies were extracted from Regions of Interests (ROIs) (Table 1) created from geo-referenced field photos and Google Earth high resolution images. These vegetation indices were evaluated by a statistical method to find an index combination with an ability to distinguish tropical forests from other land cover types.

Table 1. Regions of Interests (ROIs) created from geo-referenced field photos and Google Earth high resolution images for time-series analysis, algorithm training, and result verification in the study area.

ID	Land Cover Types	Pixels
1	Evergreen forests	26904
2	Rubber plantations	8392
3	Farmlands	268
4	Built-lands	785
5	Water bodies	538

This study's workflow for mapping tropical evergreen forests and exploring the spatial distribution is shown in Figure 2. First, annual time-series from 500-m MODIS images were analyzed to identify the key phenology phase; second, during the phenology phase, differences in 30-m Landsat NDVI, EVI, and LSWI of five typical land cover types were evaluated to select and determine the indices and relevant thresholds for extracting tropical evergreen forests; third, a 30-m map of tropical evergreen forest of the study area in 2010 was generated according to the indices and their thresholds determined above; fourth, the resultant map was verified using geo-referenced field photos and Google Earth high resolution imagery, and compared with MODIS MCD12Q1 land cover map for 2010; finally, the spatial distribution of evergreen forests at different elevation gradients was assessed based on a 90-m Digital Elevation Model (DEM) of the study area.

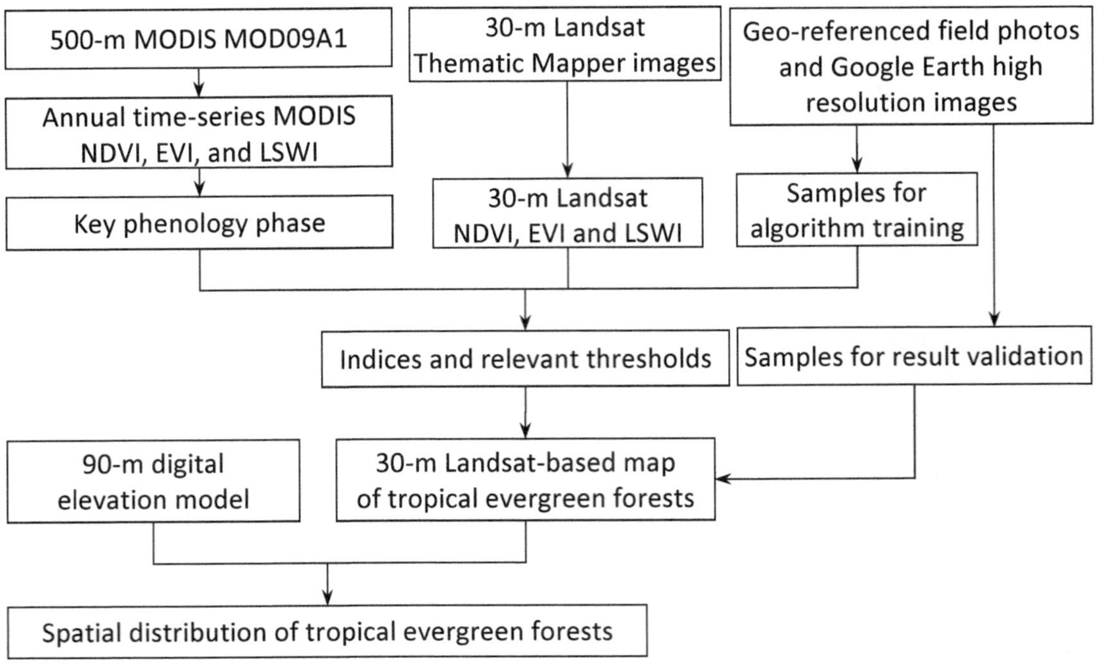

Figure 2. The technique workflow for mapping tropical evergreen forests. (1) A key phenology phase is selected based on analysis of annual time-series 500-m Moderate Resolution Imaging Spectroradiometer (MODIS) MOD09A1 Normalized Difference Vegetation Index (NDVI), Enhanced Vegetation Index (EVI), and Land Surface Water Index (LSWI) for different land cover types; (2) 30-m Landsat NDVI, EVI, and LSWI in the key phenology phase are evaluated to find an optimal combination of them for distinguishing tropical evergreen forests from other land cover types based on samples extracting from geo-referenced field photos and Google Earth high resolution images of the study area; (3) Based on (2), a 30-m map of tropic evergreen forests is generated based on Landsat Thematic Mapper (TM) imagery and verified by a confusion matrix built from samples created by the same method used above; (4) Spatial distribution of tropical evergreen forests is analyzed base on a 90-m digital elevation model.

3. Results

3.1. MODIS-Based Time-Series Analyses for Different Land Cover Types

Tropical evergreen forests can be distinguished from farmlands, built-lands, and water bodies on MODIS NDVI, EVI, and LSWI annual time series between Julian Dates 20 and 120 (Figure 3), so this time period was selected as the key phenology phase to extract evergreen forests. During this phase, the evergreen forest NDVI, EVI, and LSWI are higher than all other land cover types based on Landsat 5 TM images. This is because those evergreen forests have a strong ability to resist low air temperature in winter and spring (Julian Dates between 20 and 120), and the NDVI, EVI, and LSWI are relatively high. In addition, deciduous rubber plantations have a low canopy cover due to leaf-off conditions, and farmlands are not being cultivated at this time of year either.

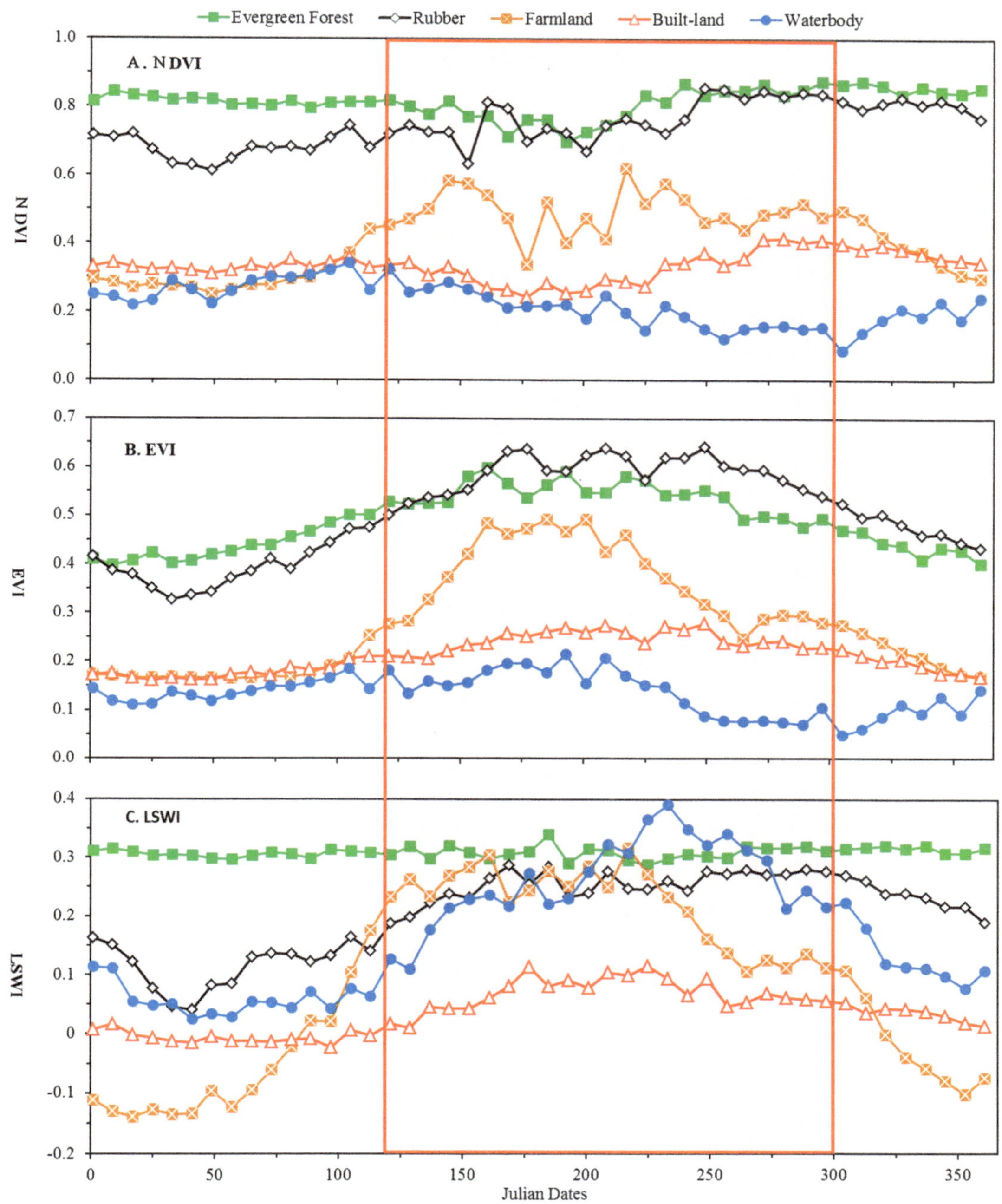

Figure 3. MODIS annual time-series analyses for different land cover types of the study area. Normalized Difference Vegetation Index (NDVI), Enhanced Vegetation Index (EVI), and Land Surface Water Index (LSWI) annual time series of five typical land cover types (evergreen forests, rubber plantations, farmlands, built-lands, and water bodies) were created from MODIS MOD09A1 between 2006 and 2014. Annual time-series profiles of NDVI (**A**); EVI (**B**); and LSWI (**C**) are separately analyzed to capture change information and rules of different land cover types of the study area. The area covered by red rectangle is the rainy season about from Julian Date 120 to 350 and the rest of the area represents the dry season in the study area.

3.2. Selection of Indices and Relevant Thresholds

NDVI of tropical evergreen forests is higher than deciduous rubber plantations, farmlands, built-lands, and water bodies, and LSWI is lower than water bodies and higher than farmlands, rubber plantations, and built-lands, respectively (Table 2). However, the evergreen forest EVI is prone to mix with rubber plantations and farmlands. Therefore, Landsat 5 TM NDVI and LSWI were selected as the index metrics to extract tropical evergreen forests in the study area. Formulas (4)–(6) were designed to calculate thresholds of NDVI and LSWI. Since NDVI of tropical evergreen forests is far higher than other land cover types, just the minimum NDVI value needs to be calculated by formula (4), and the LSWI minimum and maximum can be calculated by formula (5) and (6). According to this, the thresholds are defined as NDVI $\geq$ 0.670 and 0.447 $\geq$ LSWI $\geq$ 0.222.

$$NDVI_{low} = NDVI_r + ((NDVI_e - NDVI_{ed}) - (NDVI_r + NDVI_{rd})/2 \quad (4)$$

$$LSWI_{lo} = LSWI_c + ((LSWI_e - LSWI_{ed}) - (LSWI_c + LSWI_{cd}))/2 \quad (5)$$

$$LSWI_u = LSWI_e + ((LSWI_w - LSWI_{wd}) - (LSWI_e + LSWI_{ed}))/2 \quad (6)$$

Here, NDVI, EVI, and LSWI were respectively extracted from Landsat 5 TM on Julian Date 45 in 2010. These are the average values of samples of five typical land cover types (rubber plantations, evergreen forests, farmlands, built-lands, and water bodies) in the study area. Subscript and superscript letters stand for relevant minimum and maximum values of NDVI, EVI, and LSWI; r, e, c, w, and r separately stand for values of different land cover types, and d, ed, cd, and wd for their relevant standard deviation. Figure 4 shows the statistical comparison of 30-m Landsat 5 TM NDVI, EVI, and LSWI of five typical land cover types. NDVI and LSWI are able to discern tropical evergreen forests from other land cover types, whereas EVI is unable to do so.

Table 2. NDVI, EVI, and LSWI of different land cover types of the study area of Xishuangbanna, Yunnan, China based on Landsat 5 TM images in 2010.

Land Cover Types	NDVI	EVI	LSWI
Evergreen Forests	0.743 ± 0.044	0.426 ± 0.076	0.312 ± 0.059
Rubber Plantations	0.563 ± 0.077	0.342 ± 0.069	0.053 ± 0.095
Farmlands	0.526 ± 0.067	0.337 ± 0.051	0.113 ± 0.079
Built-lands	0.341 ± 0.089	0.203 ± 0.055	−0.029 ± 0.084
Water bodies	0.015 ± 0.078	0.008 ± 0.034	0.618 ± 0.095

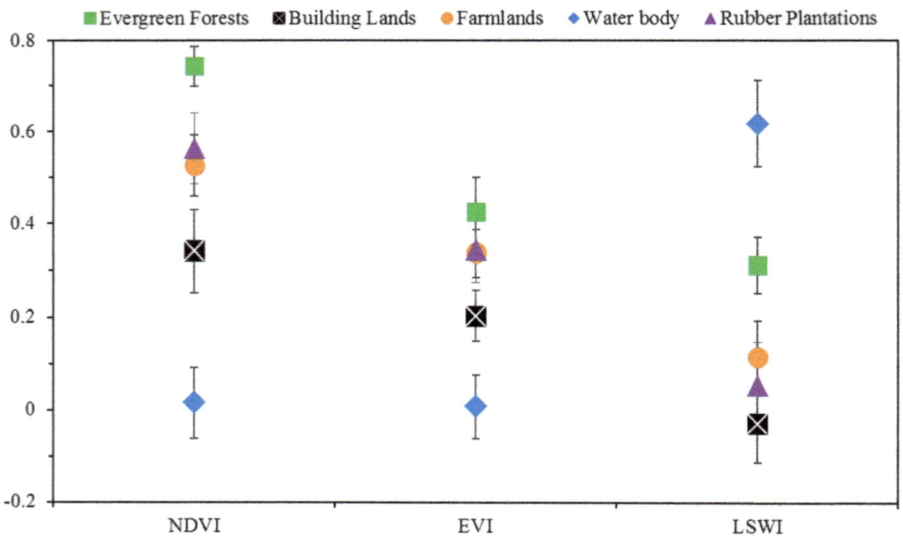

Figure 4. Signature analysis of the Normalized Difference Vegetation Index (NDVI), Enhanced Vegetation Index (EVI), and Land Surface Water Index (LSWI) for tropical evergreen forests, rubber plantations, farmlands, built-lands, and water bodies based on the Landsat Thematic Mapper (TM) 5 images between Julian Dates 20 and 120 in 2010 in the study area. Evergreen forests and other land cover types have distinctive values in NDVI and LSWI.

3.3. Landsat-Based Tropical Evergreen Forest Map of the Study Area in 2010

The output evergreen forest map (Figure 5) was verified by using the photo-interpreted locations described in the methods. To assess the map's accuracy a confusion matrix [11] was created. The samples included evergreen forests and other typical land cover types. According to the confusion matrix (Table 3), the overall accuracy is 92%; the user's and producer's accuracy of evergreen forests are 91% and 89% separately, and the accuracy of the other land cover types is 93% and 94%, respectively.

The spatial distribution of tropical evergreen forests in the output map for 2010 fits well to the false color (bands 5/4/3) composite Landsat 5 TM images-Julian Date 45 in 2010 (Figure 6). Furthermore, the resultant map (Figure 7B) was overlaid with the evergreen forests from the 500-m MODIS MCD12Q1 in 2010 (Figure 7A). The evergreen forest area of the Landsat-based output map is smaller than the one obtained from MODIS MCD12Q1. The possible reason for this is that mixed pixels are found due to the coarse spatial resolution and complex tropical mountain environment. For instance, some old rubber plantations have the similar phenology characteristics with evergreen forests and are misclassified.

Table 3. The accuracy assessment result of the tropical evergreen forest resultant map at 30-m spatial resolution in 2010 based on an error matrix. The overall accuracy is 92%.

Class		Ground Truth Samples (pixels)		Total Class pixels	Producer's Accuracy
		Evergreen Forests	Others		
Classified results	Evergreen forests	3774	353	4127	91%
	Others	373	3843	4425	93%
Total ground truth pixels		4238	6240	10,478	
User's accuracy		89%	94%		

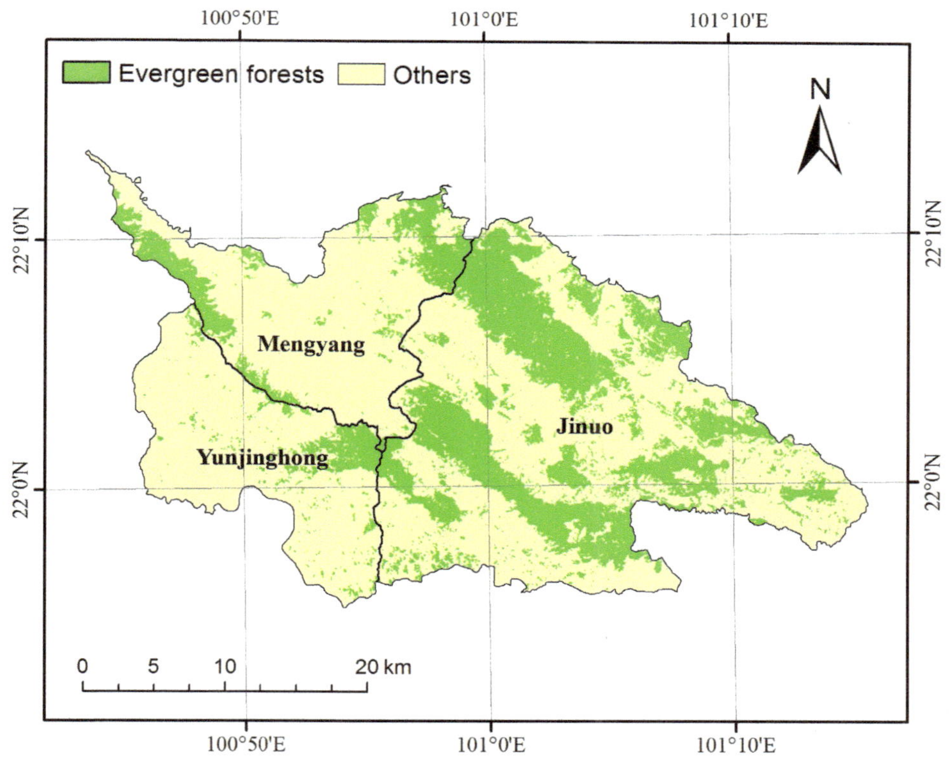

Figure 5. The 30-m Landsat-based map of tropical evergreen forest of the study area in Xishuangbanna, Yunnan, China in 2010.

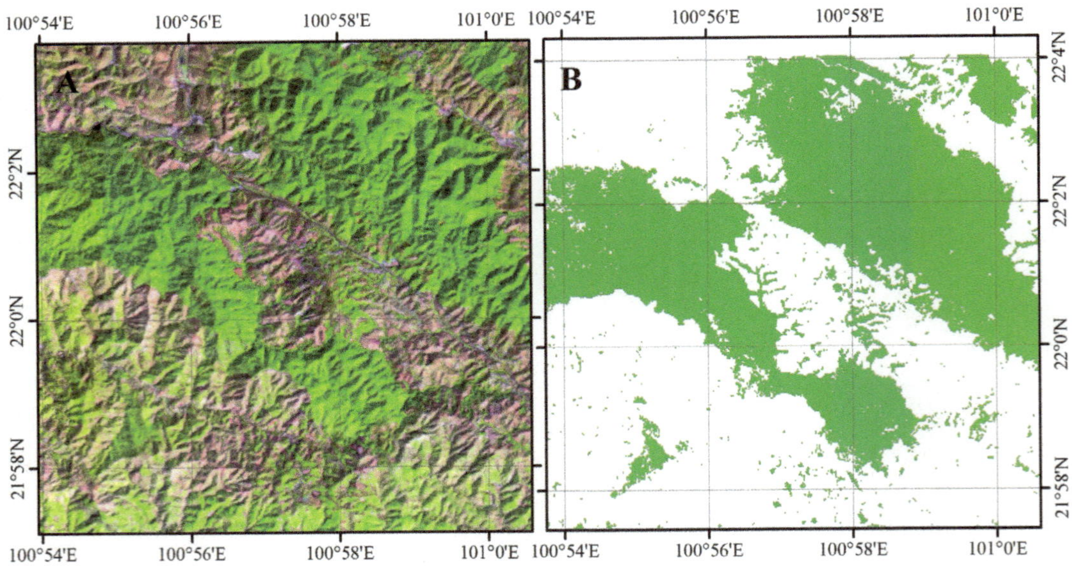

Figure 6. The resultant map is compared with the 30-m Landsat 5 Thematic Mapper (TM) image on the Julian Date 45 in 2010. The evergreen forest shows as green on the composite Landsat 5 TM image (**A**) combined as false color (bands 5/4/3); and (**B**) is the 30-m Landsat-based resultant map of tropical evergreen forests in the study area in 2010.

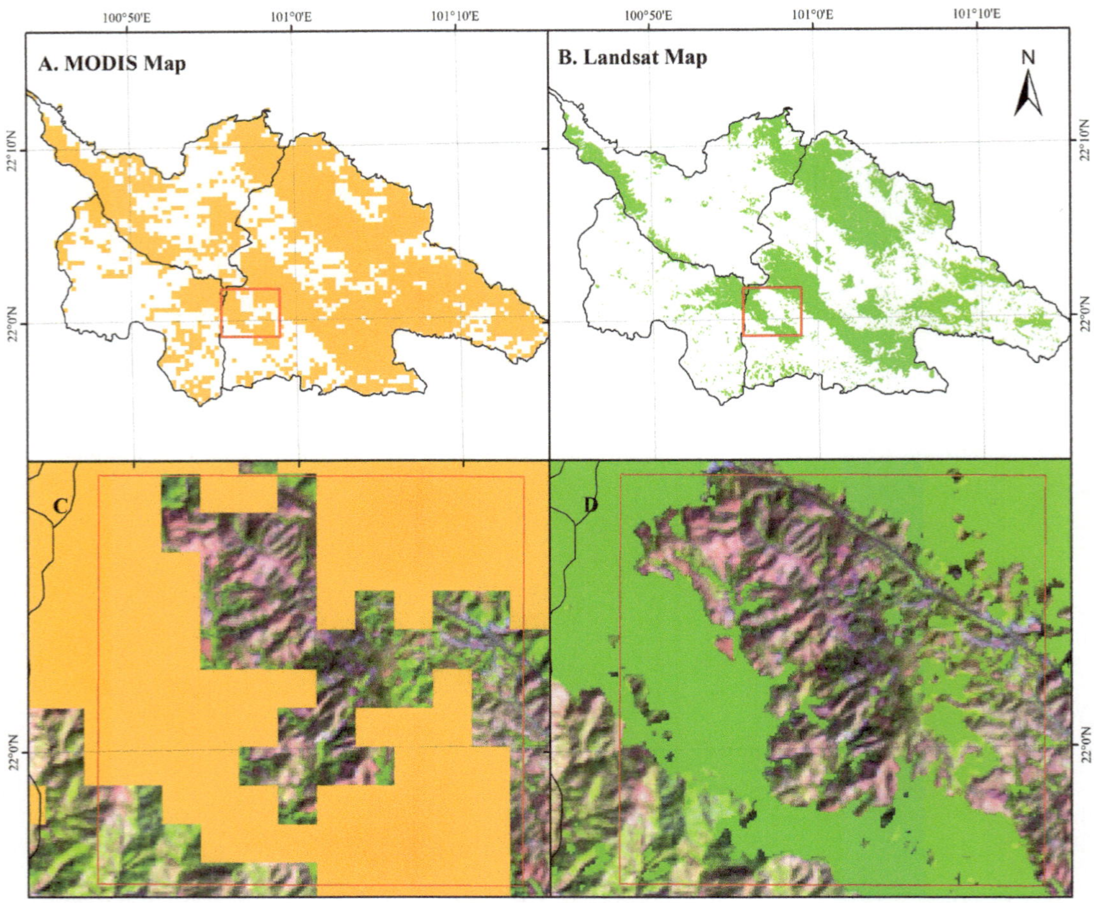

Figure 7. Comparison of eight-day 500-m MODIS MCD12Q1 (**A**) and the 30-m Landsat resultant map (**B**) in 2010. The two zoom-out windows below display the detailed information of MODIS MCD12Q1 (**C**) and the resultant map of 30-m Landsat 5 TM (**D**).

3.4. Distribution of Evergreen Forests on Different Elevation Gradients

The distribution of tropical evergreen forests of the resultant map was analyzed by resampled 30-m DEM of the study area (Figure 8). The results show that tropical evergreen forests are majorly distributed in a narrow elevation range between 900 m and 1400 m, and few forest patches are found in lower (<900 m) and higher elevations (>1400 m).

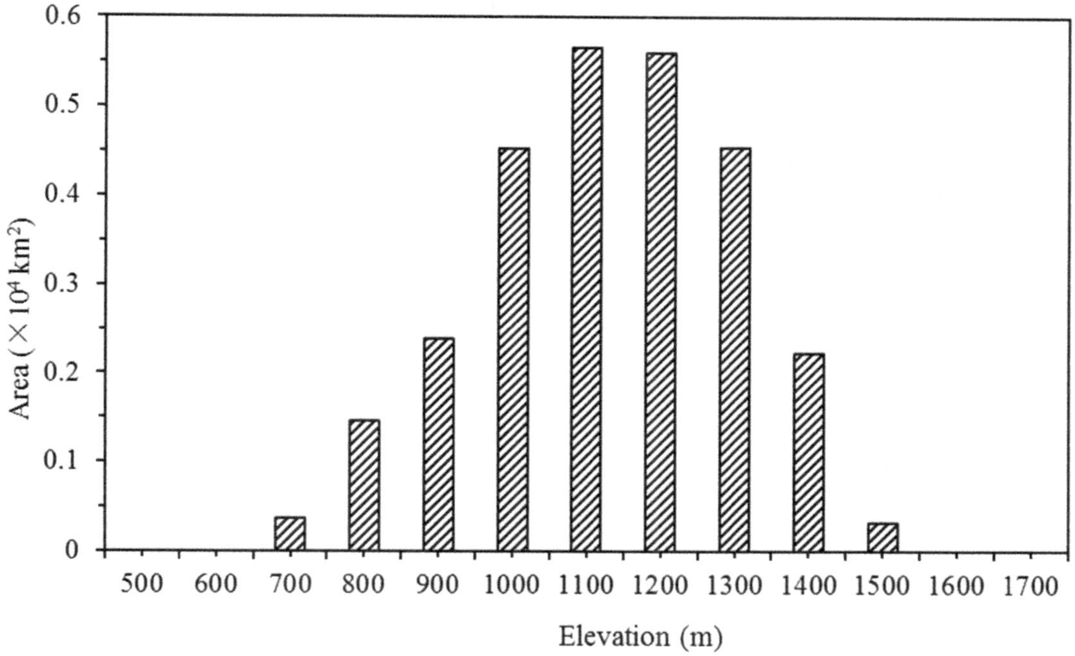

Figure 8. Distribution of tropical evergreen forests on different elevation gradients based on resampled 30-m digital elevation model in the study area of Xishuangbanna, Yunnan, China in 2010.

4. Discussion

The results from this study reinforces the notion that fine temporal resolution annual time series (eight-day MODIS) are capable of distinguishing the key phenology phase to extract tropical evergreen forests in tropical regions. These time-series can provide abundant and highly-reliable basic scientific data. In this particular case, it was established that the phenology phase can be identified from Julian Dates 20–120, and that this important phenological signature can be effectively used to extract tropical evergreen forests through analysis of NDVI, EVI, and LSWI of different land cover types. In tropical regions, cloud-free observations are difficult to get due to frequent clouds and shadows. Nevertheless, having coarser spatial resolution with high-temporal-resolution sensors can increase the possibility of getting good quality remote sensing images. This approach takes advantages of the high-revisit characteristics of the MODIS sensor, which increase the possibility of capturing good quality images from optical sensors. Annual MODIS time-series have been used in previous research [3,12,31–33], and most of them are not in tropical mountainous regions. It should be mentioned, however, that the coarse spatial resolution (500 m) may produce some inaccurate results from mixing pixels of small-area land cover types, such as narrow rivers.

Distinguishing tropical evergreen forest and deciduous forest is a challenge in Southeast Asia based on optical remote sensing images. The phenology-based method used in several studies is very useful for mapping paddy rice fields [17] and rubber plantations [11]. In this study, the key phenology phase was used to distinguish tropical evergreen forests in mountainous regions. Vegetation in different growth stages exhibits various distinct characteristics in terms of forms, color, and shape. Spectral reflectance characteristics of different land cover types are similar in some phases and different in others. For instance, as it was shown in this paper, annual time-series MODIS NDVI of evergreen forests is higher than other land cover types on Julian Dates between 20 and 120 (Figure 3). Based on this finding, evergreen forests were distinguished from other land cover types in tropical mountainous regions based on 30-m Landsat 5 TM images. This study demonstrates that highly-temporal time-series are useful in observing key phenology phase for mapping a certain land cover type in certain time periods in tropical humid regions.

The vegetation indices and related thresholds were keys for accurately mapping evergreen forests. Easy to carry out and to understand statistical methods are often used to choose indices and compute related threshold values based on samples of different land cover types. At present, there is a lack of efficient methods for choosing indices and computing related thresholds for quickly and efficiently distinguishing evergreen forests in tropical regions of Southeast Asia. This study found that 30-m Landsat NDVI and LSWI are two efficient indices for distinguishing evergreen forests in these tropical environments. A general method (formulae (4)–(6)) for calculating related thresholds by a statistical method has been described and provided in this paper. Thirty-meter Landsat NDVI and LSWI are two effective metrics ($NDVI \geq 0.670$ and $0.447 \geq LSWI \geq 0.222$) to discern tropical evergreen forests for inventories from other land cover types during the key phenology phases in tropical complex mountainous regions. In the past few decades, driven by urban and economic developments, tropical forests were substantially converted to farmlands and rubber plantations in Xishuangbanna, Yunnan China. However, an accurate spatial distribution of the remaining forests was hard to determine in this region. This study generated an accurate evergreen forest map (Figure 5) of the study area and explored the forests' distribution across an elevation gradient (Figure 8). It was found that evergreen forest is mainly distributed in a very narrow range (900 m–1400 m) of elevation. A possible reason for this is that farmlands and plantations mainly dominate the low-elevation regions, and the high-elevation regions are cultivated for planting economic crops such as tea plants. This may lead to the loss of vertical forest diversity and biodiversity, and threatens the local ecosystem, biodiversity, and environment. This implies that these ecosystems are susceptible to a land cover transition, thus local governments need to take action for their protection. In this paper, a phenology-based automated mapping algorithm was shown to perform well in tropical mountainous regions. The study explored the extent, areas, and spatial distribution of tropical evergreen forest in this region of South Asia, and evaluated the efficiency and applicability of the method.

Additionally, samples created from geo-referenced photos and high resolution images of Google Earth provide reliable basic data for algorithm training and result validation. These geo-referenced filed photos are firstly transferred into xml files and then linked with high resolution imagery of Google Earth, in which samples are visually and interactively created.

5. Conclusions

This study attempted to develop a novel phenology-based method to map tropical evergreen forests by integrating 30-m Landsat and eight-day 500-m MODIS remotely sensed images during a key phenology phase. The developed and verified protocol is simple yet effective. Integration of multi-source remote sensing data with different spatial and temporal resolutions proved satisfactory to obtain accurate estimates of the extent, areas, and spatial distribution of tropical evergreen forest in Southeast Asia. Additionally, identification of the key phenology phase is very important to distinguish evergreen forest from other land cover types based on vegetation indices including NDVI, EVI, and LSWI. The assessment of the spatial distribution of evergreen forest across the elevation gradient showed that the remaining forest is fragmented and needs to be protected. Additionally, the study suggests that the local government can compensate farmers to encourage them to farm in low-elevation gradients and to withdraw from ecosystem-weak regions in high-elevation gradients, for example, by establishing guidelines for planting rubber trees below 950 m. This study also illustrates that our output map can provide more detailed and accurate spatial information than MODIS MCD12Q1. However, there are still some uncertainties in the methods due to the relatively coarse temporal and spatial resolution of remote sensing data and the influence of frequent clouds, shadows, and aerosols. The integration with other sources (such as microwave, lidar, and optical) with finer spatial and temporal resolution may be tested in future studies to improve the presented methods for the extraction of tropical evergreen forests.

Acknowledgments: This work was supported by the National Natural Science Foundation of China (No. 31400493) and the Research Center of Kunming Forestry Information Engineering Technology (Grant No. 2015FIB04). We thank three anonymous peer reviewers for their helpful and valuable suggestions and comments to improve this manuscript.

Author Contributions: Weili Kou conceived and designed the experiments; Changxian Liang, Lili Wei, and Xuejing Yang performed the experiments; Weili Kou and Alexander J. Hernandez wrote the paper. All co-authors assisted the lead author in writing and revising the manuscript.

Conflicts of Interest: The authors declare no conflicts of interest.

References

1. Qin, Y.; Xiao, X.; Dong, J.; Zhang, G.; Shimada, M.; Liu, J.; Li, C.; Kou, W.; Moore, B., III. Forest cover maps of china in 2010 from multiple approaches and data sources: Palsar, landsat, modis, fra, and nfi. *ISPRS J. Photogramm. Remote Sens.* **2015**, *109*, 1–16. [CrossRef]
2. Pongratz, J.; Reick, C.H.; Raddatz, T.; Claussen, M. Biogeophysical versus biogeochemical climate response to historical anthropogenic land cover change. *Geophys. Res. Lett.* **2010**, *37*, 162–169. [CrossRef]
3. Qin, Y.; Xiao, X.; Dong, J.; Zhang, G.; Roy, P.S.; Joshi, P.K.; Gilani, H.; Murthy, M.S.R.; Jin, C.; Wang, J. Mapping forests in monsoon asia with alos palsar 50-m mosaic images and modis imagery in 2010. *Sci. Rep.* **2016**, *6*, 1–10. [CrossRef] [PubMed]
4. Gibbs, H.K.; Ruesch, A.S.; Achard, F.; Clayton, M.K.; Holmgren, P. Tropical forests were the primary sources of new agricultural land in the 1980s and 1990s. *Proc. Natl. Acad. Sci. USA* **2010**, *107*, 16732–16737. [CrossRef] [PubMed]
5. Kwak, Y.; Park, J.; Fukami, K. Near real-time flood volume estimation from modis time-series imagery in the indus river basin. *IEEE J. Sel. Top. Appl. Earth Obs. Remote Sens.* **2014**, *7*, 578–586. [CrossRef]
6. Hansen, M.C.; Stehman, S.V.; Potapov, P.V. From the cover: Quantification of global gross forest cover loss. *Proc. Natl. Acad. Sci. USA* **2010**, *107*, 8650–8655. [CrossRef] [PubMed]
7. Yi, Z.F.; Cannon, C.H.; Chen, J.; Ye, C.X.; Swetnam, R.D. Developing indicators of economic value and biodiversity loss for rubber plantations in Xishuangbanna, southwest China: A case study from menglun township. *Ecol. Indic.* **2014**, *36*, 788–797. [CrossRef]
8. Liang, X.; Kankare, V.; Hyyppä, J.; Wang, Y.; Kukko, A.; Haggrén, H.; Yu, X.; Kaartinen, H.; Jaakkola, A.; Guan, F.; et al. Terrestrial laser scanning in forest inventories. *ISPRS J. Photogramm. Remote Sens.* **2016**, *115*, 63–77. [CrossRef]
9. McRoberts, R.E.; Tomppo, E.O. Remote sensing support for national forest inventories. *Remote Sens. Environ.* **2007**, *110*, 412–419. [CrossRef]
10. McRoberts, R.E.; Wendt, D.G. Using a land cover classification based on satellite imagery to improve the precision of forest inventory area estimates. *Remote Sens. Environ.* **2002**, *81*, 36–44. [CrossRef]
11. Dong, J.; Xiao, X.; Chen, B.; Torbick, N.; Jin, C.; Zhang, G.; Biradar, C. Mapping deciduous rubber plantations through integration of palsar and multi-temporal landsat imagery. *Remote Sens. Environ.* **2013**, *134*, 392–402. [CrossRef]
12. Sakamoto, T.; Yokozawa, M.; Toritani, H.; Shibayama, M.; Ishitsuka, N.; Ohno, H. A crop phenology detection method using time-series modis data. *Remote Sens. Environ.* **2005**, *96*, 366–374. [CrossRef]
13. Giri, C.; Ochieng, E.; Tieszen, L.L.; Zhu, Z.; Singh, A.; Loveland, T.; Masek, J.; Duke, N. Status and distribution of mangrove forests of the world using earth observation satellite data. *Glob. Ecol. Biogeogr.* **2011**, *20*, 154–159. [CrossRef]
14. Gong, P.; Wang, J.; Yu, L.; Zhao, Y.; Zhao, Y.; Liang, L.; Niu, Z.; Huang, X.; Fu, H.; Liu, S. Finer resolution observation and monitoring of global land cover: First mapping results with landsat tm and etm+ data. *Int. J. Remote Sens.* **2012**, *34*, 2607–2654. [CrossRef]
15. Hansen, M.C.; Potapov, P.V.; Moore, R.; Hancher, M.; Turubanova, S.A.; Tyukavina, A.; Thau, D.; Stehman, S.V.; Goetz, S.J.; Loveland, T.R.; et al. High-resolution global maps of 21st-century forest cover change. *Science* **2013**, *342*, 850–853. [CrossRef] [PubMed]

16. Kou, W.; Xiao, X.; Dong, J.; Gan, S.; Zhai, D.; Zhang, G.; Qin, Y.; Li, L. Mapping deciduous rubber plantation areas and stand ages with palsar and landsat images. *Remote Sens.* **2015**, *7*, 1048–1073. [CrossRef]
17. Xiao, X.; Boles, S.; Liu, J.; Zhuang, D.; Frolking, S.; Li, C.; Salas, W.; Iii, B.M. Mapping paddy rice agriculture in southern china using multi-temporal modis images. *Remote Sens. Environ.* **2005**, *95*, 480–492. [CrossRef]
18. Liu, X.; Feng, Z.; Jiang, L.; Zhang, J. Rubber plantations in xishuangbanna: Remote sensing identification and digital mapping. *Resour. Sci.* **2012**, *34*, 1769–1780.
19. Senf, C.; Pflugmacher, D.; Linden, S.V.D.; Hostert, P. Mapping rubber plantations and natural forests in Xishuangbanna (southwest China) using multi-spectral phenological metrics from modis time series. *Remote Sens.* **2013**, *5*, 2795–2812. [CrossRef]
20. Xie, D.; Sun, P.; Zhang, J.; Zhu, X. Autumn Crop Identification Using High-Spatial-Temporal Resolution Time Series Data Generated by Modis and Landsat Remote Sensing Images. In Proceedings of the IGARSS 2014—2014 IEEE International Geoscience and Remote Sensing Symposium, Quebec City, QC, Canada, 13–18 July 2014; pp. 2118–2121.
21. Editorial Board of the Year Statistical Book of Xishuangbanna. *The Year Statistical Book of Xishuangbanna in 2010*; Dehong Nationalities Publishing House: Dehong, China, 2011.
22. Lü, X.T.; Yin, J.X.; Tang, J.W. Diversity and composition of understory vegetation in the tropical seasonal rain forest of Xishuangbanna, SW China. *Rev. Biol. Trop.* **2011**, *59*, 455–463. [CrossRef] [PubMed]
23. Masek, J.G.; Vermote, E.F.; Saleous, N.; Wolfe, R.; Hall, F.G.; Huemmrich, K.F.; Gao, F.; Kutler, J.; Lim, T.K. *Ledaps Calibration, Reflectance, Atmospheric Correction Preprocessing Code, Version 2*; ORNL Distributed Active Archive Center: Oak Ridge, TN, USA, 2013.
24. Zhu, Z.; Woodcock, C.E. Object-based cloud and cloud shadow detection in landsat imagery. *Remote Sens. Environ.* **2012**, *118*, 83–94. [CrossRef]
25. Dong, J.; Xiao, X.; Sheldon, S.; Biradar, C.; Xie, G. Mapping tropical forests and rubber plantations in complex landscapes by integrating palsar and modis imagery. *ISPRS J. Photogramm. Remote Sens.* **2012**, *74*, 20–33. [CrossRef]
26. Lu, N.; Hernandez, A.J.; Ramsey, R.D. Land cover dynamics monitoring with landsat data in kunming, china: A cost-effective sampling and modelling scheme using google earth imagery and random forests. *Geocarto Int.* **2014**, *30*, 186–201. [CrossRef]
27. Friedl, M.A.; Sulla-Menashe, D.; Tan, B.; Schneider, A.; Ramankutty, N.; Sibley, A.; Huang, X. Modis collection 5 global land cover: Algorithm refinements and characterization of new datasets. *Remote Sens. Environ.* **2010**, *114*, 168–182. [CrossRef]
28. Liu, F.J.; Huang, C.; Pang, Y.; Li, M.; Song, D.X.; Song, X.P.; Channan, S.; Sexton, J.O.; Jiang, D.; Zhang, P. Assessment of the three factors affecting myanmar's forest cover change using landsat and modis vegetation continuous fields data. *Int. J. Digit. Earth* **2016**, *9*, 1–24. [CrossRef]
29. Setiawan, Y.; Pawitan, H.; Prasetyo, L.B.; Lubis, M.I.; Parlindungan, M.; Nurdiana, A. Characterizing spatial distribution and environments of sumatran peat swamp area using 250 m multi-temporal modis data. *Procedia Environ. Sci.* **2016**, *33*, 117–127. [CrossRef]
30. Dong, J.; Xiao, X. Evolution of regional to global paddy rice mapping methods: A review. *ISPRS J. Photogramm. Remote Sens.* **2016**, *119*, 214–227. [CrossRef]
31. Wardlow, B.D.; Egbert, S.L.; Kastens, J.H. Analysis of time-series modis 250 m vegetation index data for crop classification in the U.S. Central great plains. *Remote Sens. Environ.* **2007**, *108*, 290–310. [CrossRef]
32. Dorjsuren, M.; Liou, Y.-A.; Cheng, C.-H. Time series MODIS and in situ data analysis for mongolia drought. *Remote Sens.* **2016**, *8*, 509. [CrossRef]
33. Zeng, L.; Wardlow, B.D.; Wang, R.; Shan, J.; Tadesse, T.; Hayes, M.J.; Li, D. A hybrid approach for detecting corn and soybean phenology with time-series modis data. *Remote Sens. Environ.* **2016**, *181*, 237–250. [CrossRef]

3

Buffer-Mediated Effects of Clearcutting on In-Pool Amphibian Productivity: Can Aquatic Processes Compensate for Terrestrial Habitat Disturbance?

Jessica S. Veysey Powell * and Kimberly J. Babbitt

Department of Natural Resources and the Environment, University of New Hampshire, 56 College Road, Durham, NH 03824, USA; kbabbitt@unh.edu
* Correspondence: jsveysey@gmail.com

Academic Editor: Deanna H. Olson

Abstract: Natural resource extraction and wildlife conservation are often perceived as incompatible. For wetland-dependent amphibians, forested buffers may mitigate timber-harvest impacts, but little empirical research has focused on buffers around lentic habitats. We conducted a landscape experiment to examine how spotted salamander and wood frog reproductive output (i.e., eggmass and metamorph production) respond to clearcutting mediated by buffers of different widths (i.e., uncut, 30 m buffer, 100 m buffer) at ephemeral pools in an industrial forest. We found complex interactions between buffer treatment and reproductive output, which were strongly mediated by hydroperiod. Overall, reproductive output was most sensitive at 30 m-buffer pools and for salamanders, but responses diverged across productivity metrics even within these categories. Notably, for both cut treatments over time, while salamander eggmass abundance decreased, metamorph productivity (i.e., snout-vent length [SVL] and abundance) tended to increase. For example, average metamorph SVLs were predicted to lengthen between 0.2 and 0.4 mm per year post-cut. Additionally, typical relationships between reproductive output and hydroperiod (as indicated by the reference treatment) were disrupted for both species in both cut treatments. For example, long-hydroperiod pools produced more salamander metamorphs than short-hydroperiod pools in both the reference and 30 m-buffer treatments, but the rate of increase was lower in the 30 m-buffer treatment such that a long-hydroperiod pool in the reference treatment was predicted to produce, on average, 24 more metamorphs than a similar pool in the 30 m-buffer treatment. From a conservation perspective, our results highlight the importance of evaluating both terrestrial and aquatic responses to terrestrial habitat disturbance, since responses may be reinforcing (i.e., exert similarly positive or negative effects, with the potential for amplification in the aquatic habitat) or decoupled (i.e., operate independently or be negatively correlated, with responses in the aquatic habitat potentially dampening or counteracting responses in the terrestrial habitat).

Keywords: *Ambystoma maculatum*; eggmass; ephemeral pool; hydroperiod; *Lithobates sylvaticus*; metamorph; reproductive output

---

1. Introduction

Pressure to manage forests for multiple uses is increasing globally as climate change progresses, and renewable energy demands, human populations, and per capita land consumption grow [1]. Natural resource extraction and wildlife conservation are two uses that are frequently pitted against each other during inter-user discussions over forest management plans [2–4]. Clearcuts are an effective timber harvest technique for diversifying the age structure of trees within forests and for creating habitat for wildlife species that specialize in young forest habitats, but there is evidence that clearcuts negatively impact most amphibian species (e.g., [5–7]).

For wetland-dependent amphibians, forested buffers adjacent to wetlands hold promise as a simple technique for mitigating the impacts of clearcuts. Buffers effectively protect surface water quality and mitigate flooding [8–10]. In stream systems, where most manipulative buffer work has been conducted, riparian buffers have also been successfully used to mitigate the effects of clearcuts on multiple stream-dependent amphibian species (e.g., [11–13]). Considerably less research has focused on the wildlife value of buffers around lentic habitats [14,15]. While many stream-dependent amphibian species typically remain close to streams (i.e., within 100 m), some wetland-dependent amphibians regularly range hundreds of meters into the surrounding landscape and thus may respond very differently to buffers [16–18].

To address the lack of empirical research on buffer utility for wetland-dependent amphibians, we conducted a landscape-scale experiment that examined how amphibian communities in ephemeral pools in an industrial forest respond to clearcutting mediated by buffers of different widths. Previously, we showed that even in the presence of buffers, clearcuts exerted negative effects on breeding-adult population composition and size structure, but that wider buffers mitigated some of these effects; specifically: 100-m-wide buffers were associated with larger adult recapture rates and body size than were 30-m-wide buffers [14,15]. Here, we assess reproductive output (i.e., eggmass abundance and metamorph production) at these same pools, embedded within the same buffer treatments. We focus on two species that are aquatic as embryos and larvae, but leave their natal pools after metamorphosis, only to return during annual breeding events. The spotted salamander (*Ambystoma maculatum*) is long-lived and refrains from breeding during some years [19–21]. Conversely, the wood frog (*Lithobates sylvaticus*) typically lives less than half as long and may depend on annual reproductive output for population persistence [22–24]. These contrasting life-history strategies suggest that each species might respond uniquely to the different buffer treatments.

Though the relative importance of the aquatic phase may vary by species, it is important to determine whether effects on aquatic-phase productivity confirm or contrast with adult responses to buffer treatment for several reasons. First, for both species, regional populations persist through inter-pool dispersal and metamorphs are the primary dispersers [25–27]. Metamorph productivity thus constrains the number of dispersal propagules on the landscape. Second, metamorphs are a significant vector for transfer of energy and nutrients from the aquatic into the terrestrial system [28–30]. Finally, since eggmasses and larvae are confined to the pool basin and, unlike adults, cannot escape negative habitat conditions by moving to a different location, treatment impacts may manifest sooner or be more severe in these life stages.

Given their bi-phasic life histories, clearcutting could impact eggmass and metamorph production directly by changing environmental conditions in ephemeral pools and/or indirectly by altering the surrounding terrestrial habitat, causing adult amphibians to invest less of their energetic resources in reproduction and/or reducing breeding-population size through increased terrestrial-phase mortality and/or dispersal out of the population [31,32]. Ephemeral pools are complex systems, however, with dynamic food webs [28,33,34], competitive interactions [35–37], and high intra- and inter-annual variability [23,28,38]. A species' reproductive output depends not only on embryonic condition, but also on community composition [36,39,40]; conspecific density [23,41,42]; and environmental factors, like light, vegetative structure, hydroperiod, and temperature [43–45]. Negative larval density-dependence strongly shapes the number and condition of metamorphs emerging from pools [23,46,47]. Consequently, even if negative effects from the terrestrial environment carry-over into the aquatic phase via reduced eggmass quality (e.g., clutch size, egg lipid content), it is difficult to predict the ultimate effects on metamorph production. Negative terrestrial responses could be amplified in the aquatic phase or disrupted by complex in-pool interactions. Nevertheless, we made the following initial predictions for both species.

1. Eggmass abundance would increase with buffer width because eggmass number is positively correlated with adult abundance and wider buffers provide forested habitat for more adults.

2. The number of metamorphs produced and metamorph snout-vent length would be negatively correlated with buffer width because these metamorphic traits are strongly shaped by negative density dependence in the larval stage. Thus, if eggmass abundance increased with buffer width (due to increased adult abundance), larval density should increase, and metamorph productivity decrease, with buffer width.
3. Any treatment effects on eggmass and metamorph abundance and metamorph length would diminish over time as clearcuts regenerated. However, wood frogs would respond faster to regeneration than spotted salamanders, since wood frogs breed every year, whereas individual salamanders often skip years between breeding. Further, wood frogs can generate many times more metamorphs than spotted salamanders, during productive years [48].

2. Materials and Methods

We used eleven natural ephemeral pools from a private, industrial forest in east-central Maine, USA for this study. We randomly assigned each pool to one of three treatments: reference (i.e., uncut; $N = 3$), 100 m buffer ($N = 4$), or 30 m buffer ($N = 4$). During winter 2003–2004, we used clearcutting to create experimental buffers such that post-cut, pools in the two buffer treatments had, respectively, a 100-m or 30-m-wide upland buffer encircling the pool and a 100-m-wide concentric clearcut around the buffer (Figure 1).

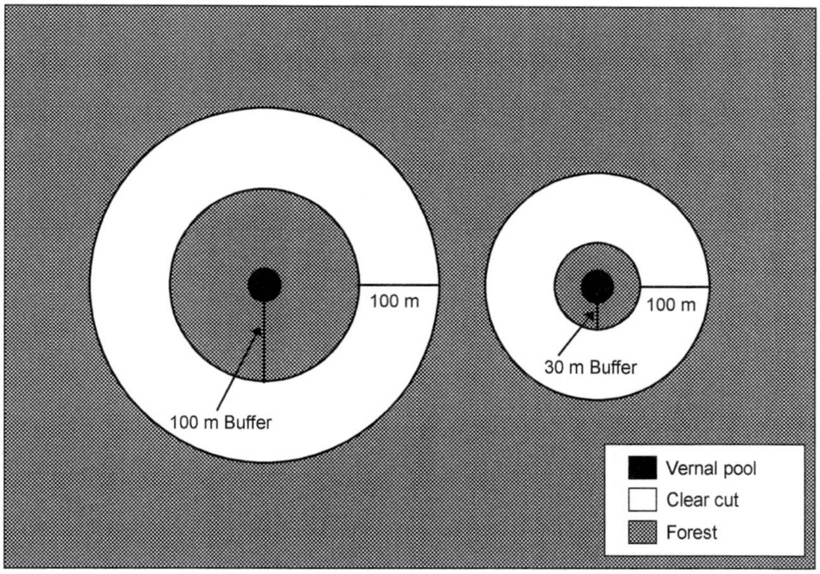

Figure 1. Experimental design implemented at 11 natural ephemeral pools in east-central Maine, USA. Undisturbed buffers of either 100 m (left; $n = 4$) or 30 m (right; $n = 4$) were left adjacent to pools and 100 m wide clearcuts were created around the buffers. Forest beyond the clearcut was undisturbed. No cutting occurred at reference pools (not shown; $n = 3$) [18].

From 2004 to 2009, we used a circular drift fence/pit fall trap array, installed within the forest/buffer and about 5 m upgradient from the high-water elevation of each pool, to capture, count, and sex all spotted salamanders and wood frogs emerging from the pools. We also measured the snout-vent/snout-urostyle (SVL) length of each emerging metamorph, except for brief periods in mid-July of 2004–2006 and 2008 when emerging metamorphs were excessively abundant at certain pools and we were unable to measure each individual. During these periods, we measured as many metamorphs as logistically possible.

We conducted annual eggmass counts for each species in each pool, starting in 2003 (i.e., pre-cut). Between 26 April and 20 May, we censused each pool for eggmasses on at least two occasions, with successive census dates for a given pool at least one week apart. Because eggmass integrity degrades relatively quickly, we were easily able to distinguish the old from the new eggmasses during the second census for each pool. For each species, we used the total number of novel eggmasses observed across the multiple sampling dates as the eggmass count for a given pool in a given year. We assume that eggmass detection approached 100% because the pools were relatively small and had clear water, allowing easy movement and high visibility throughout the pool during the censuses. Finally, we documented hydroperiod, or the number of days a pool held water between ice-out and the day the pool dried completely, for each pool in each year. For more details on the study site, experimental design, amphibian sampling, and hydroperiod determination, see Veysey Powell and Babbitt [14].

2.1. Statistical Analyses

To assess the relative impacts of forestry treatment and hydroperiod on spotted salamander and wood frog eggmass abundance and metamorph SVL, and on wood frog metamorph abundance, we developed linear mixed-effects regression models (LME) using the "lme" function in S-Plus 8.0 (Insightful Corporation, Seattle, WA, USA, 2007). Because salamander metamorph abundance was zero-inflated, we modeled this metric using a two-part process. During the first step, we used the "glme" function in the correlatedData library of S-Plus 8.0 to create a mixed-effects logistic regression model with a logit link that specified whether or not a given pool produced salamander metamorphs in a given year (hereafter: the Metamorph-Presence model). During the second step, we used LME to model metamorph abundance for those instances when abundance was greater than zero (hereafter: the Metamorph-Count model). To meet the assumptions of LME, we used $\ln(y + 0.5)$ as the outcome variable in all of the LME-based abundance analyses.

We treated year and pool ID as crossed random effects [49], except in the following cases. (1) For the wood frog SVL model, we treated pool ID and month-nested-within-year as crossed random effects. (2) If a model did not converge with crossed effects, we simplified the model to include either a random intercept for year or for pool, whichever provided a better model fit, as determined by likelihood ratio tests (LRTs). For this reason, we only used pool random intercepts in the salamander SVL and Metamorph-Presence models. We also modeled the variance-covariance structure for each regression to account for heterogeneous variance across groups and correlation among individuals from the same pool (Table S1). We used LRTs to optimize the variance-covariance structure of each model.

We included the following predictor variables in all models: buffer treatment, pool hydroperiod, an interaction between treatment and pool hydroperiod, and a pair of numeric dummy variables representing an interaction between treatment and study year. We used the first dummy variable (cut.year) to distinguish whether a pool was subjected to clearcutting or not. We used the second dummy variable (30 m.year) to indicate marginal impacts to the 30 m treatment pools. To represent hydroperiod in the eggmass models, we used both mean pool hydroperiod (i.e., the mean hydroperiod for each pool across the six study years) and the standard deviation of pool hydroperiod (as calculated for each pool across the six study years) as predictors, and included mean pool hydroperiod in the interaction. In the metamorph models, we used the current-year hydroperiod (i.e., the hydroperiod documented in a given [individual] study year) as the hydroperiod metric.

Additionally, we included the total number of metamorphs produced in a given pool in a given year as a predictor in the metamorph SVL models, and both breeding-female abundance and eggmass abundance as predictors in the metamorph-abundance models. We also wanted to include breeding-female abundance as a predictor in the eggmass-abundance models. However, we had eggmass data for 2003–2009 (the 2003 data were from the year preceding the clearcut), but only had female abundance data for 2004–2009. Thus, we present two alternative models per species for eggmass abundance. The first includes the pre-cut eggmass data from 2003 (in addition to the eggmass data from 2004 to 2009), but not female abundance (hereafter: the Pre-cut Eggmass model). The second includes female abundance and the eggmass data from 2004 to 2009, but not the 2003 eggmass data (hereafter: the Female Eggmass model).

We used analyses of variance (i.e., ANOVAs_ to determine the overall significance of each fixed effect and t tests to assess the significance of the different treatment levels ($\alpha = 0.05$). We used treatment contrasts to compare the reference treatment to each respective cut treatment. Based on the results of these initial inter-treatment comparisons, we subsequently performed a post-hoc comparison between the 100 m and 30 m treatments for the salamander metamorph SVL model. Based on an a priori decision, we removed the hydroperiod interaction from a model if the interaction was not significant and refit the model for the remaining fixed effects. Ultimately, we dropped the hydroperiod interaction from the salamander Metamorph-Presence model, and the wood frog metamorph and eggmass abundance models. In their final forms, all models satisfied their respective regression assumptions.

2.2. Compliance with Ethical Standards

We conducted all of the research in accordance with the rules of the Institutional Animal Care and Use Committee at the University of New Hampshire (IACUC-UNH). IACUC-UNH approved our research protocol, as detailed in permits: 020601 and 050604.

3. Results

Across the study period and the 11 ephemeral pools, we censused 3030 and 3877 eggmasses and counted 2645 and 26819 metamorphs for spotted salamanders and wood frogs, respectively. We provide descriptive statistics for the relevant demographic and hydroperiod data in Table 1. Two pools (one reference, one 100 m) failed to produce salamander metamorphs during the study, though salamander eggmasses were deposited in the reference pool every year and in the 100 m pool every year, except 2006. Notably, these two pools had the shortest mean hydroperiods (i.e., 45 and 66 days, respectively) of all pools. This reference pool also only produced 12 wood frog metamorphs during the study, though frog eggmasses were deposited in all years, except 2005 and 2006. Over all 65 year X pool cases (where "case" refers to the unique data generated for each species at a single pool in a single year), no metamorphs were produced in 37% of the frog and 12% of the salamander cases, and fewer than 10 metamorphs were produced in 17% of the frog and 31% of the salamander cases. All results presented in the rest of this section were statistically significant, unless otherwise indicated. (We use the phrase "tended to" to denote marginally significant results; i.e., $0.05 \leq p \leq 0.09$).

Table 1. Mean and variability for amphibian abundance and pool hydroperiod across the 11 study pools and 6 study years [a].

	Mean ± SE	**Range**		
hydroperiod (days)				
current-year	126 ± 7	0–211		
mean	126 ± 6	45–197		
standard deviation	32 ± 2	6–49		
	Spotted Salamander		Wood Frog	
	Mean ± SE	**Range**	**Mean ± SE**	**Range**
# Eggmasses				
Reference	35 ± 5	0–90	56 ± 10	23–235
100 m	52 ± 11	0–200	38 ± 6	0–105
30 m	31 ± 4	4–79	58 ± 11	2–235
# Metamorphs				
Reference	28 ± 11	0–137	587 ± 331	0–5905
100 m	60 ± 24	0–462	151 ± 73	0–1713
30 m	30 ± 10	0–195	558 ± 331	0–7691
SVL/SUL [b] (mm)				
Reference	26.3 ± 0.14	20–42	15.6 ± 0.03	8–28
100 m	29.3 ± 0.07	19–43	17.3 ± 0.04	10–32
30 m	27.6 ± 0.14	18–45	17.8 ± 0.02	10–33

[a] Eggmass data include one year of pre-cut data and six years of post-cut data; all other metrics are for six years of post-cut data; [b] Snout-vent length or snout-urostyle length, for salamander and wood frogs, respectively.

3.1. Spotted Salamanders

Salamander eggmass abundance declined over time in the cut treatments, as demonstrated by the significance of the "cut.year" parameter in the Pre-cut Eggmass model (Table S2; Figure 2). Though the mean decline rate for eggmass abundance was predicted to be about 9% per year, the most dramatic decline happened during the first year post-cut and in the 100 m treatment, a result derived by comparing results for "cut.year" across the two salamander eggmass models.

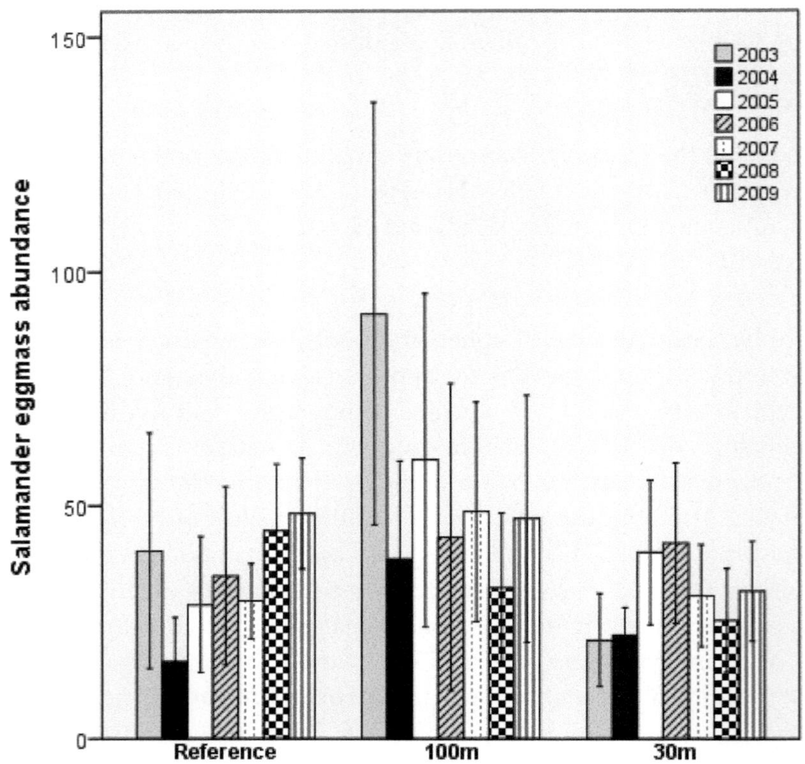

Figure 2. Mean (±1 SE) annual abundance of spotted salamander eggmasses deposited in 11 natural ephemeral pools across 3 experimental forestry treatments in east-central Maine, USA. Treatments were: reference (uncut), 100 m undisturbed buffer, and 30 m undisturbed buffer. Eggmass data from 2003 was collected prior to the experimental cut; eggmass data from 2004 through 2009 were collected post-cut.

As expected, the Female Eggmass model showed that the number of eggmasses deposited tended to increase with the number of breeding females. Otherwise, the two eggmass-abundance models were consistent: they both showed that eggmass abundance in 100 m pools was partly a function of mean-pool hydroperiod and that the standard deviation of mean-pool hydroperiod was negatively associated with eggmass abundance across all treatments. Specifically, while eggmass abundance in the reference treatment was not significantly related to mean hydroperiod, abundance in the 100 m treatment increased by about 2% for each additional day of mean hydroperiod and tended to be lower at short-hydroperiod 100 m (versus reference) pools (Figure 3). Conversely, for each additional day that hydroperiod varied inter-annually, eggmass abundance decreased by just over 3%.

Salamander metamorph abundance varied between pools and inter-annually. For instance, across all pools, we captured 1027 metamorphs in 2006, but only 44 in 2007; two pools (one reference and one 100-m buffer) did not produce metamorphs during the study. The Metamorph-Presence model, used to analyze whether or not individual pools produced salamander metamorphs in a given year, correctly classified 92% of cases. In both cut treatments, pools tended to be more likely to produce metamorphs as the cuts regenerated, such that for every additional year post-cut, the odds of a pool producing metamorphs increased by a factor of 3.2. Across all treatments, as expected, the pools were more likely to produce metamorphs when they held water for an extended period during the growing season. Specifically, for every additional day the pool held water in a given year, the odds of the pool producing any metamorphs increased by 7.4%.

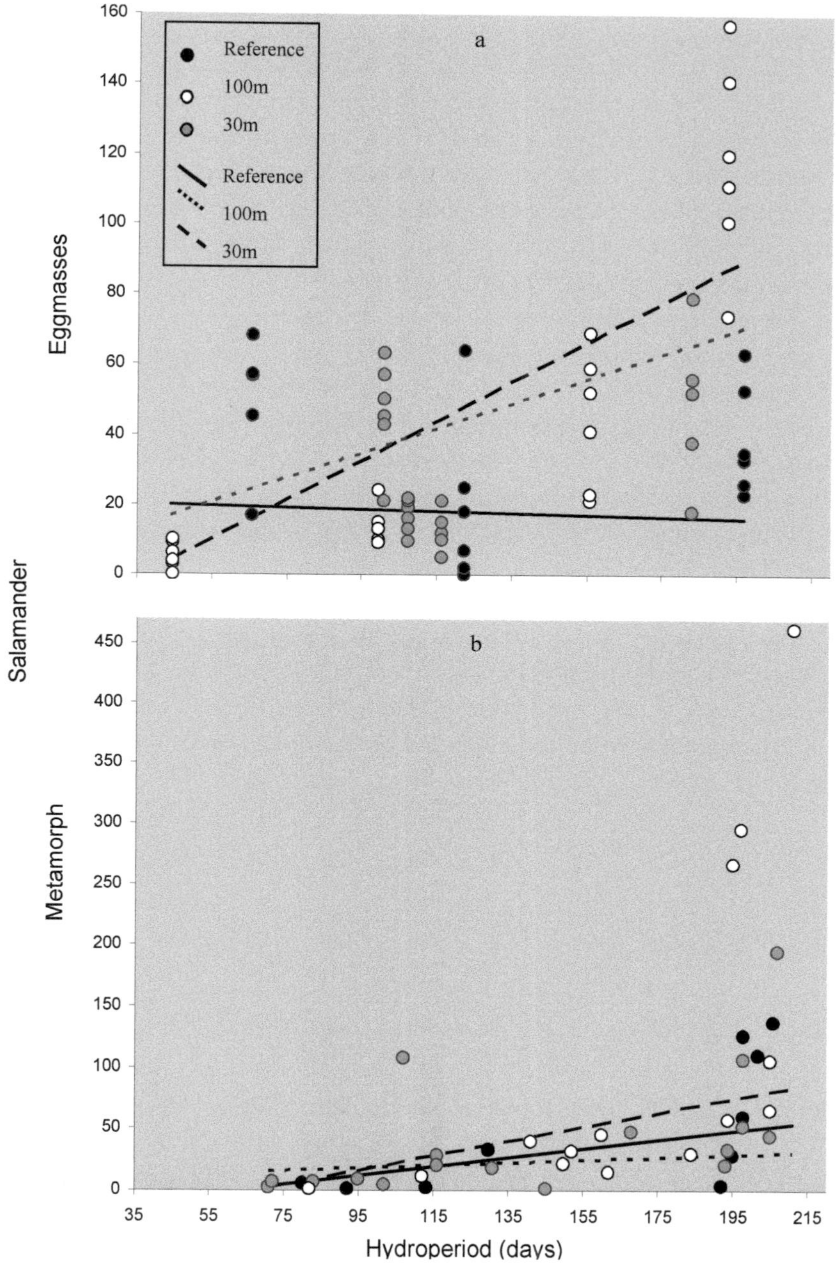

Figure 3. Spotted salamander abundance by hydroperiod (days) across 3 experimental forestry treatments at 11 natural ephemeral pools in east-central Maine, USA. Treatments were: reference (uncut), 100 m undisturbed buffer, and 30 m undisturbed buffer. (**a**) Eggmass abundance by mean-pool hydroperiod; (**b**) metamorph abundance, in pools that produced metamorphs, by current-year hydroperiod. Lines represent predicted values based on the fixed-effects regression coefficients.

Among the pools that produced metamorphs in a given year, the number of metamorphs that emerged depended, in the 30 m treatment, on both current-year hydroperiod and time since cut; and in all treatments, on eggmass abundance. On average, while metamorph abundance was predicted to increase by about 1.9% for every additional day of hydroperiod in the reference treatment, the production rate was significantly lower in the 30 m treatment, where metamorph abundance was predicted to increase by only about 0.44% per additional day that the pool held water (Figure 3). However, 30 m treatment pools with short hydroperiods tended to produce more metamorphs than short-hydroperiod reference pools. For example, and on average, we predicted that short-hydroperiod (e.g., 71 days) pools in the 30 m versus reference treatment would produce 16 versus 4 metamorphs, while long hydroperiod pools (e.g., 211 days) would produce 30 versus 54 metamorphs in the 30 m versus reference treatment. Additionally, metamorph abundance in the 30 m treatment tended to increase by about 0.45% for every year post-cut (i.e., on average, one additional metamorph was predicted at each 30 m pool over the entire study). Across all treatments, the number of metamorphs produced increased by about 1.3% for every additional eggmass deposited in the pool. For example, and on average, we predicted that a reference pool with 10 eggmasses would produce 8 metamorphs, while a similar pool with 100 eggmasses would produce 42 metamorphs. Finally, metamorph abundance was unrelated to breeding-female abundance.

Salamander metamorph SVL increased in the cut treatments over time, such that for every additional year post-cut, metamorphs were predicted to be 0.24 mm longer, on average (Figure 4). Conversely, the more metamorphs produced by a pool in a given year, the shorter each individual metamorph: for every additional metamorph emerging from the pool, the average metamorph length was predicted to decrease by 0.01 mm. In a single year in a reference pool, for instance, and on average, we predicted that a typical metamorph would have a SVL of 24.5 mm if 100 additional metamorphs were produced in the pool, but a SVL of 23.5 mm if 200 additional metamorphs were produced. Finally, while metamorph SVL in the reference treatment was unrelated to current-year hydroperiod, metamorph SVL in both cut treatments tended to increase with hydroperiod, but at a faster rate (i.e., by about 0.048 mm more per day) in the 100 m versus 30 m treatment. For example, and on average, we predicted that the SVL of a typical metamorph would increase from 24.8 mm to 28.8 mm in short (e.g., 71 days) versus long (e.g., 211 days) hydroperiod pools in the 30 m treatment, but would increase from 21.7 mm to 32.4 mm as hydroperiod extended by the same amount in the 100 m treatment.

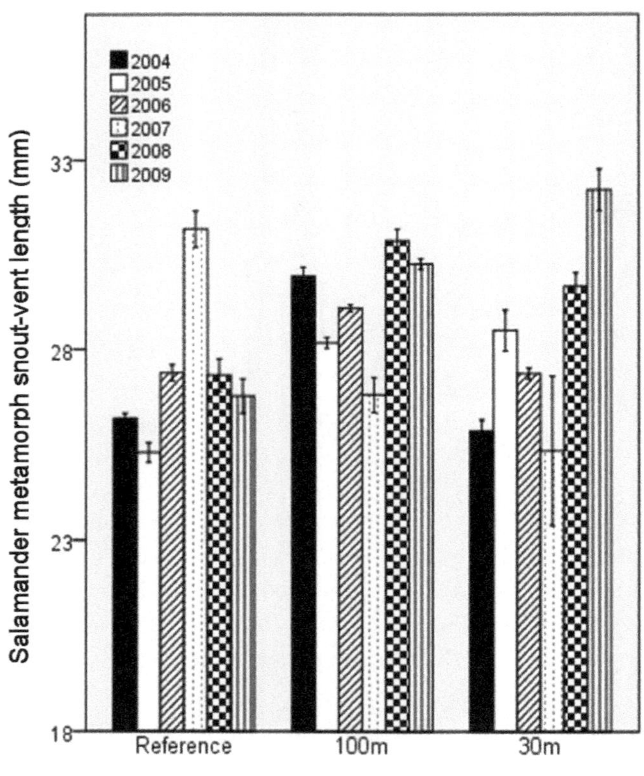

Figure 4. Mean (±1 SE) spotted salamander metamorph snout-vent length, across 3 experimental forestry treatments and 6 study years at 11 natural ephemeral pools in east-central Maine, USA. Treatments were: reference (uncut), 100 m undisturbed buffer, and 30 m undisturbed buffer.

3.2. *Wood Frogs*

Both eggmass models were consistent in that they showed no relationship between wood frog eggmass abundance and buffer treatment. According to the Female Eggmass model, the number of deposited eggmasses increased with the number of breeding females (by about 2% for each additional female), but decreased as the hydroperiod became more variable (by about 2% per additional day of variability). The Pre-cut Eggmass model, however, did not identify any significant relationships between the tested predictors and wood frog eggmass abundance.

Inter-pool and inter-annual wood frog metamorph production were highly variable. For example, during the five years that we collected metamorph data at one 30 m pool, the numbers of metamorphs produced were: 9, 51, 7691, 6, and 0. We observed a similar boom-and-bust pattern at one of the reference pools. Though abundance in the remaining pools never spiked higher than 1700 metamorphs, these pools nevertheless also had quite variable production.

In the 30 m treatment, wood frog metamorph production increased over time. On average, for every additional study year, about 71% more wood frog metamorphs were predicted from 30 m pools. It is important to stress, however, that this is the average pattern identified via the deployed linear modeling technique; this technique directly accounted for pool and year-specific heterogeneity in metamorph production, but individual pools did deviate from this average pattern in particular years. Finally, and in all treatments, wood frog metamorph production increased as current-year hydroperiod increased (by about 3.1% for every additional day that a pool held water).

Wood frog metamorph SVL varied with most of the tested predictors. Mean metamorph SVL increased over time in both cut treatments, but at a faster rate in the 100 m versus 30 m treatment (by about 0.4 mm and 0.2 mm per year, respectively; Figure 5). The relationship between frog metamorph SVL and treatment also varied with current-year hydroperiod. Whereas at reference pools, mean SVL increased by 0.01 mm per additional day of hydroperiod, mean SVL decreased with hydroperiod in both cut treatments. For every additional day a pool held water, the average metamorph was 0.024 mm and 0.003 mm shorter in the 30 m and 100 m treatments, respectively. Further, metamorphs at short-hydroperiod pools were significantly longer in the 30 m versus reference treatment. Finally, across all treatments, the more metamorphs produced by a pool in a given year, the shorter the average metamorph.

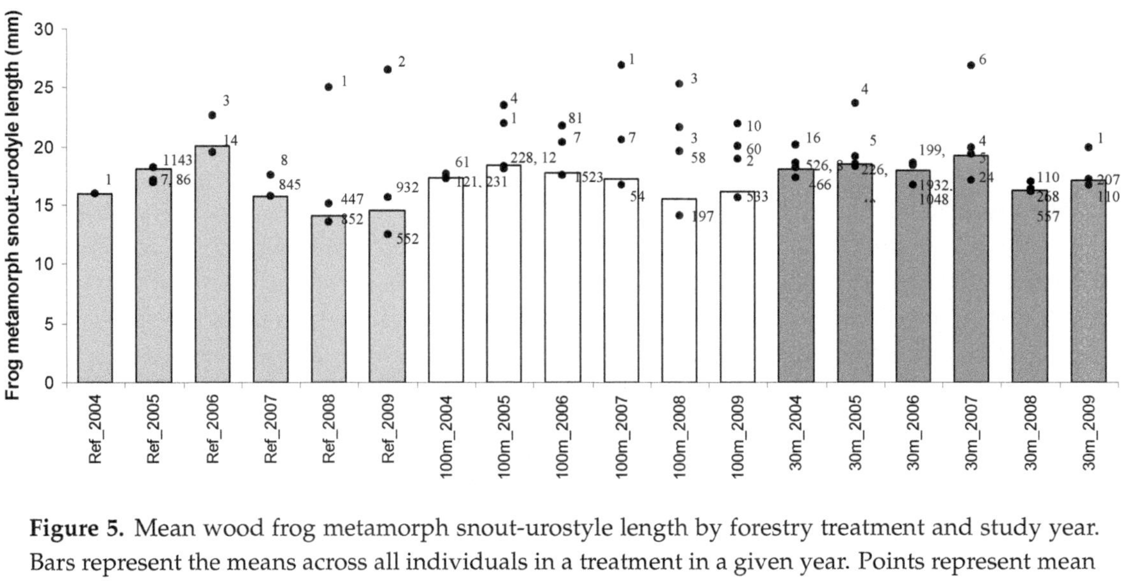

Figure 5. Mean wood frog metamorph snout-urostyle length by forestry treatment and study year. Bars represent the means across all individuals in a treatment in a given year. Points represent mean values for individual pools within each treatment in a given year, with the sample size for each pool provided by the number to the right of each point. These data are for populations at 11 natural ephemeral pools in east-central Maine, USA. Treatments were: reference (uncut), 100 m undisturbed buffer, and 30 m undisturbed buffer.

4. Discussion

Sub-adult amphibians can serve as indicators of aquatic environmental conditions, connect terrestrial and aquatic energy cycles, constitute the pool of dispersal propagules, and presage future population size. It is thus important to understand how buffer width interacts with forest harvesting to influence sub-adult amphibian productivity. Contrary to our predictions, we observed neither consistently positive relationships between buffer width and eggmass abundance nor consistently negative relationships between buffer width and metamorph productivity. Nor did we find that treatment effects neatly diminished over time. Instead, our results reveal complex interactions between buffer treatment and productivity, which were strongly mediated by hydroperiod and, potentially, other environmental conditions. Overall, productivity was most sensitive at 30 m pools and for spotted salamanders, but we found divergent responses across productivity metrics even within these categories. In particular, for both cut treatments over time, while salamander eggmass abundance decreased, metamorph productivity tended to increase. As in many amphibian systems, our data were highly variable and potentially stochastic. We acknowledge the difficulty in modeling such systems and recognize that our sample size (i.e., 11 pools) and limited study period (i.e., six years) may also have constrained our ability to model treatment effects on amphibian reproductive output. Despite these potential limitations, we note that mixed-effects regression is a powerful statistical technique that specifically allowed us to model heterogeneity across individual pools and years. Further, we emphasize that our models describe average conditions. Thus, while the raw data are variable across individual pools and years, we use the models to understand mean responses to experimental treatments. That we were able to discover robust, significant mean trends within highly variable data actually indicates the strength of those trends. In the following discussion, we explore the mechanisms that potentially explain these mean, modeled results.

4.1. Eggmass Abundance

Eggmass abundance is a useful productivity measure for two principle reasons. Foremost, it directly reflects the terrestrial conditions experienced by breeding adults and is a pathway for those conditions to influence the aquatic community. In terrestrial life-stages, both species exhibit plastic resource allocation between growth, reproduction, storage, and maintenance [19,50,51]. In response to stressful terrestrial conditions, like clearcuts, individuals may prioritize allocation to maintenance and growth, at the expense of reproduction, leading to delayed maturity or, for salamanders, an extended inter-breeding interval [15,52,53]. In turn, breeding-population and eggmass abundance (which are typically tightly and positively correlated; [54–56]) can decline. Second, eggmass abundance provides a rough approximation of the number of embryos deposited in a pool and thus signifies constraints on maximum metamorph production.

Buffer treatment was not a significant predictor of wood frog eggmass abundance, suggesting that the terrestrial impacts of clearcutting on adult frogs were not severe enough to be perpetuated in the aquatic environment, at least not in the form of fewer deposited eggmasses. In Veysey Powell and Babbitt [14], we showed that fewer mature wood frogs were recaptured in the 30 m treatment (compared to the 100 m treatment), but that this was supplemented by immigration, which buoyed the breeding population. The eggmass results imply that adult immigration offset any potential decline in eggmass productivity otherwise associated with the loss of mature, resident breeders.

By contrast, buffer treatment was a strong predictor of spotted salamander eggmass abundance. Eggmass abundance decreased in both cut-treatments over time, with this average trend driven primarily by a marked drop in the 100 m treatment immediately post-cut. Decreased eggmass abundance was likely a direct consequence of declining breeding-female abundance in the cut treatments, which was more severe in the 100 m treatment—this may be explained by a combination of reduced philopatry and increased mortality and temporary emigration—and is explored in Veysey Powell and Babbitt [14]. In the 100 m treatment, recaptured-breeding-female body condition also declined over time [15], suggesting that 100 m treatment females may have restricted reproductive investment by limiting per capita eggmass production.

The relationship between hydroperiod and salamander eggmass abundance was also altered in the 100 m treatment. Whereas hydroperiod was not a significant predictor in the reference treatment, in the 100 m treatment, eggmass abundance was positively correlated with mean-pool hydroperiod and, among short-hydroperiod pools, there tended to be fewer eggmasses. We observed a similar pattern for multiple metrics across the broader experiment, including for salamanders: metamorph SVL and breeding-adult abundance; and for breeding, female frogs: SVL of recaptured individuals and biomass of new-captured individuals [14]. Such repetition suggests impaired terrestrial habitat quality in the 100 m treatment that triggered a tight link between hydroperiod and fitness.

Hydroperiod typically exerts strong controls on salamander development during aquatic life stages and subsequently on metamorph production, with long-hydroperiod pools generally producing greater numbers of more robust metamorphs [41,57,58]. The lack of a significant relationship between eggmass abundance and mean-pool hydroperiod in the reference treatment, however, suggests that in undisturbed conditions, the terrestrial environment provided sufficient resources to support reproductive maturation, despite disparities in metamorph starting point, and/or facilitated sufficient immigration to offset such disparities. Comparatively, terrestrial conditions in the 100 m treatment may not have been conducive to compensatory growth of metamorphs and juveniles originating from short-hydroperiod pools, leading to reduced abundance of breeding adults and, subsequently, eggmasses. Further research is needed to understand whether hydroperiod effects on metamorph productivity influence subsequent adult resource allocation and eggmass production. Addressing the potential for such carry-over effects is beyond the scope of the present study. Finally, we cannot satisfactorily explain why we did not find a similarly altered hydroperiod relationship in the 30 m treatment. Since eggmass abundance largely mirrors breeding-adult abundance, however, we refer to Veysey Powell and Babbitt [14], for hypotheses that could explain the hydroperiod-abundance patterns for both breeding-adults and eggmasses.

As one of two aquatic life-stages, eggmasses are an essential link in understanding how terrestrial forestry treatments could influence population size. Our results indicate that the negative effects of cutting on terrestrial adults were perpetuated in the aquatic environment in salamander, but not frog, eggmass abundance. For salamanders, this could limit metamorph productivity and population size. Given the complexity of in-pool interactions and variability in embryonic maternal investment, however, other outcomes are possible, as explored in the next section. Ultimately, understanding how eggmass abundance responds to buffer treatment may be most valuable as confirmation of adult-abundance patterns.

4.2. Metamorph Productivity

Metamorphs represent population potential: they are the newest generation of amphibians and their abundance and condition can strongly influence (meta) population dynamics [25,59,60]. They also embody a significant transfer of energy and nutrients from wetland to upland systems [28–30]. These roles are especially pronounced for wood frog metamorphs, which are produced by the thousands in boom years, and because wood frogs, having a relatively short life-span, depend on recruitment for persistence [23,61,62]. While metamorph productivity is a function of eggmass quantity and egg condition [23,63,64], it is also influenced by in-pool conditions, like intra- and inter-specific competition, temperature, hydroperiod, predation, and refuge and food availability [43,63,65]. Consequently, buffer treatment can influence metamorph productivity via numerous pathways. We expected metamorph abundance and size to decrease with increasing buffer width, but our results were more complex. For both species and cut treatments, some results showed increased metamorph productivity over time, while others indicated altered hydroperiod dynamics.

4.2.1. Productivity over Time

Over time, wood frog metamorph abundance increased in the 30 m treatment, while metamorph SVL increased in both cut treatments (SVL increased more slowly in the 30 m versus 100 m treatment). For spotted salamanders, metamorph SVL increased with time in both cut treatments. Our salamander results also indicated marginally significant trends of increased metamorph-production probability in both cut treatments and greater metamorph abundance in the 30 m treatment, over time. In general, these patterns did not represent a recovery following post-cut reductions in abundance or SVL, but rather a genuine increase beyond metamorph productivity observed at the reference pools. While logistics inhibited complete examination of potential explanatory mechanisms, we used adult demographic data from the larger experiment to isolate the most likely mechanisms.

Among the adult data, the only relevant metrics that increased over time were: (a) SVL and mass of new-captured female wood frogs in both cut treatments; and in the 30 m treatment; (b) mass and body condition of recaptured female spotted salamanders; (c) mass of recaptured male salamanders; and (d) SVL, mass, and body condition of new-captured male salamanders [14,15]. Larger females could contribute to increased metamorph abundance and SVL by producing more eggs per eggmass and/or eggs with higher nutrient content [66–68]. Similarly, larger males could enhance metamorph production through increased egg-fertilization rates [69,70]. However, these adult metrics increased only after a post-cut decline and were predicted to recover to reference levels between four and 11 years post-cut. Thus, increased adult size and condition may partially, but not completely, explain metamorph productivity patterns.

For cut-treatment salamanders, declining eggmass abundance could help explain increasing metamorph SVL. For all treatments, the SVL model indicated negative density dependence among larval salamanders, whereas the Metamorph-Count model showed that eggmass and metamorph abundance declined in tandem. Thus, in the cut treatments, the decline in eggmass abundance could lead to fewer metamorphs, a release from intra-specific competition, enhanced conditions for metamorph growth, and longer metamorphs. We also attribute inter-treatment differences in the rate of increase for wood frog SVLs over time to variation in density dependence. Though negative density dependence was indicated for all treatments, only in the 30 m treatment did metamorph abundance increase with time. Thus, intra-specific competition was likely greater, and metamorph SVL more constrained, over time in the 30 m versus 100 m treatment. Given the relatively large mean annual predicted increase in wood frog metamorph abundance at 30 m treatment pools (i.e., 71% more metamorphs per year), it is surprising that competition did not, apparently, limit SVLs even more severely in the 30 m treatment. This may be attributed to the fact that predicted metamorph abundance at the start of the study was low (e.g., <25 metamorphs) for short-to-medium hydroperiod (e.g., 40–130 days) pools; thus, even an annual increase in abundance of 71% would not generate great numbers of competing larvae and metamorphs. Only at pools with longer hydroperiods did we predict large numbers of metamorphs (e.g., >200) throughout the study. Overall, competitive effects on SVL in long-hydroperiod pools may have been offset by a lack of competition in the short-medium hydroperiod pools.

4.2.2. Productivity and Hydroperiod

For both species, reaching metamorphosis requires balancing desiccation and predation risks. If a pool dries before larvae metamorphose, high larval mortality results [23,71]. If the hydroperiod is too long, larvae risk predation from an increasingly abundant and rich predator pool [65,72,73]. Inter and intra-specific competition also influence development and metamorph productivity across the hydroperiod spectrum [23,63,74]. In the absence of competition, both species typically achieve higher rates of metamorphosis, and salamanders attain greater metamorphic size, in pools with intermediate hydroperiods (i.e., >14 weeks, but non-permanent; [71,75,76]). (For wood frogs, minimal research explores how size varies with hydroperiod). Our reference-treatment metamorph-productivity results generally mirrored typical patterns observed in the absence of competition. In the cut treatments, by contrast, relationships between metamorph productivity and hydroperiod differed from those observed typically in the literature and in our reference treatment. Specifically, as hydroperiod increased: salamander metamorph abundance increased at a slower rate in the 30 m versus reference treatment; salamander metamorph SVL increased in the two cut treatments, but was unaltered in the reference treatment; and frog metamorph SVL decreased in the two cut treatments, but increased in the reference treatment. Salamander metamorph SVL also differed between the two cut treatments: SVL increased more quickly with hydroperiod in the 100 m versus 30 m-buffer treatment. Cumulatively, these productivity-hydroperiod results suggest altered in-pool dynamics between predation, competition, and hydroperiod in the cut treatments.

Though metamorph productivity is tightly linked with in-pool dynamics, we were logistically constrained from collecting sufficient data to test for inter-treatment differences in in-pool conditions. Nevertheless, we suggest that the following mechanisms, both linked to in-pool dynamics, most plausibly explain inter-treatment differences in metamorph productivity and merit further investigation. First, to distinguish reference from cut-treatment patterns: use of the buffers as refuge by larval-amphibian predators and competitors. Second, to distinguish between the cut-treatments: easier predator/competitor detection of, and stronger edge effects at, pools with the narrower 30 m buffer. We briefly explore each of these mechanisms below.

Previous research suggests that various larval-amphibian predators and competitors would seek refuge in the buffers, in response to stressful conditions in the clearcuts (e.g., reduced water availability, restricted movement permeability, increased wind velocities; [48,77,78]). If this led to increased predator/competitor colonization of the embedded pools, then metamorph productivity would decline in the pools. Predation could limit metamorph abundance via direct consumption of larval amphibians and could constrain both SVL and abundance by triggering anti-predator behavior, expressed as reduced larval activity and subsequently, reduced foraging and growth rate [55,79,80]. Competition, by comparison, could reduce productivity by limiting resources available for growth. We expect this refuge-effect to be strongest in long-hydroperiod pools because (a) predator abundance normally increases with hydroperiod [72,73,81]; and (b) key competitors may thrive with longer hydroperiods (at least across the hydroperiods recorded in our study; e.g., midges; [82,83]).

For both species, the relationship between metamorph productivity and hydroperiod likely deviated more strongly from reference conditions in the 30 m versus 100 m treatment for two reasons. First, it may have been easier for dispersing predators and competitors to detect and colonize pools across the narrower 30 m buffer [84,85]. Second, cut-associated edge effects could extend across the buffer and into pools in the 30 m, but not 100 m, treatment. Edge effects, including increased light and temperature, can extend 30–60+ m into forest patches [86–88]. Higher in-pool light and temperature can lead to greater food availability, through enhanced periphyton [and subsequently herbivorous invertebrate] growth, and faster growth rates for amphibian larvae [89–91]. Faster larval growth rates could result in reduced mortality, leading to higher larval densities and competition. Ultimately, however, more intense competition in the 30 m treatment could slow growth rates and limit SVLs, with potentially negative consequences for terrestrial-stage survival [23,92,93].

5. Conclusions and Conservation Implications

Given the negative effects that clearcuts exert on terrestrial, adult spotted salamanders and wood frogs [14,15], it is important to evaluate whether these effects are perpetuated or amplified during their aquatic life stages. In this study, we examined how buffer width affects eggmass and metamorph productivity. Overall, we found mixed effects of terrestrial habitat disturbance on these aquatic stages and species-specific responses.

The negative effects of cutting were perpetuated in the aquatic stages for spotted salamanders (i.e., declining eggmass abundance over time), but not wood frogs. Nevertheless, for both species, certain metamorph-productivity measures improved with time. This potentially indicates a degree of decoupling between terrestrial and aquatic habitats and compensation within the aquatic environment for disturbance in the terrestrial environment. It also suggests that the aquatic stages provide some insurance against terrestrial habitat disturbance for these species' populations. This key conclusion provides further insight into the mechanisms by which breeding-populations remained relatively resilient, despite poor adult response to terrestrial-habitat alteration [14,15]. The extent to which increased metamorph productivity subsequently enhances population fitness or individual dispersal success, thereby truly buffering populations against terrestrial habitat disturbance, merits further investigation, however. Similarly, it is important to determine whether increased metamorph productivity resulted from reduced intra-specific competition (i.e., a negative consequence of reduced eggmass abundance) or improved in-pool conditions (e.g., from edge effects and increased solar radiation in pools).

Despite the potentially beneficial decoupling of aquatic and terrestrial life-stages in terms of metamorph production over time, we emphasize that typical relationships between aquatic production and hydroperiod were disrupted for both species in both cut treatments. This disruption was especially prominent for salamanders and in the 30 m treatment. The insurance potential of the aquatic life stages may be weakened by the dependent relationship between reproductive output and hydroperiod, which is a changeable, landscape-specific factor.

Furthermore, the implications of decoupled metamorph production may vary with species. Spotted salamanders may be less dependent on metamorph production, being somewhat insulated from population threats by having long-lived adults capable of both temporary emigration and iteroparity [52,61]. Wood frogs may benefit more from decoupled production, since adults live just a few years and often only breed once, making population persistence more dependent on births and immigration [23,61]. Enhanced metamorph production may not always benefit wood frogs, however. If poor terrestrial habitat conditions, such as those created by cuts, inhibit metamorph growth and survival and/or instigate dispersal to other pools, then a pool might experience limited recruitment into its breeding population, despite abundant production of metamorphs. From Veysey Powell and Babbitt [14], we know that in the 30 m treatment, where wood frog metamorph abundance increased over time, the number of recaptured adults was persistently low. This implies that the metamorph surplus was not recruited into the breeding population, but instead died off and/or emigrated to other pools.

From a conservation perspective, the eggmass and metamorph results suggest three important objectives for these and other wetland-dependent species with complex life cycles. First, when developing conservation plans, managers must evaluate both terrestrial and aquatic responses to habitat disturbance, as responses may be either decoupled or reinforcing. Similarly, planners must assess both the effects of management actions (e.g., cutting) and how other key drivers mediate those effects. In undisturbed ephemeral pools, hydroperiod strongly structures community composition and reproductive output [65,72,94]. Our study demonstrates that hydroperiod remains a key driver in disturbed pool-systems, mediating the impacts of cutting on amphibian fecundity. Second, the focal species exhibited distinct responses to experimental treatments, with spotted salamanders displaying greater sensitivity, and wood frogs potentially more resilient, to cutting. Conservation planners should consider tailoring management strategies to accommodate such interspecies differences. Likewise, planners may need to customize strategies to reflect present and predicted hydroperiod regimes of individual pools. Finally, to minimize the negative effects of clearcuts on amphibian reproductive output, planners should consider retaining buffers >30 m wide, especially for spotted salamanders. Though we observed detrimental cutting effects in both buffer treatments, negative effects occurred more frequently in the 30 m versus 100 m treatment.

Acknowledgments: We acknowledge International Paper Company and Sustainable Forestry Technologies for their cooperation and permission to conduct research on their property. We thank Matt Baber, Andrew Cooper, Mark Ducey, and Jennifer Purrenhage for their research advice, our graduate students and field technicians for their dedicated service, and one anonymous reviewer for thoughtful comments. This project was supported by the: National Research Initiative of the USDA Cooperative State Research, Education, and Extension Service (grant numbers: 2003-35101-12922 and 2007-35101-18281); Northeastern States Research Cooperative; Robert and Patricia Switzer Foundation; University of New Hampshire Graduate School; University of New Hampshire Cooperative Extension; New Hampshire Agricultural Experiment Station (Scientific Contribution number: **2701**); and USDA National Institute of Food and Agriculture McIntire-Stennis Project (accession number: 0226124).

Author Contributions: K.J.B. conceived and designed the experiments; K.J.B. and J.S.V.P. performed the experiments; J.S.V.P. analyzed the data; K.J.B. contributed materials and analysis tools; J.S.V.P. wrote the paper.

Conflicts of Interest: The authors declare no conflict of interest. The funding sponsors had no role in the design of the study; in the collection, analyses, or interpretation of data; in the writing of the manuscript, or in the decision to publish the results.

References

1. United Nations Forum on Forests. Conclusions and recommendations for addressing key challenges of forests and economic development. In Proceedings of the Tenth Session of the United Nations Forum on Forests, Istanbul, Turkey, 8–19 April 2013; p. 9.
2. Wondolleck, J. *Public Lands Conflict and Resolution: Managing National Forest Disputes*, 2nd ed.; Springer: New York, NY, USA, 1991.
3. Winkel, G. When the pendulum doesn't find its center: Environmental narratives, strategies, and forest policy change in the US Pacific Northwest. *Glob. Environ. Chang.* **2014**, *27*, 84–95. [CrossRef]
4. Ringold, P.L.; Mulder, B.; Alegria, J.; Czaplewski, R.L.; Tolle, T.; Burnett, K. Design of an ecological monitoring strategy for the Forest Plan0020in the Pacific Northwest. In *Monitoring Ecosystems: Interdisciplinary Approaches for Evaluating Ecoregional Initiatives*; Busch, D.E., Trexwell, J.C., Eds.; Island Press: Washington, DC, USA, 2003; pp. 73–99.
5. Tilghman, J.M.; Ramee, S.W.; Marsh, D.M. Meta-analysis of the effects of canopy removal on terrestrial salamander populations in North America. *Biol. Conserv.* **2012**, *152*, 1–9. [CrossRef]
6. Morris, D.L.A.; Porneluzi, P.; Haslerig, J.; Clawson, R.L.; Faaborg, J. Results of 20 years of experimental forest management on breeding birds in Ozark forests of Missouri, USA. *For. Ecol. Manag.* **2013**, *310*, 747–760. [CrossRef]
7. Semlitsch, R.D.; Todd, B.D.; Blomquist, S.M.; Calhoun, A.J.K.; Gibbons, J.W.; Gibbs, J.P.; Graeter, G.J.; Harper, E.B.; Hocking, D.J.; Hunter, M.L.; et al. Effects of timber harvest on amphibian populations: Understanding mechanisms from forest experiments. *BioScience* **2009**, *59*, 853–862. [CrossRef]
8. Broadmeadow, S.; Nisbet, T.R. The effects of riparian forest management on the freshwater environment: A literature review of best management practice. *Hydrol. Earth Syst. Sci.* **2004**, *8*, 286–305. [CrossRef]
9. Mayer, P.M.; Reynolds, S.K.; McCutchen, M.D.; Canfield, T.J. Meta-analysis of nitrogen removal in riparian buffers. *J. Environ. Qual.* **2007**, *36*, 1172–1180. [CrossRef] [PubMed]
10. Castelle, A.J.; Johnson, A.W.; Conolly, C. Wetland and stream buffer size requirements—A review. *J. Environ. Qual.* **1994**, *23*, 878–882. [CrossRef]
11. Dupuis, L.; Steventon, D. Riparian management and the tailed frog in northern coastal forests. *For. Ecol. Manag.* **1999**, *124*, 35–43. [CrossRef]
12. Johnston, B.; Frid, L. Clearcut logging restricts the movements of terrestrial Pacific giant salamanders (*Dicamptodon tenebrosus* Good). *Can. J. Zool.* **2002**, *80*, 2170–2177. [CrossRef]
13. Vesely, D.G.; McComb, W.C. Salamander abundance and amphibian species richness in riparian buffer strips in the Oregon coast range. *For. Sci.* **2002**, *48*, 291–297.
14. Veysey Powell, J.S.; Babbitt, K.J. An experimental test of buffer utility as a technique for managing pool-breeding amphibians. *PLoS ONE* **2015**, *10*, e0133642. [CrossRef] [PubMed]
15. Powell, J.S.V.; Babbitt, K.J. Despite buffers, experimental forest clearcuts impact amphibian body size and biomass. *PLoS ONE* **2015**, *10*, e0143505.
16. Semlitsch, R.D.; Bodie, J.R. Biological criteria for buffer zones around wetlands and riparian habitats for amphibians and reptiles. *Conserv. Biol.* **2003**, *17*, 1219–1228. [CrossRef]

17. Olson, D.H.; Anderson, P.D.; Frissell, C.A.; Welsh, H.H., Jr.; Bradford, D.F. Biodiversity management approaches for stream-riparian areas: Perspectives for Pacific Northwest headwater forests, microclimates, and amphibians. *For. Ecol. Manag.* **2007**, *246*, 81–107. [CrossRef]
18. Veysey, J.S.; Babbitt, K.J.; Cooper, A. An experimental assessment of buffer width: Implications for salamander migratory behavior. *Biol. Conserv.* **2009**, *142*, 2227–2239. [CrossRef]
19. Wilbur, H.M. Propagule size, number, and dispersion pattern in *Ambystoma* and *Asclepias*. *Am. Nat.* **1977**, *111*, 43–68. [CrossRef]
20. Kinkead, K.E.; Otis, D.L. Estimating superpopulation size and annual probability of breeding for pond-breeding salamanders. *Herpetologica* **2007**, *63*, 151–162. [CrossRef]
21. Flageole, S.; Leclair, R., Jr. Etude démographique d'une population de salamandres (*Ambystoma maculatum*) à l'aide de la méthode squeletto-chronologique. *Can. J. Zool.* **1992**, *70*, 740–749. [CrossRef]
22. Redmer, M.; Trauth, S.E. *Rana sylvatica* LeConte, 1825, Wood Frog. In *Amphibian Declines: The Conservation Status of United States Species*; Lannoo, M.J., Ed.; University of California Press: Berkeley, CA, USA, 2005.
23. Berven, K.A. Factors affecting population fluctuations in larval and adult stages of the wood frog (*Rana sylvatica*). *Ecology* **1990**, *71*, 1599–1608. [CrossRef]
24. Berven, K.A. Population regulation in the wood frog, *Rana sylvatica*, from three diverse geographic localities. *Aust. J. Ecol.* **1995**, *20*, 385–392. [CrossRef]
25. Semlitsch, R.D. Differentiating migration and dispersal processes for pond-breeding amphibians. *J. Wildl. Manag.* **2008**, *72*, 260–267. [CrossRef]
26. Harper, E.B.; Patrick, D.A.; Gibbs, J.P. Impact of forestry practices at a landscape scale on the dynamics of amphibian populations. *Ecol. Appl.* **2015**, *25*, 2271–2284. [CrossRef] [PubMed]
27. Peterman, W.E.; Rittenhouse, T.A.G.; Earl, J.E.; Semlitsch, R.D. Demographic network and multi-season occupancy modeling of *Rana sylvatica* reveal spatial and temporal patterns of population connectivity and persistence. *Landsc. Ecol.* **2013**, *28*, 1601–1613. [CrossRef]
28. Regester, K.J.; Whiles, M.R.; Lips, K.R. Variation in the trophic basis of production and energy flow associated with emergence of larval salamander assemblages from forest ponds. *Freshw. Biol.* **2008**, *53*, 1754–1767. [CrossRef]
29. Regester, K.J.; Lips, K.R.; Whiles, M.R. Energy flow and subsidies associated with the complex life cycle of ambystomatid salamanders in ponds and adjacent forest in Southern Illinois. *Oecologia* **2006**, *147*, 303–314. [CrossRef] [PubMed]
30. Capps, K.A.; Berven, K.A.; Tiegs, S.D. Modelling nutrient transport and transformation by pool-breeding amphibians in forested landscapes using a 21-year dataset. *Freshw. Biol.* **2015**, *60*, 500–511. [CrossRef]
31. Pittman, S.E.; Semlitsch, R.D. Habitat type and distance to edge affect movement behavior of juvenile pond-breeding salamanders. *J. Zool.* **2013**, *291*, 154–162. [CrossRef]
32. Osbourn, M.S.; Connette, G.M.; Semlitsch, R.D. Effects of fine-scale forest habitat quality on movement and settling decisions in juvenile pond-breeding salamanders. *Ecol. Appl.* **2014**, *24*, 1719–1729. [CrossRef]
33. Wilbur, H.M. Experimental ecology of food webs: Complex systems in temporary ponds. *Ecology* **1997**, *78*, 2279–2302. [CrossRef]
34. Schriever, T.A.; Williams, D.D. Ontogenetic and individual diet variation in amphibian larvae across an environmental gradient. *Freshw. Biol.* **2013**, *58*, 223–236. [CrossRef]
35. Burley, L.A.; Moyer, A.T.; Petranka, J.W. Density of an intraguild predator mediates feeding group size, intraguild egg predation, and intra- and interspecific competition. *Oecologia* **2006**, *148*, 641–649. [CrossRef] [PubMed]
36. Relyea, R.A.; Auld, J.R. Predator- and competitor-induced plasticity: How changes in foraging morphology affect phenotypic trade-offs. *Ecology* **2005**, *86*, 1723–1729. [CrossRef]
37. Mott, C.L.; Maret, T.J. Species-specific patterns of agonistic behavior among larvae of three syntopic species of Ambystomatid salamanders. *Copeia* **2011**, *2011*, 9–17. [CrossRef]
38. Pechmann, J.H.K.; Scott, D.E.; Semlitsch, R.D.; Caldwell, J.P.; Vitt, L.J.; Gibbons, J.W. Declining amphibian populations: The problem of separating human impacts from natural fluctuations. *Science* **1991**, *253*, 892–895. [CrossRef] [PubMed]
39. Brodman, R. Intraguild predation on congeners affects size, aggression, and survival among Ambystoma Salamander larvae. *J. Herpetol.* **2004**, *38*, 21–26. [CrossRef]

40. Rubbo, M.J.; Shea, K.; Kiesecker, J.M. The influence of multi-stage predation on population growth and the distribution of the pond-breeding salamander, *Ambystoma jeffersonianum*. *Can. J. Zool.* **2006**, *84*, 449–458. [CrossRef]
41. Brodman, R. Effects of intraguild interactions on fitness and microhabitat use of larval Ambystoma salamanders. *Copeia* **1996**, *1996*, 372–378. [CrossRef]
42. Davis, A.K.; Maerz, J.C. Effects of larval density on hematological stress indices in salamanders. *J. Exp. Zool.* **2009**, *311A*, 697–704. [CrossRef] [PubMed]
43. Berven, K. The genetic basis of altitudinal variation in the wood frog *Rana sylvatica*. I. An experimental analysis of life history traits. *Evolution* **1982**, *36*, 962–983. [CrossRef]
44. Werner, E.E.; Glennemeier, K.S. Influence of forest canopy cover on the breeding pond distributions of several amphibian species. *Copeia* **1999**, *1999*, 1–12. [CrossRef]
45. Rowe, C.L.; Dunson, W.A. Impacts of hydroperiod on growth and survival of larval amphibians in temporary ponds of central Pennsylvania, USA. *Oecologia* **1995**, *102*, 397–403. [CrossRef]
46. Semlitsch, R.D.; Walls, S.C. Competition in two species of larval salamanders—A test of geographic-variation in competitive ability. *Copeia* **1993**, *1993*, 587–595. [CrossRef]
47. Anderson, T.L.; Whiteman, H.H. Non-additive effects of intra- and interspecific competition between two larval salamanders. *J. Anim. Ecol.* **2015**, *84*, 765–772. [CrossRef] [PubMed]
48. Popescu, V.D.; Patrick, D.A.; Hunter, M.L., Jr.; Calhoun, A.J.K. The role of forest harvesting and subsequent vegetative regrowth in determining patterns of amphibian habitat use. *For. Ecol. Manag.* **2012**, *270*, 163–174. [CrossRef]
49. Pinheiro, J.C.; Bates, D.M. *Mixed-Effects Models in S and S-PLUS*; Springer: New York, NY, USA, 2000.
50. Crespi, E.J.; Warne, R.W. Environmental conditions experienced during the tadpole stage alter post-metamorphic glucocorticoid response to stress in an amphibian. *Integr. Comp. Biol.* **2013**, *53*, 989–1001. [CrossRef] [PubMed]
51. Homan, R.N.; Regosin, J.V.; Rodrigues, D.M.; Reed, J.M.; Windmiller, B.S.; Romero, L.M. Impacts of varying habitat quality on the physiological stress of spotted salamanders (*Ambystoma maculatum*). *Anim. Conserv.* **2003**, *6*, 11–18. [CrossRef]
52. Bull, J.J.; Shine, R. Iteroparous animals that skip opportunities for reproduction. *Am. Nat.* **1979**, *114*, 296–303. [CrossRef]
53. Spotila, J.R. Role of temperature and water in the ecology of lungless salamanders. *Ecol. Monogr.* **1972**, *42*, 95–125. [CrossRef]
54. Crouch, W.B.; Paton, P.W.C. Using egg-mass counts to monitor wood frog populations. *Wildl. Soc. Bull.* **2000**, *28*, 895–901.
55. Savage, W.K.; Zamudio, K.R. *Ambystoma maculatum* (Shaw, 1802), Spotted Salamander. In *Amphibian Declines: The Conservation Status of United States Species*; Lannoo, M.J., Ed.; University of California Press: Berkeley, CA, USA, 2005; pp. 621–627.
56. Veysey, J.S.; Mattfeldt, S.D.; Babbitt, K.J. Comparative influence of isolation, landscape, and wetland characteristics on egg-mass abundance of two pool-breeding amphibian species. *Landsc. Ecol.* **2011**, *26*, 661–672. [CrossRef]
57. Scott, D.E. Effects of larval density in *Ambystoma opacum*: An experiment in large-scale field enclosures. *Ecology* **1990**, *71*, 296–306. [CrossRef]
58. Taylor, B.E.; Scott, D.E.; Gibbons, J.W. Catastrophic reproductive failure, terrestrial survival, and persistence of the marbled salamander. *Conserv. Biol.* **2006**, *20*, 792–801. [CrossRef] [PubMed]
59. Berven, K.A.; Grudzien, T.A. Dispersal in the wood frog (*Rana sylvatica*): Implications for genetic population-structure. *Evolution* **1990**, *44*, 2047–2056. [CrossRef]
60. Gamble, L.R.; McGarigal, K.; Compton, B.W. Fidelity and dispersal in the pond-breeding amphibian, *Ambystoma opacum*: Implications for spatio-temporal population dynamics and conservation. *Biol. Conserv.* **2007**, *139*, 247–257. [CrossRef]
61. Harper, E.B.; Rittenhouse, T.A.G.; Semlitsch, R.D. Demographic consequences of terrestrial habitat loss for pool-breeding amphibians: Predicting extinction risks associated with inadequate size of buffer zones. *Conserv. Biol.* **2008**, *22*, 1205–1215. [CrossRef] [PubMed]
62. Berven, K.A. Density dependence in the terrestrial stage of wood frogs: Evidence from a 21-year population study. *Copeia* **2009**, *2009*, 328–338. [CrossRef]

63. Wilbur, H.M. Competition, predation, and structure of *Ambystoma-Rana-sylvatica* community. *Ecology* **1972**, *53*, 3–21. [CrossRef]
64. Bennett, A.M.; Murray, D.L. Maternal body condition influences magnitude of anti-predator response in offspring. *Proc. R. Soc. B Biol. Sci.* **2014**, *281*, 20141806. [CrossRef] [PubMed]
65. Wellborn, G.A.; Skelly, D.K.; Werner, E.E. Mechanisms creating community structure across a freshwater habitat gradient. *Annu. Rev. Ecol. Evolut. Syst.* **1996**, *27*, 337–563. [CrossRef]
66. Scott, D.E.; Fore, M.R. The effect of food limitation on lipid levels, growth, and reproduction in the marbled salamander, *Ambystoma opacum*. *Herpetologica* **1995**, *51*, 462–471.
67. Berven, K.A. Factors affecting variation in reproductive traits within a population of wood frogs (*Rana sylvatica*). *Copeia* **1988**, *1988*, 605–615. [CrossRef]
68. Woodward, B.D. Local intraspecific variation in clutch parameters in the spotted salamander (*Ambystoma maculatum*). *Copeia* **1982**, *1982*, 157–160. [CrossRef]
69. Smith-Gill, S.J.; Berven, K.A. In vitro fertilization and assessment of male reproductive potential using mammalian gonadotropin-releasing hormone to induce spermiation in *Rana sylvatica*. *Copeia* **1980**, *1980*, 723–728. [CrossRef]
70. Chandler, C.H.; Zamudio, K.R. Reproductive success by large, closely related males facilitated by sperm storage in an aggregate breeding amphibian. *Mol. Ecol.* **2008**, *17*, 1564–1576. [CrossRef] [PubMed]
71. Shoop, C.R. Yearly variation in larval survival of *Ambystoma maculatum*. *Ecology* **1974**, *55*, 440–444. [CrossRef]
72. Batzer, D.P.; Wissinger, S.A. Ecology of insect communities in nontidal wetlands. *Annu. Rev. Entomol.* **1996**, *41*, 75–100. [CrossRef] [PubMed]
73. Tarr, T.; Baber, M.; Babbitt, K. Macroinvertebrate community structure across a wetland hydroperiod gradient in southern New Hampshire, USA. *Wetl. Ecol. Manag.* **2005**, *13*, 321–334. [CrossRef]
74. Holbrook, C.T.; Petranka, J.W. Ecological interactions between *Rana sylvatica* and *Ambystoma maculatum*: Evidence of interspecific competition and facultative intraguild predation. *Copeia* **2004**, *2004*, 932–939. [CrossRef]
75. Karraker, N.E.; Gibbs, J.P. Amphibian production in forested landscapes in relation to wetland hydroperiod: A case study of vernal pools and beaver ponds. *Biol. Conserv.* **2009**, *142*, 2293–2302. [CrossRef]
76. Phillips, C.A.; Johnson, J.R.; Dreslik, M.J.; Petzing, J.E. Effects of hydroperiod on recruitment of mole salamanders (genus *Ambystoma*) at a temporary pond in Vermilion County, Illinois. *Trans. Ill. State Acad. Sci.* **2002**, *95*, 131–139.
77. Semlitsch, R.D.; Conner, C.A.; Hocking, D.J.; Rittenhouse, T.A.G.; Harper, E.B. Effects of timber harvesting on pond-breeding amphibian persistence: Testing the evacuation hypothesis. *Ecol. Appl.* **2009**, *18*, 283–289. [CrossRef]
78. Whitaker, D.M.; Carroll, A.L.; Montevecchi, W.A. Elevated numbers of flying insects and insectivorous birds in riparian buffer strips. *Can. J. Zool.* **2000**, *78*, 740–747. [CrossRef]
79. Relyea, R.A. Morphological and behavioral plasticity of larval anurans in response to different predators. *Ecology* **2001**, *82*, 523–540. [CrossRef]
80. Schoeppner, N.M.; Relyea, R.A. Phenotypic plasticity in response to fine-grained environmental variation in predation. *Funct. Ecol.* **2009**, *23*, 587–594. [CrossRef]
81. Babbitt, K.J.; Baber, M.J.; Tarr, T.L. Patterns of larval amphibian distribution along a wetland hydroperiod gradient. *Can. J. Zool.* **2003**, *81*, 1539–1552. [CrossRef]
82. Brooks, R.T. Annual and seasonal variation and the effects of hydroperiod on benthic macroinvertebrates of seasonal forest ("vernal") ponds in central Massachusetts, USA. *Wetlands* **2000**, *20*, 707–715. [CrossRef]
83. Hanson, M.A.; Church, J.O.; Miller, A.T.; Palik, B.J.; Butler, M.G. Testing the efficacy of harvest buffers on the invertebrate communities in seasonal forest wetlands. In *Summaries of Wildlife Research Findings*; Wingate, P.J., Kimmel, R.O., Lawrence, J.S., Lenarz, M.S., Eds.; Minnesota Department of Natural Resources, Division of Fish and Wildlife, Wildlife Populations and Research Unit: St. Paul, MN, USA, 2005; pp. 164–179.
84. Miller, A.T. Aquatic Communities in Northern Forest Wetlands: Characteristics and Influences of Timber Harvest. Master's Thesis, North Dakota State University of Agriculture and Applied Science, Fargo, ND, USA, 2001.
85. Hanson, M.; Palik, B.; Church, J.; Miller, A. Influences of upland timber harvest on aquatic invertebrate communities in seasonal ponds: Efficacy of forested buffers. *Wetl. Ecol. Manag.* **2010**, *18*, 255–267. [CrossRef]

86. Chen, J.; Saunders, S.C.; Crow, T.R.; Naiman, R.J.; Brosofske, K.D.; Mroz, G.D.; Brookshire, B.L.; Franklin, J.F. Microclimate in forest ecosystem and landscape ecology. *BioScience* **1999**, *49*, 288–297. [CrossRef]
87. De Maynadier, P.G.; Hunter, M.L. Effects of silvicultural edges on the distribution and abundance of amphibians in Maine. *Conserv. Biol.* **1998**, *12*, 340–352. [CrossRef]
88. Hocking, D.J.; Babbitt, K.J.; Yamasaki, M. Comparison of silvicultural and natural disturbance effects on terrestrial salamanders in northern hardwood forests. *Biol. Conserv.* **2013**, *167*, 194–202. [CrossRef]
89. Halverson, M.A.; Skelly, D.K.; Kiesecker, J.M.; Freidenburg, L.K. Forest mediated light regime linked to amphibian distribution and performance. *Oecologia* **2003**, *134*, 360–364. [CrossRef] [PubMed]
90. Skelly, D.K.; Freidenburg, L.K.; Kiesecker, J.M. Forest canopy and the performance of larval amphibians. *Ecology* **2002**, *83*, 983–992. [CrossRef]
91. Batzer, D.P.; George, B.M.; Braccia, A. Aquatic invertebrate responses to timber harvest in a bottomland hardwood wetland of South Carolina. *For. Sci.* **2005**, *51*, 284–291.
92. Altwegg, R.; Reyer, H.-U. Patterns of natural selection on size at metamorphosis in water frogs. *Evolution* **2003**, *57*, 872–882. [CrossRef] [PubMed]
93. Scott, D.E. The effect of larval density on adult demographic traits in *Ambystoma opacum*. *Ecology* **1994**, *75*, 1383–1396. [CrossRef]
94. Herrmann, H.L.; Babbitt, K.J.; Baber, M.J.; Congalton, R.G. Effects of landscape characteristics on amphibian distribution in a forest-dominated landscape. *Biol. Conserv.* **2005**, *123*, 139–149. [CrossRef]

4

Natural Regeneration in a Multi-Layered *Pinus sylvestris-Picea abies* Forest after Target Diameter Harvest and Soil Scarification

Lars Drössler [1,*], Nils Fahlvik [1], Natalia K. Wysocka [1], Karin Hjelm [2] and Christian Kuehne [3]

1 Southern Swedish Forest Research Centre, Swedish University of Agricultural Sciences, Box 49, 230 53 Alnarp, Sweden; nils.fahlvik@slu.se (N.F.); natalia_wysocka@tlen.pl (N.K.W.)

2 Skogforsk; Ekebo 2250, 268 90 Svalöv, Sweden; karin.hjelm@skogforsk.se

3 School of Forest Resources, University of Maine; 5755 Nutting Hall, Orono, ME 04469, USA; christian.kuehne@maine.edu

* Correspondence: Lars.drossler@slu.se

Abstract: Forest management in Sweden can be characterized by even-aged silviculture heavily relying on three established harvest regimes: clearcutting, the seed-tree method, and the shelterwood system. Less intense, small-scale retention harvest systems such as single tree and group selection harvest are rarely used. In addition, natural regeneration dynamics without enrichment planting have barely been studied. Consequently, this study examined natural regeneration establishment in a multi-layered *Pinus sylvestris-Picea abies* forest stand in southwest Sweden after target diameter harvesting and soil scarification. The creation of forest canopy gaps had a positive effect on total seedling density five years after harvest, mainly due to a significantly higher number of *Betula pendula* individuals. Seedling density of more desirable tree species suitable for continuous cover forestry such as *Fagus sylvatica*, *Quercus petraea* and *Picea abies* also increased substantially in gaps when compared to pre-harvest conditions or the unharvested plots. In contrast, soil scarification did not increase the number of seedlings of desired tree species due to a significant decrease in *Picea abies* abundance. Soil moisture and gap size significantly improved *Betula pendula* seedling establishment while a larger number of *Quercus petraea* seedlings were observed in *Vaccinium myrtillus* patches. We conclude that canopy gaps are beneficial under the encountered stand conditions to initiate forest regeneration, and that soil scarification without the timely occurrence of a mast year of desired tree species is not effective in the type of forest studied.

Keywords: Sweden; continuous cover forestry; heterogeneous stand structure; regeneration density; seedling height growth

---

1. Introduction

Managed forests in northern Europe are dominated by even-aged, single-species stands with trees forming a single canopy layer [1,2]. In order to promote more heterogeneous stand structures and to obtain more resilient, multifunctional forest ecosystems, management approaches are required that complement the prevailing systems of even-aged silviculture [3]. Some alternative management approaches such as single-tree selection in boreal Norway spruce stands and diameter limited cutting with enrichment planting in mature coniferous stands have been examined in northern and central Sweden [4–7] while the options to apply natural regeneration in multi-layered stands have rarely been studied in southern Sweden. Potential stand development trajectories including regeneration and ingrowth dynamics thus are difficult to predict. The effect of silvicultural treatments on natural regeneration has to be estimated from shelterwood trials in Sweden [8–11] and selection cuttings in multi-layered stands in Norway and Finland [12–14]. On poor sites in northern Sweden or Finland, patch cuttings can complement alternative approaches to continuous cover forestry in mature pine stands [15–17]. In northern Germany, Tremer [18] estimated the ingrowth of new trees in mature pine stands that have been managed by target diameter harvest for the last three decades. Drössler [19] provided an overview of seedling densities, height growth and ingrowth rates reported in these studies. In addition, there are other relevant management strategies to promote more heterogeneous stand structures in central Europe and Denmark. Duncker et al. [20] summarizes them as close-to-nature forest management. They use natural forest development as an important reference for forest management and involve the transformation to more irregular forest structures mainly by target diameter harvest or group selection cutting in mature stands [21–24].

The most common managed forest types with multi-layered stand structures in southern Sweden are mature Scots pine (*Pinus sylvestris* L.) stands with advanced undergrowth of Norway spruce (*Picea abies* (L.) H. Karst.) and uneven-aged Norway spruce stands [25,26]. Mixed Scots pine-Norway spruce stands are the most frequent type of multi-layered forest in south Sweden [27], with fewer single-layered than multi-layered forest stands. Population density in southern Sweden is substantially higher compared with other regions in northern Europe, potentially affecting forest management of municipality woodlands and forested recreational areas. In fact, driven by public debate on common forestry practices, interest in the aforementioned mixed and structurally heterogeneous forest types as well as their sustainable management has increased in recent years. However, managing more heterogeneous stands is complex and alternative approaches within the concepts of continuous cover forestry or disturbance-based forest management [2,28] are thus needed to address the more diversified objectives of Swedish forest land owners [29,30]. These less intense, small-scale management approaches should be able to sustainably provide a variety of forest goods and services apart from timber, such as recreation and biodiversity values. In general, alternative methods to clearfelling and the seed-tree method are increasing in importance in the region.

By using an experiment laid out in a multi-layered Scots pine-Norway spruce forest with varying small-scale stand structure, we aimed to analyze the effects of soil scarification and partial, low intensity tree removal on natural regeneration patterns. The two main research questions were (i) How do varying site and forest conditions (e.g., stand structure, ground vegetation) influence seedling occurrence and density? and (ii) How do the implemented management measures affect regeneration characteristics?

2. Materials and Methods

2.1. Site and Initial Stand Characteristics

The study area is located 20 km east of Halmstad (56°42′02′′ N, 13°7′56′′ E, 115–140 m ASL) on a productive site in the northern temperate forest zone [31]. Mean annual precipitation is 1050 mm, mean annual air temperature is 7 °C and the number of days with a mean air temperature >5 °C (growing season) is 215. The soil type is a podzol developed over sandy moraine. The ground cover vegetation is mainly characterized by *Vaccinium myrtillus* and *Deschampsia flexuosa*. Site index (dominant height base age 100) is 32 m for Norway spruce and 28 m for Scots pine. Stand establishment was initiated by seeding Scots pine in 1912 to replace the *Calluna* spp. heathland while other tree species established naturally over time. Today, the stand is dominated by mature Scots pine trees in the overstory and Norway spruce trees of all size classes. Other species admixed in the under- and mid-story include sessile oak (*Quercus petraea*), silver birch (*Betula pendula*), European beech (*Fagus sylvatica*), rowan (*Sorbus aucuparia*), and aspen (*Populus tremula*).

The study stand was thinned several times before the establishment of the experiment. A detailed description of the site, the study stand, the forest structure, and the implemented experimental design can be found in [32].

2.2. Experimental Design and Data Collection

In spring 2009, the total study area of 19 ha was divided into three blocks, and four treatment areas of approximately 1 ha were randomly applied within each block. Treatments consisted of (1) control (CON) without any tree removal; (2) target diameter harvesting (TDH) where trees with a diameter at 1.3 m breast height (DBH) equal to or larger than listed in Table 1 were removed; (3) TDH complemented by soil scarification in canopy openings (TDH+); and (4) target diameter harvesting with slightly altered target diameters compared to TDH (TDH2) (Table 1). Target diameters applied in TDH and TDH+ were chosen according to economic criteria based on timber quality and average timber prices per tree species and product (e.g., sawlogs, pulpwood) published annually by the south Swedish forest owner association "Södra" for 2006. Modified target diameters in TDH2 were chosen in order to promote broadleaf tree species and to cut more Norway spruce trees. Harvesting took place in spring 2009, and soil scarification was carried out in autumn 2010 by disc-trenching (double row scarifier Bracke T21 mounted on a forwarder) resulting in approximately 0.5 m wide rows of exposed mineral soil ca. 2 m apart from each other. About one third of the area within each canopy gap was affected by this measure. The driver was instructed to scarify all canopy openings but to exclude areas with advanced natural regeneration. Detailed post-harvest stand characteristics for each treatment are provided in Table 2.

Table 1. Treatment-specific target diameter thresholds (DBH, cm) applied per tree species and timber quality class. Quality class 1 included trees with branches thinner than 6 cm while quality class 2 comprised low quality trees with thicker branches, spike-knots or forks.

Tree Species	TDH/TDH+		TDH2
	Class 1	Class 2	Class 1&2
Scots pine	40	30	40
Norway spruce	36	26	26
Silver birch	30	20	30
Sessile oak	60	30	60
European beech	50	30	50

Table 2. Mean and standard errors of post-harvest stand characteristics for the four experimental treatments.

Treatment	CON	TDH	TDH+	TDH2
N before harvest (trees·ha^{-1})	1151 ± 79	1045 ± 56	1082 ± 98	849 ± 69
N (trees·ha^{-1})	1151 ± 79	815 ± 64	905 ± 113	698 ± 81
BA (m^2·ha^{-1})	36 ± 3.6	19 ± 2.0	23 ± 2.4	20 ± 1.9
BA pine (%)	41	35	44	50
BA spruce (%)	44	52	40	28
BA oak (%)	7	18	12	14
BA birch (%)	7	1	3	5
BA others (%)	1	15	1	3
Vol (m^3·ha^{-1})	325 ± 27	160 ± 19	193 ± 17	164 ± 19
Removal (m^3·ha^{-1})	0	133 ± 24	155 ± 28	130 ± 24

Species and DBH of all living overstory trees (DBH $\geq$ 5 cm) were recorded on four systematically distributed circular overstory measurement plots (radius = 10 m) per treatment area before the harvest took place in 2009 and five years after the harvest in 2014. In addition, all canopy openings (termed gaps hereafter) were measured for the entire study area in 2009 and 2014. Gaps were defined as stand areas larger than 20 m^2 without overstory tree canopy cover in these pre- and post-harvest gap surveys. Gap size was calculated as the area of an ellipse using measured gap length and width while the size of non-ellipsoid gaps was estimated by several ellipse shapes or using polygons along the crowns of edge trees [33]. If a gap was divided by different treatments, each ellipsoid canopy opening was assigned to the dominant treatment.

Aiming at a total sampling area of 1000 m^2 to describe the initial regeneration of the stand, seedlings (height $\geq$ 20 cm and < 200 cm) and saplings (height $\geq$ 200 cm and DBH < 5 cm) were sampled in four systematically distributed circular permanent regeneration subplots (5 m^2) within each overstory measurement plot in summer 2009 (Figure 1a). The tree regeneration was resampled at the same locations but on larger subplots (10 m^2) five years later in summer 2014 (Figure 1b, Table 3). Subplot size was increased in 2014 in order to increase the total number of sampled seedlings and saplings necessary for species-specific treatment-wise comparisons. Additional information recorded for each subplot and inventory year included location specifics (skid road, scarified area, and gap ID if the plot center was located within a gap), ground vegetation type ("blueberry", "fern", "moss", and "grass"; the latter includes thin- and broad-leaved grasses according to [34]) and soil moisture class. The number of seedlings per tree species and height class (20–50, 50–100, 100–150, 150–200, and >200 cm) was recorded in summer 2009 and summer 2014. Also, browsing damage during the previous year, and annual terminal shoot length of the previous three years of the largest individual per tree species were measured on every regeneration subplot.

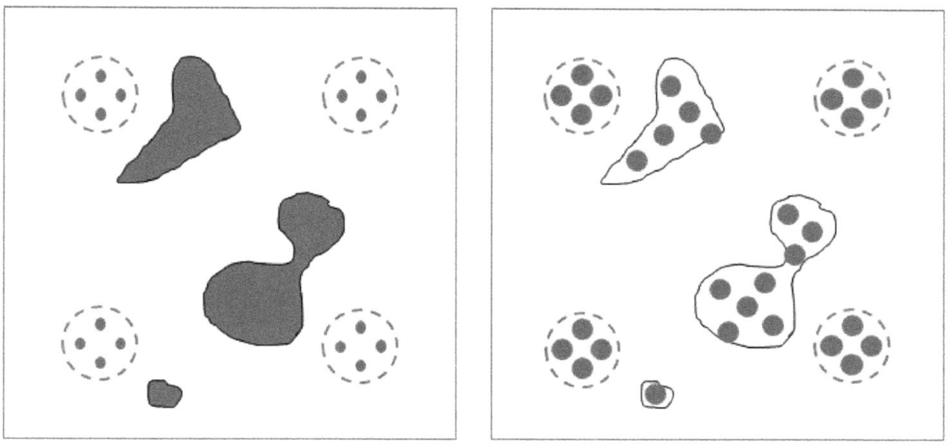

(a) (b)

Figure 1. Schematic depiction of different regeneration sampling schemes used in this study (grey = sampling area): (**a**) 5 m^2 circular sample plots systematically distributed within overstory measurement plots in 2009 (dashed line) and 100% inventory of all seedlings and saplings in canopy gaps 2009 (black line); (**b**) 10 m^2 sample plots systematically distributed within overstory measurement plots in 2014 (dashed line) and 10 m^2 sample plots systematically distributed within canopy gaps in 2014 (black line).

Table 3. Sampling method and number of regeneration plots with different sizes.

	2009		2014	
	Stand	Gaps	Stand	Gaps
Number of sample plots	192	-	192	636
Sampling area (m^2)	960	28 620	1 920	6 360
Number of sampled seedlings	29	2 862	351	5 215

In addition to the regeneration surveys conducted in overstory measurement plots, seedlings and saplings were also recorded within every canopy gap in 2009 and 2014 (Figure 1). Sampling in 2009 was a complete enumeration, i.e., all gap seedlings and saplings were measured to capture initial variation and diversity within the forest regeneration layer in canopy openings. Due to a significant increase in gap regeneration density as well as to harmonize the surveying schemes, gap sampling in

2014 followed the procedure described above for regeneration surveys in overstory measurement plots using systematically distributed circular permanent regeneration subplots (10 m^2). Therefore, a central regeneration subplot was established at the intersection point of the previously located length and width transects within each gap. Additional permanent regeneration subplots were established every 5 m along the length and width transect lines, respectively. As a result, the number of plots per gap varied depending on gap size. Regeneration surveys conducted in 2009 were considered pre-harvest, as they reflect conditions prior to tree cutting operations.

On average, 42%–45% of the 320 $m^3 \cdot ha^{-1}$ initial standing volume was removed by the first cutting in spring 2009. The pre-harvest total gap area was 3%, and the post-harvest total gap area amounted to 15% of the total stand area in 2009. About 6% of the total gap area of the stand was located in treatment CON, while gap area and sizes were evenly distributed in the three management treatments. Windthrow of individual trees was rare throughout the five-year study period and thus influenced gap characteristics only marginally. Gap size distribution and a more detailed description of the stand are given in [20]. The regeneration within the two experimental treatments TDH and TDH2 is jointly presented in this study, as preliminary analysis revealed very small differences in regeneration density and composition between the two target diameter harvest treatments. TDH+ with the additional soil scarification treatment, however, is presented separately.

Other information recorded for every stand and gap regeneration subplot comprised soil moisture class [35], vegetation type [34] and canopy closure class (0–33%, 34%–66%, 67%–100%) estimated using a spherical densiometer [36].

2.3. Data Analysis

Regeneration establishment was first quantified using seedlings 20–200 cm in height to evaluate overall density per treatment and in gaps. In addition, tree species composition and height growth were compared between treatments and in gaps. The regeneration sampled 2009 in overstory measurement plots was analyzed as a single stratum without the consideration of treatments applied in that year. Student's two-tailed *t*-test was applied to detect statistical differences in species-specific annual seedling height growth between gaps and treatment TDH/TDH+. The test was also used to compare seedling densities between non-scarified and scarified gaps in 2014, and to compare the 2014 gap regeneration subplots with the following subplot subsamples of the same inventory: (1) gaps >100 m^2; (2) scarified gaps >100 m^2; and (3) individual canopy closure classes.

To further gain insight into the factors influencing seedling establishment and relative species proportion, regeneration models were developed and evaluated using all regeneration subplot data. Due to the hierarchical structure (spatial correlation: regeneration subplots within overstory measurement plots within treatment areas within blocks) and the zero-inflation (i.e., a large number of subplots with no seedlings or saplings) of the combined regeneration data set, generalized linear mixed effects modelling was applied using the R package "glmmADMB" [37,38]. In a first step, a model analyzing total subplot seedling density was developed using standardized seedling density (# of seedlings per 10 m^2) because of the varying regeneration subplot size used in surveys during 2009 and 2014 [39]. In a second step, we developed species-specific relative proportion models using the subsample of stocked regeneration subplots only [40]. A species-specific relative proportion was calculated for each analyzed species as the subplot-specific percentage of the total number of seedlings and saplings. Zero-inflated negative binomial and binomial error structures were used in modeling steps 1 and 2, respectively [41]. Various nested random effect structures on the intercept (e.g., plots within treatment areas within blocks) were tested [42]. However, best model performance evaluated using Akaike's information criterion (AIC) was achieved with the inclusion of random block terms only. Potential predictor variables were first tested for collinearity. Subsequent preliminary analyses examining the predictive power of explanatory variables within each modeling step showed that gap size was a better predictor to quantify overstory tree density than canopy closure class. Total basal area could not be included because overstory tree data were not collected for gaps. Consequently, considered predictor variables included gap size (m^2; 0–1155 with 0 for regeneration subplots under closed canopy), time since harvest (years; 0 or 5), and various binomial indicator variables (0 or 1) namely soil scarification, skid road, soil moisture classes "moist" and "wet", as well as the vegetation types "blueberry", "fern", "grass", and "moss". Starting with a null model containing an intercept only, predictor variables were added stepwise and retained in the model if the inclusion showed a significant effect ($p < 0.05$) and resulted in improved model performance which was evaluated using AIC [42].

3. Results

3.1. Regeneration Establishment

Only 10% of the regeneration subplots contained seedlings in 2009. On average, the number of seedlings was 300 individuals per hectare in the entire stand, but increased in all treatments in 2014. After cutting, the number increased considerably to 1400 (TDH+) and 1800 (TDH) individuals ha^{-1} (Table 4). While Norway spruce dominated the seedlings initially, its relative proportion decreased substantially in the harvest treatments five years after cutting. The decrease was mainly caused by the large number of new birch and rowan seedlings. Dividing the regeneration into two height classes, the proportion of spruce in year 2014 was 12% of small seedlings (20–100 cm height) and 30% of larger seedlings (101–200 cm).

Table 4. Number of seedlings ha^{-1} and standard error by tree species (height 20–200 cm) before and after treatment.

Tree Species	2009		2014			
	Stand	Gaps	Stand			Gaps
			TDH	TDH+	CON	
Spruce	135 ± 42	422 ± 65	229 ± 144	104 ± 106	354 ± 427	929 ± 93
Oak	10 ± 69	36 ± 8	167 ± 160	83 ± 80	21 ± 41	263 ± 23
Beech	63 ± 53	37 ± 16		21 ± 41	125 ± 173	41 ± 10
Birch	10 ± 10	167 ± 31	771 ± 619	500 ± 400		5031 ± 271
Rowan	42 ± 21	220 ± 70	458 ± 243	375 ± 192	83 ± 80	925 ± 59
Pine		5 ± 2				149 ± 25
Willow		14 ± 11	42 ± 82	83 ± 80		357 ± 45
Aspen				21 ± 41		94 ± 18
Others	52 ± 34	98 ± 43	146 ±148	188 ± 183		472 ± 48
Total N·ha^{-1}	300 ± 80	1000 ± 158	1800 ± 440	1400 ± 270	600 ± 250	8300 ± 320
N of plots	192	174 gaps	48	48	48	636

In canopy gaps, on average 1000 seedlings ha^{-1} were found in 2009. The proportion of spruce was 50%. In total, the proportion of birch and rowan was higher in gaps than in the entire stand already and similar to the proportion of spruce (Table 4). Five years after harvest, an eightfold increase of total seedling density was observed, mainly due new birch seedlings (Table 4). The more shade-tolerant tree species spruce, oak and beech, increased in total from 208 to 306 individuals ha^{-1} in the stand on the overstory measurement plots. The seedling density after soil scarification in treatment TDH+ was lower with 224 individuals ha^{-1}. In gaps, the absolute density of these three tree species together increased from 450 to 1200 individuals ha^{-1} while their relative proportion decreased (Table 4).

3.2. Seedling Height Growth

While seedling number was large enough for comparisons between the stand and gaps in 2014, only Norway spruce provided enough seedlings to describe the height growth in the stand in 2009 and in the CON treatment in 2014 (Table 5). Annual terminal shoot length of spruce was 3 cm on average in 2009, and 5 cm in the CON treatment in 2014. In contrast, spruce height growth in gaps and the treatments TDH or TDH+ substantially increased to 10 cm per year in 2014. More pronounced differences between gaps in 2014 and the entire stand were found for broadleaf tree species and Scots pine (Table 5). Leader shoot growth in beech and pine was on average 5 cm greater in gaps compared to the treatments TDH/TDH+. A small difference was also found for oak, accompanied with the highest browsing damage (Table 5). Over all tree species, no effect of scarification in TDH+ on seedling height growth could be found.

Table 5. Top shoot length with standard error (cm per year) of the tallest individual per tree species and per gap or subplot of different inventories (if available; see Table 3) and initial browsing damage of the top shoot per tree species.

Tree Species	2009		2014			Browsing Damage
	Stand	Gaps	Stand		Gaps	
			TDH/TDH+	CON		2009
Spruce	3 ± 0.5	8 ± 0.6	10 ± 1.9	5 ± 0.7	10 ± 1.0	2%
Oak		8 ± 0.8	6 ± 0.8		8 ± 0.6	51%
Beech		11 ± 0.9	11 ± 2.7		16 ± 3.0	37%
Birch		18 ± 1.2	14 ± 1.3		17 ± 0.8	42%
Pine		8 ± 0.8	5.5 ± 0.7		11 ± 0.7	33%

3.3. Influence of Soil Scarification, Soil Moisture, Vegetation Type and Gap Size

Table 4 pointed out a somewhat lower total regeneration density in treatment TDH+ with soil scarification (1400 seedlings) compared with treatment TDH (1800 seedlings). In canopy gaps, however, there was still a reduction in the number of seedlings after scarification but this difference diminished with 8600 seedlings ha^{-1} in non-scarified gaps and 8200 seedlings ha^{-1} in scarified gaps. A statistically significant difference was found only for spruce ($p = 0.03$), when comparing the 900 spruce seedlings ha^{-1} in non-scarified gaps with 500 seedlings ha^{-1} in gaps with soil scarification. Pine and willow increased their small number by 25%–35% after scarification, but the differences were not statistically significant.

In gaps larger than 100 m^2, seedling density increased to 9200 ± 430 (SE) individuals ha^{-1} in 2014, but no significant difference compared with the total regeneration in gaps was found. The density in scarified gaps larger than 100 m^2 was 8400 ± 880 individuals ha^{-1}. Also in these gaps, a significant decrease was revealed for spruce with 490 individuals ha^{-1} (less than half the value in non-scarified larger gaps).

A lower total density of 6100 ± 730 seedlings was found on mesic micro-sites in large, scarified gaps (compared with the average in all gaps). On such sites, the number of birch seedlings (3200 ± 520) was lower than on moister sites (7300 ± 830). The same trend was found for all gaps surveyed in 2014, where fewer birch seedlings occurred on mesic sites than on moist sites. Norway spruce, on the contrary, was more abundant on mesic subplots.

Comparing the regeneration density in gaps in 2014 between the two dominant ground vegetation types "blueberry" and "grass" with total gap regeneration density, a higher seedling density of 9100 (±470) individuals ha^{-1} was found on the vegetation type "grass". Total seedling density on regeneration subplots dominated by blueberry was slightly lower than the average with 7400 (± 460) individuals ha^{-1}. As suggested in the respective proportion model, a higher number of oak seedlings was found in the blueberry type with 380 ± 40 individuals ha^{-1}, compared to 249 oak seedlings ha^{-1} across all vegetation types.

In Table 6, three different classes of light conditions in gaps were compared. In 2014, total seedling density was significantly higher in subplots with 0–33% canopy coverage (11000 ind. ha^{-1}), and significantly lower in subplots with 67%–100% coverage (6500 ind. ha^{-1}). At the species level, the number of birch, pine and oak seedlings was significantly lower in subplots in the highest canopy coverage class, while the number of spruce seedlings was higher in these subplots (Table 6).

Table 6. Effect of canopy closure on seedling density in forest gaps. *p*-values refer to the comparison with all 636 regeneration plots in gaps 2014 (Table 3).

Tree Species	Canopy Coverage											
	0–33%				34%–66%				67%–100%			
	Plot No.	N·ha^{-1}	SE	*p*	Plot No.	N·ha^{-1}	SE	*p*	Plot No.	N·ha^{-1}	SE	*p*
Spruce	121	**620**	110	0.04	298	970	140	0.83	217	1100	180	0.52
Oak	121	370	60	0.10	298	280	40	0.68	217	**180**	30	0.03
Birch	121	**7600**	770	<0.01	298	5100	400	0.94	217	**3600**	360	<0.01
Pine	121	170	50	0.79	298	220	50	0.16	217	**40**	20	<0.01
Total	121	**11,000**	890	<0.01	298	8500	470	0.67	217	**6500**	440	<0.01

Generalized linear mixed effects modelling revealed that time since harvest, gap size, and the vegetation type "grass" were among the most influential predictors for total seedling density. All three predictors exhibited a positive relationship (Table 7). In addition, skid road and soil moisture class "moist" also significantly increased regeneration density. No statistically significant effect was found for soil scarification on regeneration density. A significantly negative effect on the relative proportion of spruce was observed for gap size, soil scarification, skid road and vegetation type "grass". Increasing gap size also significantly decreased the relative proportions of beech and rowan. In contrast, gap size as well as soil moisture were positively correlated with the relative proportion of birch. Proportion models for pine, aspen and willow could not be derived due to their low total seedling number.

Table 7. Parameter estimates and significance levels of predictor variables including various soil moisture classes (SMC) and vegetation types (VT) of generalized linear mixed effects models for regeneration subplot-level total seedling density and relative species-specific proportion.

Predictor	Seedling Density		Relative Proportion									
			Beech		Birch		Oak		Rowan		Spruce	
Intercept	−0.7066	***	−4.2552	***	−2.7897	***	−3.8160	***	−0.9641	*	−1.1238	***
Time since harvest	0.4165	***							−0.2767	**		
Gap size	0.0007	***	−0.0218	*	0.0011	*			−0.0035	**	−0.0038	**
Soil scarification											−0.8925	**
Skid road	0.3894	***									−2.0437	*
SMC 'moist'	0.3521	***			0.9051	**						
SMC 'wet'					2.9427	***						
VT 'blueberry'					−0.6955	*	0.9130	*				
VT 'grass'	0.4002	***									−1.3305	***
VT 'moss'												
VT 'fern'			7.2066	**	1.6109	*			2.0296	*		

[1] Significance: *, $p \leq 0.05$; **, $p \leq 0.01$; ***, $p \leq 0.001$.

4. Discussion

This study is one of the first to analyze regeneration establishment without enrichment planting in managed multi-layered stands under transition towards an uneven-aged forest structure on fertile forest sites in southern Sweden. Target diameter harvesting, and the creation of gaps in particular, significantly increased seedling density over the five-year study period. This was mostly due to an increase in the number of birch seedlings. However, seedling density of other, more desirable species such as spruce, oak, and beech also significantly increased in studied canopy gaps. Soil scarification, in contrast, did not prove to be an effective management practice to improve regeneration establishment under the studied forest conditions. Gap size, soil moisture, and ground vegetation type were most influential in predicting species-specific relative proportions of the forest regeneration layer. The creation of gaps also increased annual height growth rates of the regeneration, in line with other regeneration studies in the boreal and temperate region of Europe [43–45].

4.1. Regeneration Establishment

Many ecological studies have emphasized the importance of canopy gaps for initiating regeneration processes and maintaining stand structural complexity in natural forests, including their role for pioneer tree species [2,43–47]. Our case-study highlighted the importance of canopy gaps for forest regeneration establishment in a managed forest, which has not been reported in previous studies in southern Sweden. The total regeneration density was on average approximately four times greater in gaps than elsewhere in the stand. Despite covering only 15% and 11% of the total stand area in 2009 after cutting and 2014, respectively, half of the seedlings were found in canopy gaps. Consequently, gap size was found to be crucial in initiating forest regeneration. Earlier work in boreal Scots pine forests in northern Sweden found that competition for nutrients is crucial [16,48], while studies in temperate Norway spruce stands highlighted the importance of light for successful regeneration [49]. In boreal Norway spruce single-tree selection forests of central Sweden, no relationship between seedling occurrence and canopy gaps was found [50]. Hence, the role of above- and below-ground competition remains unclear under southern Swedish conditions and for the study stand in particular.

Strong relationships between harvest removal levels and seedling occurrence as well as seedling height growth were reported for managed pine- and spruce-dominated stands in previous experimental trials across Sweden, in line with the findings for undisturbed soil in the present study. For example, Nilsson et al. [8] recorded approximately 2000 seedlings ha^{-1} five years after shelterwood cutting (95 and 155 shelter trees ha^{-1}) in a spruce-dominated stand on a similar site. However, the number of seedlings increased 2- to 3-fold after soil preparation. Irrespective of the soil disturbance, Norway spruce seedlings dominated the newly established regeneration layer [8]. Also, within an experimental shelterwood trial containing 22 mature pine stands across Sweden, an average 10,000 seedlings ha^{-1} were counted five to eight years after harvest, and almost 20,000 seedlings ha^{-1} with an additional soil preparation treatment [9]. Here, the difference in seedling density was mainly caused by a substantially higher number of birch seedlings, as observed in our study. In addition, positive significant effects of shelter and scarification on pine seedling density in comparison to clearcutting were revealed for the whole trial in Sweden, but not for the southern or northern region with a large natural variation between stands [9]. Even in a single stand, seedling density can range from 14,000 to 321,000 ha^{-1} depending on stand density [51]. Regeneration occurrence and survival are highly stochastic processes compared with individual tree growth dynamics following seedling establishment and are influenced by various factors apart from management-generated variation in canopy closure and soil disturbance [52]. However, in the majority of the above-mentioned studies, the number of seedlings was significantly lower in non-harvested areas, a finding that was verified in our study (Table 4). Only for spruce and beech did the absolute number remain constant.

4.2. Seedling Height Growth

A substantial increase in annual height growth was observed for spruce seedlings after harvest. The silvicultural intervention in general caused the difference as height growth was the same in gaps and post-harvest in the stand (Table 5). High browsing pressure by ungulates may have confounded the results for other tree species, especially oak.

Compared to the stand average, beech, birch and pine exhibited accelerated height growth in gaps. However, the absolute height growth of all seedlings was less than previously reported for advanced regeneration under shelterwood and in stands under transformation towards more heterogeneous stand structure [10,13,14,51]. On the other hand, the magnitude was larger than observed for newly established seedlings under shelterwood [8,9] and in boreal Norway spruce single-tree selection stands [4,5,53]. Only a southern site in central Sweden [4] showed comparable annual height growth rates that ranged from 1–19 cm on average. In addition, further increases in height growth rates can be expected in subsequent years given results reported from older experiments [12,13].

4.3. *Influence of Soil Scarification, Soil Moisture, Vegetation Type and Gap Size*

Soil scarification did not increase seedling density but instead had a significant negative effect on the number of spruce seedlings (Tables 4 and 7). While rather unexpected, [9] reported similar findings for spruce regeneration in shelterwood cuttings. No frost heaving damage was observed after scarification and thus can be excluded as a potential cause for the decline [54]. We therefore consider unintended mechanical damage from soil scarification as the main reason for the reduction of spruce seedling density. As the less browsed, initially more abundant, and shade-tolerant Norway spruce can be considered as one of the important tree species to constitute the future forest, the additional soil scarification treatment was not effective in improving conditions for future ingrowth. Beside mechanical damage to existing regeneration, the delayed timing and insufficient seed crops following soil preparation may explain the poor results of this measure [55]. Given more promising findings in earlier experimental shelterwood trials using various methods of soil preparation [8,56,57], additional scarification measures should be planned only when a lack of regeneration becomes evident and a mast year of desired species is expected.

As shown in previous studies, gap size increased seedling density in birch and pine [58,59]. Despite the small number of pine seedlings found in our study, seed years provide opportunities for successful Scots pine regeneration, especially on drier sites, in larger gaps or by reducing overstory canopy levels rapidly [56].

In contrast and likely in addition to the lack of seed trees, high frost sensitivity of the late successional beech is likely to explain the negative relationship with gap size as temperature extremes occur, especially in larger canopy openings [60]. In a beech shelterwood experiment in southern Sweden, a higher density of beech seedlings was found on scarified soil than on undisturbed soil [61]. Karlsson and Nilsson [57] concluded that scarification and shelterwood cutting are efficient tools to promote natural regeneration under conditions similar to this study. The authors highlighted the importance of sufficient seed supply and concluded that wind-dispersed tree species including birch and willow were favored by scarification, while animal-dispersed tree species such as beech and oak were more favored by shelterwood than by clearcutting [57].

Concerning other factors, a significantly higher density of birch seedlings was found with increasing soil moisture, while spruce tended to be less frequent under such conditions (Table 7). Although Norway spruce is well adapted to moist sites, the competition with birch or other vegetation can be a reason for the observed lower density in spruce. Furthermore, a significantly higher proportion of oak was observed in blueberry patches. This could be related to the seed dispersal activities of the Eurasian jay (*Garrulus glandarius*), known to preferably hide acorns under dwarf shrub vegetation [62].

5. Conclusions

The observed increase in post-harvest seedling density to around 1500 ha^{-1} after target diameter harvest provides sufficient ingrowth potential for sustainable management if future target diameter harvests will have similar positive effects on regeneration. Subsequent inventories in coming years will show if the current, rather undesired high proportion of birch will decrease substantially as a result of the observed and ongoing rapid gap closure [20,58,63]. Consequently, future management activities should promote recruitment of late-successional species as phytosociological and silvicultural experiences point to the long-term disappearance of light-demanding tree species under close-to-nature management practices [60,64].

We conclude that heterogeneously structured stands like the one studied here appear to provide feasible silvicultural options to develop uneven-aged stand structures [24]. Under such a management goal, a more explicit spatial consideration of advanced regeneration and the creation of canopy gaps become crucial silvicultural tools to enhance stand regeneration processes [65,66]. Schütz [24] highlighted that transformation to more irregular stand structures needs time and frequent interventions. To develop the study stand into an uneven-aged single-tree selection forest with a sustainable supply of smaller trees [32,67] therefore, will take many decades. Aiming at the transformation to a spruce-dominated single-tree selection forest, more attention should be paid to tree species with vital growth under varying levels of canopy closure. Given our findings and current management conditions in the study region, Norway spruce, European beech and sessile oak appear to be the most suitable species in this regard [26,68]. To balance removal and ingrowth of trees over time, both gaps to initiate or release regeneration and closed canopy sections without regeneration are necessary, as this will assure the continuous supply of regeneration and mature, harvestable trees.

Acknowledgments: The study was supported by the Swedish research foundations Skogssällskapet and Ekfrämjandet, the research program TC4F at SLU, and the Southern Swedish Forest Research Centre. We thank two anonymous reviewers for their valuable comments and Ulf Johansson and the staff at the experimental forest in Tönnersjöheden for their work and care pertaining to this experiment.

Author Contributions: L.D. and N.K.W. carried out the regeneration and gap inventories. L.D., N.K.W. and C.K. analyzed the data and all co-authors collaborated in writing and revising the manuscript.

Conflicts of Interest: The authors declare no conflict of interest.

References

1. Axelsson, R. Forest Policy, Continuous Tree Cover Forest and Uneven-Aged Forest Management in Sweden's Boreal Forest. Master's Thesis, Department of Forest Products, Swedish University of Agricultural Sciences, Uppsala, Sweden, 2008.
2. Kuuluvainen, T. Forest management and biodiversity conservation based on natural ecosystem dynamics in northern Europe: The complexity challenge. *Ambio* **2009**, *38*, 309–315. [CrossRef] [PubMed]
3. Puettmann, K.J.; Wilson, S.; Baker, S.; Donoso, P.; Drössler, L.; Amente, G.; Harvey, B.D.; Knoke, T.; Lu, Y.; Nocentini, S.; et al. Silvicultural alternatives to conventional even-aged forest management—What limits global adoption? *For. Ecosyst.* **2015**, *2*, 8. [CrossRef]
4. Nilson, K.; Lundqvist, L. Effect of stand structure and density on development of natural regeneration in two *Picea abies* stands in Sweden. *Scand. J. For. Res.* **2001**, *16*, 253–259. [CrossRef]
5. Chrimes, D.; Nilson, K. Overstorey density influence on the height of *Picea abies* regeneration in northern Sweden. *Forestry* **2005**, *78*, 433–442. [CrossRef]
6. Hagner, M. Naturkultur. Ekonomiskt Skogsbruk kännetecknat av Befriande Gallring och Berikande Plantering (Naturkultur—Forest Management with Single Tree Release and Enrichment Planting). UBICON: Umeå, Sweden, 2004; ISBN: 91-631-5010-7. (In Swedish)
7. Wikberg, P.-E.; Lundmark, T. *Naturkultur—Utvecklingen i Försöksserien de 10 Första åren (Naturkultur—The Development of the Study Trial after 10 Years)*; Rapport 23; Skogsstyrelsen: Jönköping, Sweden, 2008. (In Swedish)
8. Nilsson, U.; Gemmel, P.; Johansson, U.; Karlsson, M.; Welander, T. Natural regeneration of Norway spruce, Scots pine and birch under Norway spruce shelterwoods of varying densities on a mesic-dry site in southern Sweden. *For. Ecol. Manag.* **2002**, *161*, 133–145. [CrossRef]
9. Nilsson, U.; Örlander, G.; Karlsson, M. Establishing mixed forests in Sweden by combining planting and natural regeneration—Effects of shelterwoods and scarification. *For. Ecol. Manag.* **2006**, *237*, 301–311. [CrossRef]
10. Glöde, D. Survival and growth of Picea abies regeneration after shelterwood removal with single and double-grip harvester systems. *Scand. J. For. Res.* **2002**, *17*, 417–426. [CrossRef]
11. Holgén, P.; Hånell, B. Performance of planted and naturally regenerated seedlings in *Picea abies* dominated shelterwood stands and clearcuts in Sweden. *For. Ecol. Manag.* **2000**, *127*, 129–138. [CrossRef]
12. Granhus, A.; Brække, F.H. Nutrient status of Norway spruce stands subjected to different levels of overstorey removal. *Trees* **2001**, *15*, 393–402. [CrossRef]
13. Eerikäinen, K.; Valkonen, S.; Saksa, T. Ingrowth, survival and height growth of small trees in uneven-aged Picea abies stands in southern Finland. *For. Ecosyst.* **2014**, *1*, 10. [CrossRef]
14. Lin, C.J.; Laiho, O.; Lähde, E. Norway spruce (*Picea abies* L.) regeneration and growth of understory trees under single-tree selection silviculture in Finland. *Eur. J. For. Res.* **2012**, *131*, 683–691. [CrossRef]
15. Erefur, C. Regeneration in Continuous Cover Forestry Systems. Doctoral Thesis, Swedish University of Agricultural Sciences, Acta Universitatis Agriculturae Sueciae, Uppsala, Sweden, 2010; p. 42.

16. Axelsson, E.P.; Lundmark, T.; Högberg, P.; Nordin, A. Belowground competition directs spatial patterns of seedling growth in boreal pine forests in Fennoscandia. *Forests* **2014**, *5*, 2106–2121. [CrossRef]
17. Valkonen, S.; Siitonen, J. Tree regeneration in patch cutting in Norway spruce stands in northern Finland. *Scand. J. For. Res.* **2016**, *31*, 271–278. [CrossRef]
18. Tremer, N. Untersuchungen zur Verjüngung von Waldbeständen in Nordwestdeutschland (Study on Regeneration of Forest Stands i Nortwestern Germany. Ph.D. Thesis, University of Göttingen, München, Germany, 2008.
19. Drössler, L.; Ekö, P.M.; Balster, R. Short-Term development of a multi-layered forest stand after target diameter harvests in southern Sweden. *Can. J. For. Res.* **2015**, *45*, 1198–1205. [CrossRef]
20. Duncker, P.S.; Barreiro, S.M.; Hengeveld, G.M.; Lind, T.; Mason, W.L.; Ambrozy, S.; Spiecker, H. Classification of forest management approaches: A new conceptual framework and its applicability to European forestry. *Ecol. Soc.* **2012**, *17*, 51. [CrossRef]
21. Spellmann, H. Zielstärkennutzung: Waldbauliche und ertragskundliche Aspekte. (Target diameter cutting: Silvicultural and yield aspects.). Berichte zur Jahrestagung der Sektion Ertragskunde, Deutscher Verband Forstlicher Forschungsanstalten: Grünberg, Germany, 1997; pp. 186–198. (In German)
22. Sterba, H.; Zingg, A. Target diameter harvesting—A strategy to convert even-aged forests. *For. Ecol. Manag.* **2001**, *151*, 95–105. [CrossRef]
23. Schütz, J.-P. Opportunities and strategies of transforming regular forests to irregular forests. *For. Ecol. Manag.* **2001**, *151*, 87–94. [CrossRef]
24. Larsen, J.B. Close-to-Nature Forest Management: The Danish Approach to Sustainable Forestry. In *Sustainable Forest Management - Current Research*; Garcia, J.M., Casero, J.J.D., Eds.; InTech: Rijeka, Croatia, 2012.
25. Lundqvist, L. Blädning i Granskog—Strukturförändringar, Volymtillväxt, Inväxning och Föryngring på Försöksytor Skötta med Stamvis blädning. Ph.D. Thesis, Department of Silviculture, Swedish University of Agricultural Sciences, Umeå, Sweden, 1989. (English Chapters).
26. Drössler, L.; Fahlvik, N.; Elfving, B. Application and limitations of growth models for silvicultural purposes in heterogeneously structured forest in Sweden. *J. For. Sci.* **2013**, *59*, 458–473.
27. Drössler, L. Tree species mixtures—A common feature of southern Swedish forests. *Forestry* **2010**, *83*, 433–441. [CrossRef]
28. Bauhus, J.; Puettmann, K.J.; Kühne, C. Close-to-Nature forest management in Europe. In *Managing Forests as Complex Adaptive Systems*; Messier, C., Puettmann, K.J., Coates, K.D., Eds.; Routledge: New York, NY, USA, 2013.
29. Ingemarson, F.; Lindhagen, A.; Eriksson, L. A typology of small-scale private forest owners in Sweden. *Scand. J. For. Res.* **2006**, *21*, 249–259. [CrossRef]
30. Nordström, E.-M.; Holmström, H.; Öhman, K. Evaluating continuous cover forestry based on the forest owner's objectives by combining scenario analysis and multiple criteria decision analysis. *Silva Fenn.* **2013**, *47*. [CrossRef]
31. Bohn, U.; Weber, H. *Karte der Natürlichen Vegetation Europas (Map of Natural Vegetation in Europe)*; Bundesamt für Naturschutz: Bonn, Germany, 2000. (In German)
32. Drössler, L.; Fahlvik, N.; Ekö, P.M. *Stand Structure and Future Development of a Managed Multi-Layered Forest in Southern Sweden*; Report 44; SLU, Southern Swedish Forest Research Centre: Alnarp, Sweden, 2012.
33. Runkle, J.R. *Guidelines and Sample Protocol for Sampling Forest Gaps*; General Technical Report PNW-GTR-283; U.S. Department of Agriculture, Forest Service, Pacific Northwest Research Station: Portland, OR, USA, 1992.
34. Hägglund, B.; Lundmark, J.-E. Handledning i Bonitering Med Skogshögskolans Boniteringssystem—Del 3 Markvegetationstyper—Skogsmarksflora (Manual of the Swedish Site Quality Assessment – Part 3 Floor Vegetation Sites). Skogsstyrelsen: Jönköping, Sweden, 1994.
35. Riksinventering av skog (RIS). Manual of the National Forest Inventory in Sweden. Institutionen för skoglig resurshushållning, SLU: Umeå, Sweden, 2012. (In Swedish)
36. Lemon, P.E. A spherical densiometer for estimating forest overstorey density. *For. Sci.* **1956**, *2*, 313–320.
37. Bolker, B.; Skaug, H.; Magnusson, A.; Nielsen, A. Getting Started with the glmmADMB Package. R Foundation for Statistical Computing: Vienna, Austria. Available online: http://glmmadmb.r-forge.r-project.org/glmmADMB.pdf (accessed on June 2016).
38. Moreno-Fernández, D.; Cañellas, I.; Barbeito, I.; Sánchez-González, M.; Ledo, A. Alternative approaches to assessing the natural regeneration of Scots pine in a Mediterranean forest. *Ann. For. Sci.* **2015**, *72*, 569–583. [CrossRef]

39. Fortin, M.; DeBlois, J. Modeling tree recruitment with zero-inflated models: The example of hardwood stands in southern Québec, Canada. *For. Sci.* **2007**, *53*, 529–539.
40. Schweiger, J.; Sterba, H. A model describing natural regeneration recruitment of Norway spruce (*Picea abies* (L.) Karst.) in Austria. *For. Ecol. Manag.* **1997**, *97*, 107–118. [CrossRef]
41. Li, R.; Weiskittel, A.R.; Kershaw, J.A., Jr. Modeling annualized occurrence, frequency, and composition of ingrowth using mixed-effects zero-inflated models and permanent plots in the Acadian Forest Region of North America. *Can. J. For. Res.* **2011**, *41*, 2077–2089. [CrossRef]
42. Zuur, A.F.; Leno, E.N.; Walker, N.J.; Saveliev, A.A.; Smith, G.M. *Mixed Effects Models and Extensions in Ecology with R*; Springer: New York, NY, USA, 2009.
43. Leemans, R. Canopy gaps and establishment patterns of spruce (*Picea abies* (L.) Karst.) in two old-growth coniferous forests in central Sweden. *Vegetatio* **1991**, *93*, 157–165. [CrossRef]
44. Liu, Q.; Hytteborn, H. Gap structure, disturbance and regeneration in a primeval *Picea abies* forest. *J. Veg. Sci.* **1991**, *2*, 391–402. [CrossRef]
45. Kathke, S. The Role of Gap Dynamics for the Regeneration of the Natural Spruce Forests in the Harz Mountains. Doctoral Thesis, Faculty of Biosciences, Martin-Luther-Universität Halle-Wittenberg, Stuttgart, Germany, 2010.
46. Kuuluvainen, T.; Aakala, T. Natural forest dynamics in boreal Fennoscandia: A review and classification. *Silva Fenn.* **2011**, *45*, 823–841. [CrossRef]
47. Garbarino, M.; Mondino, E.B.; Lingua, E.; Nagel, T.A.; Dukić, V.; Govedar, Z.; Motta, R. Gap disturbances and regeneration patterns in a Bosnian old-growth forest: A multispectral remote sensing and ground-based approach. *Ann. For. Sci.* **2012**, *69*, 617–625. [CrossRef]
48. Strand, M.; Ottoson-Löfvenius, M.; Bergsten, U.; Lundmark, T.; Rosvall, O. Height growth of planted conifer seedlings in relation to solar radiation and position in Scots pine shelterwood. *For. Ecol. Manag.* **2006**, *224*, 258–265. [CrossRef]
49. Rumpf, H.; Ditges, J. Jugendwachstum von Fichtennaturverjüngung in Abhängigkeit von Überschirmungsdichte und Pflegestrategie (Early growth of spruce regeneration in relation to canopy density and tending strategy. *Forst. Holz.* **2008**, *63*, 20–25. (In German with English Summary)
50. Lundqvist, L. Simulation of sapling population dynamics in uneven-aged Picea abies forests. *Ann. Bot.* **1995**, *76*, 371–380. [CrossRef]
51. Örlander, G.; Karlsson, C. Influence of shelterwood density on survival and height increment of *Picea abies*. *Scand. J. For. Res.* **2000**, *15*, 20–29. [CrossRef]
52. Weiskittel, A.R.; Hann, D.W.; Kerhsaw, J.A., Jr.; Vanclay, J.K. *Forest Growth and Yield Modeling*; Wiley: Chichester, UK, 2011.
53. Lundqvist, L. Some notes on the regeneration on six permanent plots managed with single-tree selection. *For. Ecol. Manag.* **1991**, *46*, 49–57. [CrossRef]
54. De Chantal, M.; Holt Hansen, K.; Granhus, A.; Bergsten, U.; Ottosson Löfvenius, M.; Grip, H. Frost heaving damage to one-year-old Picea abies increases with soil horizon depth and canopy size. *Can. J. For. Res.* **2007**, *37*, 1236–1243. [CrossRef]
55. Johansson, K.; Nilsson, U.; Örlander, G. A comparison of long-term effects of scarification methods on the establishment of Norway spruce. *Forestry* **2013**, *86*, 91–98. [CrossRef]
56. Beland, M.; Agestam, E.; Ekö, P.M.; Gemmel, P.; Nilsson, U. Scarification and seedfall affects natural regeneration of Scots pine under two shelterwood densities and a clear-cut in southern Sweden. *Scand. J. For. Res.* **2000**, *15*, 247–255. [CrossRef]
57. Karlsson, M.; Nilsson, U. The effects of scarification and shelterwood treatments on naturally regenerated seedlings in southern Sweden. *For. Ecol. Manag.* **2005**, *205*, 183–197. [CrossRef]
58. Kuuluvainen, T. Gap disturbance, ground microtopography, and the regeneration dynamics of boreal coniferous forests in Finland: A review. *Ann. Zool. Fennici.* **1994**, *31*, 35–51.
59. Huth, F. Untersuchungen zur Verjüngungsökologie der Sand-Birke (Betula pendula Roth) (Investigations about the Regeneration Ecology of Sandy Birch). Ph.D. Thesis, Faculty of Forest-, Geo-and Hydrological Sciences, TU Dresden, Germany, 2009. (In German with English Summary)
60. Röhrig, E.; Bartsch, N.; Lüpke, B. *Waldbau auf ökologischer Grundlage (Silviculture Based on Ecology)*; Ulmer: Stuttgart, Germany, 2006. (In German)

61. Agestam, E.; Ekö, P.M.; Nilsson, U.; Welander, N.T. The effects of shelterwood density and soil preparation on natural regeneration of Fagus Sylvatica in southern Sweden. *For. Ecol. Manag.* **2003**, *176*, 61–73. [CrossRef]
62. Frost, I. Dispersal and Establishment of Quercus Robur. Ph.D. Thesis 305, Faculty of Science and Technology, Uppsala University, Uppsala, Sweden, 1997.
63. Gimingham, C.H. Ecological aspects of birch. *Proceedings of the Royal society of Edinburgh* **1984**, *85B*, 65–72. [CrossRef]
64. Engelmark, O.; Hytteborn, H. Coniferous forests. *Acta Phytogeogr. Suec.* **1999**, *84*, 55–74.
65. Greis, I.; Kellomäki, S. Crown structure and stem growth of Norway spruce undergrowth under varying shading. *Silva Fenn.* **1981**, *15*, 306–322. [CrossRef]
66. Wagner, S.; Fischer, H.; Huth, F. Canopy effects on vegetation caused by harvesting and regeneration treatments. *Eur. J. For. Res.* **2011**, *130*, 17–40. [CrossRef]
67. Drössler, L.; Nilsson, U.; Lundqvist, L. Simulated transformation of even-aged Norway spruce stands to multi-layered forests: An experiment to explore the potential of tree size differentiation. *Forestry* **2014**, *87*, 239–248. [CrossRef]
68. Hickler, T.; Vohland, K.; Feehan, J.; Miller, P.A.; Smith, B.; Costa, L.; Giesecke, T.; Fronzek, S.; Carter, T.R.; Cramer, W.; et al. Projecting the future distribution of European potential natural vegetation zones with a generalized, tree species based dynamic vegetation model. *Glob. Ecol. Biogeog.* **2012**, *21*, 50–63. [CrossRef]

5

Clearcutting and Site Preparation, but Not Planting, Promoted Early Tree Regeneration in Boreal Alaska

Miho Morimoto [1,*], Glenn Juday [1] and Brian Young [2]

[1] School of Natural Resources and Extension, University of Alaska Fairbanks, Fairbanks, AK 99775, USA; gpjuday@alaska.edu

[2] Natural Science, Landmark College, Putney, VT 05346, USA; brianyoung@landmark.edu

* Correspondence: mmorimoto@alaska.edu

Abstract: The stand initiation stage decisively influences future forest structure and composition, particularly in the boreal forest which is a stand replacement disturbance driven system. In boreal Alaska, the conventional forest management paradigm has focused on the production of large-dimension timber, particularly white spruce (*Picea glauca*). However, energy generation and heating from wood is increasing, and is likely to significantly expand total forest harvest, further shifting management focus to fuelwood production. We evaluated the effects of forest harvest management practices on post-harvest regeneration by examining whether harvest type, site preparation method, and reforestation technique resulted in differences in forest regeneration in terms of species presence, dominance, basal area, and total stem biomass using a stochastic gradient boosting (TreeNet algorithm). We recorded diameter at breast height and height of white spruce, birch (*Betula neoalaskana*), and aspen (*Populus tremuloides*) in 726 plots from 30 harvest units, distributed across the various harvest and treatment types, harvest years, harvest sizes, and geographical locations. Our results indicate that management practices suitable/acceptable for woody biomass production differ from the more traditional dimensional timber production from white spruce-focused management. Artificial reforestation does not differ from natural regeneration in obtaining more stems or producing greater biomass. Clearcutting and site preparation increased tree regeneration, basal area, and woody biomass when compared to a partial harvest with no site preparation. Planting of white spruce in the Alaskan boreal forest may only be necessary in some specific circumstances, such as years with no/low white spruce seed crop, or in landscapes depleted of seed trees.

Keywords: low-input management; post-harvest regeneration; clearcutting; site preparation; TreeNet (stochastic gradient boosting)

---

1. Introduction

In central Interior Alaska, a large portion of the most productive boreal forest sites, especially near transportation and population centers, was transferred from federal to state ownership, beginning at statehood in 1959. On these lands the conventional forest management paradigm in the second half of the 20th century focused on the production of large-dimension timber, particularly white spruce (*Picea glauca* (Moench) Voss) [1]. As a result, several silvicultural systems designed to maximize regeneration and production of white spruce have been studied, which have included both clearcutting (even-aged) and partial harvest (uneven-aged) methods combined with various site preparation treatments and assisted tree regeneration [2–5].

During ecological succession, the stand initiation stage decisively influences future forest structure and composition, particularly in the boreal forest, which is primarily a stand replacement disturbance driven system [6–8]. Fire is the dominant natural disturbance in North American boreal forest, although various fire suppression policies are in place that modify or limit its effects, particularly near communities [9–11]. Pre-fire vegetation and fire severity greatly influence post-fire tree regeneration [8,12,13]. In particular, the depth of the organic layer is one of the most important variables determining the post-fire regeneration trajectory [12,14]. In Interior Alaska, a thick organic layer tends to accumulate because the rate of organic matter decomposition in soils, particularly in spruce stands, is very slow due to cold temperatures [15]. In Alaska, post-fire plant regeneration pathways vary from self-replacement to initial or relay floristics, depending on the depth of organic layer remaining following fire [12,16]. When fire consumes a small amount of the organic layer, regeneration is dominated by previous vegetation that survived belowground and can regenerate asexually from the remaining parts. In contrast, when fire largely consumes the organic layer, burned sites promote establishment of early-successional species that germinate from seeds on exposed mineral soil [16,17].

The same principles of organic layer thickness that affect post-fire succession also influence post-harvest stand development. Although wildfire has been and continues to be the dominant disturbance in Interior Alaska, forest harvest can be one of the most important forms of disturbance in areas dedicated to sustained yield. Numerous studies examining post-fire forest succession are available for Alaska and northern boreal Canada [14,16,18–21]. However, studies of post-harvest tree regeneration are limited [3,5,22].

Within the study region, the predominant operational harvest method used from the start of industrial wood harvest in the early 20th century up to the late 1990s was clearcutting of the mature white spruce-dominated stands, and partial harvest of larger diameter white spruce trees within mixed stands [23]. However, since the 1990s, when demand decreased for large-dimension white spruce from the Asian market, the proportion of all harvesting that used the clearcutting silvicultural system decreased [24]. In the North American boreal forest, there is increasing concern over some effects observed in large clearcuts. For example, in a study in Northern Alberta, clearcuts that exceeded 100 ha experienced low spruce recruitment due to the limited seed dispersal ability of white spruce [25]. Increasingly clearcutting is used in conjunction with mitigating measures, such as variable retention and carefully planned harvest distribution and layout across the landscape [26].

Natural regeneration of white spruce as a silvicultural system is strongly affected by the masting behavior of the species. White spruce typically produces a large seed crop roughly every 11 years in Interior Alaska [27,28]. White spruce regenerates almost exclusively from seeds, thus seedbed quality is a critical factor for adequate regeneration. Mineral soil substrate is ideal for white spruce regeneration [29]. As a result, in order to achieve an adequate component of white spruce in regenerating stands, on managed forest lands in the study area between the years 1972–2012 various site preparation (16% of harvested area) and assisted tree regeneration (predominantly planting of white spruce seedlings; 49% of harvested area) treatments were applied [23].

In addition to issues of timing of seed production and seedbed receptivity, successful white spruce regeneration must overcome strong early competition. The early growth of white spruce is slower than specialized early-successional species established by seed and asexually regenerated species such as Alaska birch (*Betula neoalaskana* Sarg.), quaking aspen (*Populus tremuloides* Michx.), and bluejoint (*Calamagrostis canadensis* (Michx.) P. Beauv.) [3,30,31]. *Calamagrostis* is the major species of concern as a competitor of white spruce regeneration because it spreads rapidly by belowground rhizomes and restricts white spruce seedlings establishment and growth [32]. As a result, when *Calamagrostis* is present prior to harvest, site preparation is typically applied following harvest to remove the organic layer and belowground rhizomes of competitive vegetation [2,5]. Additionally, when state minimum stocking standards (1112 seedlings (diameter at breast height < 2.5 cm) ha^{-1}) are not met or not expected to be met, forest managers often require planting of white spruce seedlings [33].

In Interior Alaska, the use of wood for home heating and energy generation is expected to increase to meet fossil fuel reduction goals and to provide relief from high fossil fuel prices, especially in remote communities off the road system [34]. As of 2015, nine wood biomass energy facilities have been built in Interior Alaska with another ten under construction, and more than eleven are in design or feasibility status [35]. In contrast to many other forest regions, wood biomass for energy can be, and often is, the entire goal of a harvest, not simply as a byproduct generated after large dimension trees are removed. Energy demands for woody biomass are both expanding total forest harvest and changing the conventional forest management paradigm from production of large-dimension white spruce timber (the strongly dominant management goal) to a slowly expanding harvest of other species and size classes in addition. The harvest cycle is also likely to be shorter for biomass harvest than for large-dimension wood products, requiring more frequent regeneration over a given period of time [36]. Due to limited resources for management, including research, and until recently low demand for timber harvest overall, a clear, empirically based understanding of the effects of harvest has not been available. As a result, in order to meet the needs of this evolving forest management situation while ensuring sustained yield wood production, it is crucial to identify the effects of forest harvest and post-harvest management on the regeneration of not just white spruce, but other species as well.

The total area harvested in this part of the boreal forest is relatively small when compared to boreal Canada, Fennoscandia, and Russia [25,37,38]. However, during the past 40 years of regeneration under systematic silvicultural practices the total area harvested has accumulated to be more than 7000 ha. Although unconstrained logging occurred in the early 1900s near a few early population centers, river banks, and gold mines [24,39], in this study we analyze only the last 40 years of post-harvest regeneration because of the lack of records before that time period [39].

The objective of this study is to evaluate the effects of harvest methods over the past 40 years and the subsequent management practices on post-harvest tree regeneration in central Interior Alaska. To achieve this objective, we evaluated the harvest type, site preparation method, and reforestation techniques used in order to assess the differences in forest regeneration outcomes in terms of species presence, dominance, basal area, and total woody biomass. There are a few studies examining the effects of various harvest and reforestation practices on post-harvest regeneration in Interior Alaska, but those have occurred on small experimental plots [2,3,5,40]. This study is the first landscape-scale study in central Interior Alaska to examine both temporal and spatial effects of mature forest harvest on regeneration in an operational context.

2. Materials and Methods

2.1. Study Area

The study area is located in the greater Interior Alaska boreal forest which stretches from the Alaska Range in the south, the Brooks Range to the north, the Canadian border in the east, and the Chukchi Sea to the west, covering approximately 47 million ha in total (Figure 1). The study was conducted within the Fairbanks and Kantishna Management Areas of the Tanana Valley State Forest and adjacent "forest classified" lands ("state forest lands"; Figure 1) which cover 578,575 ha in Alaska, USA. These lands are available for multiple uses, including sustained yield forest management. The Interior Alaska boreal forest is composed primarily of white spruce, black spruce (*Picea mariana* (Mill.)), Alaska birch, quaking aspen, with minor amounts of balsam poplar (*Populus balsamifera* L.), and tamarack (*Larix laricina (Du Roi)* K. Koch) [41]. The most extensive forest cover types in the Tanana Valley State Forest and adjacent state lands are black spruce and mixed white spruce-hardwood types [42]. Although black spruce forest type is the most extensive, it generally occurs on low-productive, permafrost underlain soils, resulting in low productivity [43]. In contrast, white spruce types often occur on the most productive sites resulting in high productivity [43]. Soils are mostly silt loams formed from loess parent material [44] and elevations range from 100 m to 600 m.

Climate of the study area is strongly continental and varies substantially across topographic factors, including elevation and aspect [45]. The mean annual temperature for the region is −2 °C, with extreme temperatures ranging from −50 °C to 35 °C and annual precipitation of 270 mm, with about one third falling during the cold season (October through April) months. The area currently experiences a growing season of approximately 123 days [46]. However, meso-climate in the region varies substantially according to factors such as elevation and aspect [45]. Temperature inversion is a major factor that creates great temperature variation across elevation, particularly in winter [45]. Aspect also affects temperature variability because of the low-angle of the sun [45]. South-facing slopes are generally warmer and drier compared to north-facing slopes that are cold and wet, and often underlain by permafrost [45].

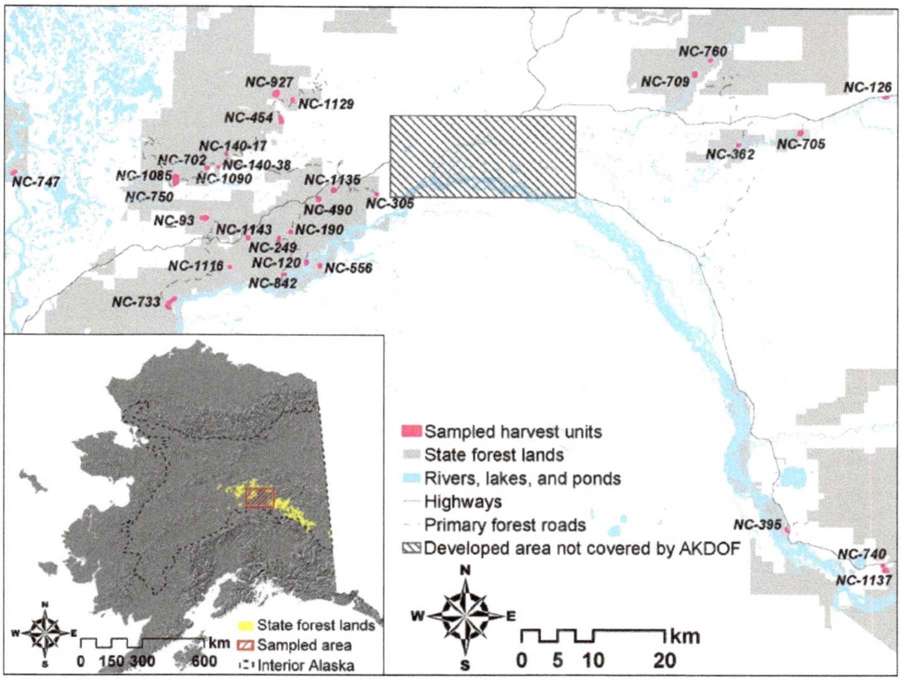

Figure 1. Map of study area. (**a**) Study area is on state forest lands within Interior Alaska boreal region [47]; (**b**) Sampled harvest units are distributed within Kantishna and Fairbanks areas of Tanana Valley State Forest and forest classified lands. NC- followed by number are sampled harvest units number.

2.2. *Silvicultural Systems*

The two primary harvesting methods used within the study region have been clearcutting and various partial cutting systems. Both of these systems have been applied for both green wood and post-fire salvage harvests. The clearcut system, as used in the study area, ranged from a conventional clearcut, to a "clearcut with reserves", as defined by Society of American Foresters [48]. Partial cuts, however, typically involved one of two types: the removal of a single species from mixed stands, predominantly white spruce, or an intermediate harvest with diameter limits [33]. Regardless of the harvest system, only whole tree harvesting was performed. All sampled regeneration units examined in this study were dominated by mature white spruce before harvest, and all stands originated from either wildfire or primary succession following flooding [23]. Although harvest of other species is likely to increase in the future, white spruce has been the major harvested species in the study area. As a result, this study examined only units that were either clearcut or partial cut for white spruce. Partial harvest units of hardwood types or post-fire salvage logging were excluded. All harvest units sampled experienced a single entry harvest between the years 1975 and 2004 and then were not subsequently disturbed by wildfire to the year sampled (2013 or 2014). The site preparation treatments used most frequently in the region and captured in this study involved mechanical scarification using either a bulldozer blade or a disk trencher [23]. Although different site preparation methods may affect regeneration slightly differently, we categorized harvest units that received any site preparation as scarified regardless of the method used. We believe this approach was justified because the main effect produced by these methods is to break up the thick organic layer of mature spruce stands and produce areas of seedbed suitable for white spruce seedling establishment. The combination of the various site disturbance methods was also necessary to have adequate replication of management treatments given the limited number of units that received site preparation.

All harvest units relied either on natural regeneration, white spruce artificial regeneration, or small amounts of planted exotic conifers. Within the study area, the most common artificial regeneration technique was the planting of container stock using locally sourced seeds [49]. In this study, we included only harvest units that experienced natural regeneration or planting of white spruce seedlings from container stock. Planting densities were generally either 2.4 m spacing (1483 trees·ha^{-1}) or 3 m spacing (1075 trees·ha^{-1}).

Within the study region, historical harvest units vary in size from ~1 to a few hundred ha. The size distribution of the harvest units was positively skewed, with a median of 4.66 ha and a mean of 10.89 ha. In this study, we excluded the smallest units (<1 ha) and extremely large units (>40 ha) because of the small number of units in these size ranges. As a result, the harvest units included in this study ranged from 1.4 to 30.3 ha in size. The total area sampled was 269 ha, accounting for approximately 3.5% of the total 7000 ha harvested in the Fairbanks and Kantishna areas of state forest lands in Interior Alaska since 1972 [23].

2.3. Sampling Design

We investigated 30 separate harvest units located in the Fairbanks and Kantishna areas of state forest lands from the Fairbanks office of the Alaska Department of Natural Resources Division of Forestry (AKDOF) Forest Management Database (FMD; Tables S2 and S3) [23]). The FMD is a file geodatabase which includes the location and types of all forest management activities that has occurred on state lands within the Fairbanks and Kantishna areas (see Figure 1) since 1972. Using this database, we selected representative harvest units that were evenly distributed across harvest types (16 clearcut and 14 partial cut units), site preparation methods (11 scarified and 19 unscarified units), reforestation techniques (16 planted and 14 naturally regenerated units), year of timber sale, and size of harvest units. Sample harvest units were also selected to achieve wide geographical coverage across the study region (Figure 1, Table 1).

To quantify tree regeneration, we used 1.69 m radius circular plots, the same plot size as operational AKDOF reforestation surveys [33]. We determined plot sampling intensity based on a preliminary test of sampling efficiency using a censused population of white spruce located in the study region [50]. Based on this analysis, we used four 1.69 m radius circular plots·ha^{-1} as our sampling intensity. To determine the placement of plots, we created a virtual 50 m $\times$ 50 m grid with points at the center of each cell over the entire study area using the Fishnet tool (ArcGIS 10.2, Esri, Redlands, CA, USA). The points falling within the selected harvest units represented the center of the plots. The number of plots in each unit varied between 7 and 120 due to the size and geographic configuration of the harvest units (Table 1). We prioritized sampling a large number of harvest units over intensive sampling in a single harvest unit to cover a greater geographic area, and allow more replications of management practices and years. Because of this strategy, when the calculated number of plots was greater than 50, the sampling intensity was truncated to 50 or fewer by sampling every other or every third plot. In units where only every other plot was sampled, sampled plots were evenly distributed starting from the first plot. The coordinates of the plots ($\pm$1 m) were exported to Trimble Pro XT GPS unit (Trimble Navigation, Sunnyvale, CA, USA) and were used to navigate to the sample plot centers in the field.

2.4. Data Collection and Preparation

2.4.1. Response Variables

Field sampling was conducted during the summers of 2013 and 2014. Within each plot, we counted all live white spruce regardless of diameter at breast height (DBH), and live birch, aspen, balsam poplar, and black spruce <1 cm in DBH. When a live white spruce crossed the DBH plane, we measured total height and DBH. We measured DBH and height if live birch, aspen, balsam poplar, and black spruce were 1 cm or greater in DBH. Residual stems were distinguished from regeneration based on estimated age of the tree, and excluded from analysis. The sampling protocol for this study will be made available at Bonanza Creek Long Term Ecological Research Site website [51].

Table 1. List of sampled harvest units. The location of the units is displayed in Figure 1.

Unit	Size (ha)	Number Plots Sampled	Logged Year	Harvest Type	Site Preparation	Reforestation
NC-120	10.4	41	1975	Partial cut	None	Plant
NC-93	17.9	35	1975	Partial cut	None	Natural
NC-190	5.1	22	1977	Clearcut	Scarify	Natural
NC-126	5.7	22	1978	Partial cut	None	Natural
NC-140-17	2.5	8	1979	Clearcut	None	Natural
NC-249	5.0	22	1980	Clearcut	Scarify	Natural
NC-362	4.4	15	1981	Partial cut	None	Natural
NC-140-38	1.5	7	1982	Clearcut	Scarify	Natural
NC-395	5.1	21	1983	Clearcut	None	Natural
NC-490	8.4	32	1985	Clearcut	None	Natural
NC-556	6.6	26	1986	Clearcut	None	Plant
NC-305	3.5	11	1987	Partial cut	Scarify	Plant
NC-705	11.0	44	1989	Clearcut	Scarify	Plant
NC-454	20.4	44	1991	Clearcut	Scarify	Plant
NC-740	1.9	8	1991	Clearcut	None	Plant
NC-709	17.2	35	1991	Clearcut	Scarify	Plant
NC-842	2.1	7	1992	Partial cut	None	Natural
NC-733	30.3	44	1992	Clearcut	Scarify	Plant
NC-702	2.0	9	1993	Clearcut	None	Plant
NC-747	8.0	31	1994	Clearcut	None	Plant
NC-750	9.8	41	1995	Clearcut	Scarify	Plant
NC-1085	22.6	47	1996	Partial cut	Scarify	Plant
NC-1137	13.5	29	1997	Clearcut	None	Plant
NC-927	22.5	43	1998	Partial cut	None	Plant
NC-760	3.4	13	1998	Partial cut	None	Natural
NC-1129	6.0	22	1999	Partial cut	None	Plant
NC-1090	1.4	7	1999	Partial cut	None	Natural
NC-1135	11.7	49	2002	Partial cut	None	Plant
NC-1116	2.4	9	2003	Partial cut	Scarify	Natural
NC-1143	6.7	28	2004	Partial cut	None	Natural

Although our response variables were initially collected as continuous variables, we categorized the variables into binary classes to obtain high accuracy for robust and reliable inferences. We predicted presence/absence for an "all size" group and for a sapling group (DBH > 2.5 cm) of white spruce, birch, and aspen. We identified dominant species in each plot in terms of stem density, and assigned 1 (dominant) or 0 (not dominant) to each species by the size class. If the stem density of different species was the same, and a third species was either absent or present at lower density, both species were classified as dominant (assigned 1). Basal area and biomass were categorized into 1 (high) or 0 (low). For basal area, we set thresholds of 0.5, 1, and 0 m^2 for white spruce, birch, and aspen, respectively, so that the binary classes were well balanced. For woody biomass calculation, we used biomass equations established using samples from Interior Alaska [52]. The equations follow the form:

$$Y = \alpha_1 \cdot DBH + \alpha_2 \cdot DBH2 + \alpha_3 \cdot height \quad (1)$$

where Y is the total above ground biomass (grams), and α_1, α_2, and α_3 are specific empirically determined coefficients for each species. We combined biomass accumulation of each woody species (white spruce, birch, aspen, balsam poplar, and black spruce) and the aggregated biomass accumulation was categorized into high ($\geq$5 t·ha^{-1}) and low (<5 t·ha^{-1}).

2.4.2. Predictors

All the predictors used in our TreeNet classifications are listed in Table 2. We obtained the values of field predictors at the center of a 50 m × 50 m lattice grid. The predictors were either publicly available or obtained from AKDOF files. We examined data and tested reliability where possible (e.g., alternative digital elevation model (DEM) data), and chose the most reliable of the publicly available data. Although the coarsest resolution was 771 m for the climate variables, we did not rescale the other predictors which had finer resolution to 771 m. We believe this approach is valid because it can be assumed that climate is largely uniform within the grid cells and topographic variables account for the mesoscale climate variations if they exist. Type and year of harvest, site preparation, and reforestation, and size of harvest unit were obtained from the FMD [23]. Elevation (m), aspect, slope (degree), and topographic position index (TPI) were obtained from a 5-m resolution DEM created by Geographic Information Network of Alaska (GINA) in ArcGIS. This DEM has 90% probability of 3 m vertical accuracy, and 90% probability of 12.2-m horizontal accuracy. The GIS data and metadata for the DEM are available at http://ifsar.gina.alaska.edu/. Aspect was transformed using the following equation:

$$1 - \cos((2\pi \times \text{aspect}/360))/2 \tag{2}$$

where aspect is measured in degrees. Slope was considered flat when it was smaller than 5 degrees. TPI was calculated using Land Facet Corridor Designer, v. 1.2.884 tool [53] in ArcGIS.

We used the AKDOF forest type map (see details in [42,54]) to calculate distances (m) from each plot within a harvest unit to various features with the "Generate Near Table" tool in ArcGIS. The features include stands of white spruce forest, birch forest, aspen forest, water features, highways, forest roads, developed area, and urban area [54]. In some cases, the sampled harvest unit might have had a white spruce stand closer than indicated on the current forest type layer because of harvest in the landscape surrounding the sampled unit. In such cases, harvests nearest to sample units were considered white spruce forests if they were harvested eight years or more following the harvest of the sample units. We used eight years because white spruce most likely produces medium to large seed crops every seven years [27]. We assigned soil subgroups to each plot in ArcGIS using soil maps obtained from the US Department of Agriculture Natural Resources Conservation Service [55].

Table 2. List of response and predictor variables.

Variable	Description	Unit	Data Source
	Response variables		
Presence/absence	Presence/absence of white spruce, birch, and aspen	category	Field sampling
Species dominance	Species dominance of white spruce, birch, and aspen	category	Field sampling
Basal area	Basal area of white spruce (high $\geq$ 0.5 m^3, low < 0.5 m^3), birch (high $\geq$ 1 m^3, low < 1 m^3), and aspen (high $\geq$ 0 m^3, low < 0 m^3)	category	Field sampling
Biomass	Biomass accumulation (high $\geq$ 5 t, low < 5 t)	category	Field sampling
	Predictor variables		
Harvest type	Harvest type: clearcut/partial cut	category	AKDOF FMD
Site preparation	Ground treatment type: none/mechanical site preparation	category	AKDOF FMD
Reforestation	Reforestation type: natural/planting white spruce seedlings	category	AKDOF FMD
Year	Year since harvest: 10–39	continuous	AKDOF FMD
Size	Size of harvest unit	hectare	AKDOF FMD
Edge	Distance to edge of harvest unit	km	AKDOF FMD
White spruce	Distance to white spruce forest	km	AKDOF vegetation map
Birch	Distance to birch forest	km	AKDOF vegetation map
Aspen	Distance to aspen forest	km	AKDOF vegetation map
Water	Distance to water	km	AKDOF vegetation map
Highway	Distance to highway	km	AKDOF vegetation map
Forest road	Distance to forest road	km	AKDOF vegetation map
Urban	Distance to urban area	km	AKDOF vegetation map

Table 2. *Cont.*

Variable	Description	Unit	Data Source
Development	Distance to development (power line, mine etc.)	km	AKDOF vegetation map
Elevation	Elevation	m	GINA DEM
Slope	Slope	degree	GINA DEM
Aspect	Aspect	category	GINA DEM
TPI	Topographic Position Index	continuous	GINA DEM
Soils	Soil subgroup	category	NRCS
May temp	Average temperature of May	°C	SNAP
June temp	Average temperature of June	°C	SNAP
July temp	Average temperature of July	°C	SNAP
Aug temp	Average temperature of August	°C	SNAP
May precip	Precipitation sum of May	mm	SNAP
June precip	Precipitation sum of June	mm	SNAP
July precip	Precipitation sum of July	mm	SNAP
Aug precip	Precipitation sum of August	mm	SNAP

AKDOF = Alaska Division of Forestry; FMD = Forest Management Database; GINA DEM = Geographic Information Network of Alaska Digital Elevation Model; NRCS = Natural Resources Conservation Service; SNAP = Scenarios Network for Alaska + Arctic Planning.

Downscaled historical average monthly temperature and monthly precipitation from 1975–2009 were obtained from the Scenarios Network for Alaska + Arctic Planning [56]. The resolution of downscaled climate data is 771 m. We used climate data of the growing season (May–August) because tree growth is greatly affected by climate variables of these summer months [57–60]. We averaged mean monthly temperatures and total monthly precipitation of twenty years post-harvest, which is the most critical time period for tree regeneration [61].

2.5. Statistical Analysis

Due to the complex and multivariate nature of the data, we used the non-parametric TreeNet algorithm to predict post-harvest regeneration [62] as implemented in the Salford Predictive Modeler version 7 (Salford System, San Diego, CA, USA). This type of model does not require the same set of assumptions as frequency statistics, such as normality and independence, which are frequently violated when analyzing ecological data [63,64]. TreeNet is known to produce highly accurate predictions even with noisy data [65]. The TreeNet algorithm is often used for prediction, but it is also a powerful tool to mine data and identify relationships between a response variable and predictors by creating partial dependence plots in multivariate settings [63,66]. As a consequence, we chose the TreeNet algorithm to identify the effects of management practices, including harvest, site preparation, and reforestation, on post-harvest regeneration. Although our focus was on the effects of management practices, we built predictive models using all available environmental variables (Table 2) to improve predictive accuracy and to place our results in a greater ecological context for a robust inference.

To construct the decision trees used by TreeNet, a balanced option which rebalances unequal class sizes was selected [67]. We decided to grow 1000 trees, but the actual number of trees generated was optimized by the program for each predictive model [67]. For validation purposes, we used the testing method of cross-validation with a randomly selected 10% sample. All other options were set at program default values [67] which are known to perform well. The model performances were evaluated by applying the predictive model to the complete data set, and obtaining average accuracy and area under the receiver operating characteristic (ROC) curve (AUC). A perfect model will score an AUC of 1, while random guessing will score an AUC of around 0.5 [68].

In order to examine the effects of management practices on post-harvest regeneration, we evaluated relative variable importance and created partial dependence plots. The importance value for any predictor is determined by averaging the number of times it is selected as a tree node over all trees and squaring improvements in error rate resulting from these nodes [69]. A relative importance value of 100 is assigned to the most important predictor.

3. Results

The TreeNet algorithm predicted species presence/absence, species dominance, basal area, and biomass of regeneration in harvest units at high accuracy using the 27 predictors evaluated (Table 3). AUC values for each prediction mostly exceeded 0.8. The main exception to this result was for the presence/absence of white spruce and birch for the "all size" group (Table 3). Predictive models for aspen displayed the highest accuracy, as demonstrated by the high AUC values among the three species evaluated here. For aspen, AUC values were all above 0.9 (Table 3). The prediction of white spruce was the least successful for all variables except for the presence/absence of the "all size" group. However, even for spruce the AUC values were above 0.8 except in one case (Table 3). Overall, the model performances were more accurate for saplings than the "all size" group for both presence/absence and dominance of all species (Table 3). For each species, the prediction was most successful for basal area among all the variables (Table 3). Predictive accuracies were well balanced between two classes (presence/absence, dominant/not dominant, and high/low; Table 3). The TreeNet algorithm provided reproducible and robust models, and findings, in our data set.

Table 3. The model performances, including contingency table, specificity (true negative rates), sensitivity (true positive rate), mean accuracy (mean of sensitivity and specificity), and receiver operating characteristic (ROC) curve AUC.

		Predictions		Specificity Sensitivity	Mean Accuracy	AUC
"All Size" Group Presence/Absence		**Absent**	**Present**			
Aspen	Absent	491	92	84.22%	0.84	0.92
	Present	22	121	84.62%		
Birch	Absent	176	91	65.92%	0.68	0.74
	Present	138	321	69.93%		
White spruce	Absent	196	74	72.59%	0.73	0.78
	Present	123	333	73.03%		
Saplings presence/absence		Absent	Present			
Aspen	Absent	653	43	93.82%	0.94	0.98
	Present	2	28	93.33%		
Birch	Absent	394	84	82.43%	0.82	0.90
	Present	47	201	81.05%		
White spruce	Absent	422	108	79.62%	0.79	0.88
	Present	44	152	77.55%		
"All size" group dominance		Low	High			
Aspen	Low	580	79	88.01%	0.88	0.95
	High	9	58	86.57%		
Birch	Low	296	84	77.89%	0.78	0.85
	High	78	268	77.46%		
White spruce	Low	377	115	76.63%	0.76	0.83
	High	58	176	75.21%		
Sapling dominance		Low	High			
Aspen	Low	656	44	93.71%	0.94	0.97
	High	2	24	92.31%		
Birch	Low	423	83	83.60%	0.84	0.90
	High	36	184	83.64%		
White spruce	Low	472	111	80.96%	0.81	0.88
	High	27	116	81.12%		
Basal area		Low	High			

Table 3. *Cont.*

"All Size" Group Presence/Absence		Predictions		Specificity Sensitivity	Mean Accuracy	AUC
		Absent	Present			
Aspen	Low	103	6	94.50%	0.94	0.98
	High	2	32	94.12%		
Birch	Low	194	30	86.61%	0.86	0.93
	High	33	201	85.90%		
White spruce	Low	190	43	81.55%	0.81	0.88
	High	45	177	79.73%		
		Low	High			
Biomass	Low	319	87	78.57%	0.78	0.85
	High	72	261	78.38%		

The different management practices, including harvest type, site preparation, and reforestation, were generally not found to make as great a contribution to accurate predictions as the environmental variables (Figure 2a–f). In particular, the reforestation method was one of the least important predictors for the three different species studied here (Figure 2a–f). This indicates a smaller effect due to the management practices on post-harvest regeneration than that of the environmental variables. In general, management practices had greater effects on saplings than on stems of the "all size" group for white spruce and aspen, as indicated by the higher relative importance values for saplings (Figure 2a–d). The importance of management practices for white spruce and aspen varied, with harvest type and site preparation contributing relatively high importance in predictions for these species. The importance of management practices was generally low for birch (Figure 2a–e). Year of harvest was one of the most important predictors for presence and dominance of the sapling group, basal area, and biomass (Figure 2a–f).

The model presented here depicted the trend in species presence/absence, species dominance, basal area, and biomass in response to harvest type, site preparation method, reforestation technique, and year of harvest (Figure 3). White spruce presence, dominance, and basal area tended to be greater in clearcut, scarified, and planted units than in partial cut, unscarified, and naturally regenerated units, except that site preparation did not appreciably contribute to the dominance of the "all size" group (Figure 3a–e). White spruce presence in the "all size" group was lower in units that were logged within 25 years of sampling and higher in units that were logged 25–35 years before, but became lower again in units in which harvest occurred 35 years or more before sampling (Figure 3a). White spruce sapling presence was higher in units with longer time since harvest (Figure 3b). White spruce dominance of the "all size" group was greater in units with shorter time since harvest, while white spruce sapling dominance was greater in units with longer time since harvest (Figure 3c–d). White spruce basal area was low until 15 years after harvest, and became high after that time period (Figure 3e).

Clearcutting resulted in greater birch presence, birch sapling dominance, but lower birch "all size" group dominance and birch basal area (Figure 3a–e). Birch sapling presence and "all size" group dominance were greater in scarified than in unscarified units, while birch sapling dominance and birch basal area were greater in unscarified than in scarified units (Figure 3b–e). Birch dominance and basal area were greater in planted units than in naturally regenerated units (Figure 3c–e). Birch presence, sapling dominance, and basal area were low in units harvested 15 or fewer years before our sampling, and much greater in units harvested more than 15 years before sampling (Figure 3a,d,e). Dominance of birch in the "all size" group was greatest in units harvested 20–25 years or earlier (Figure 3c).

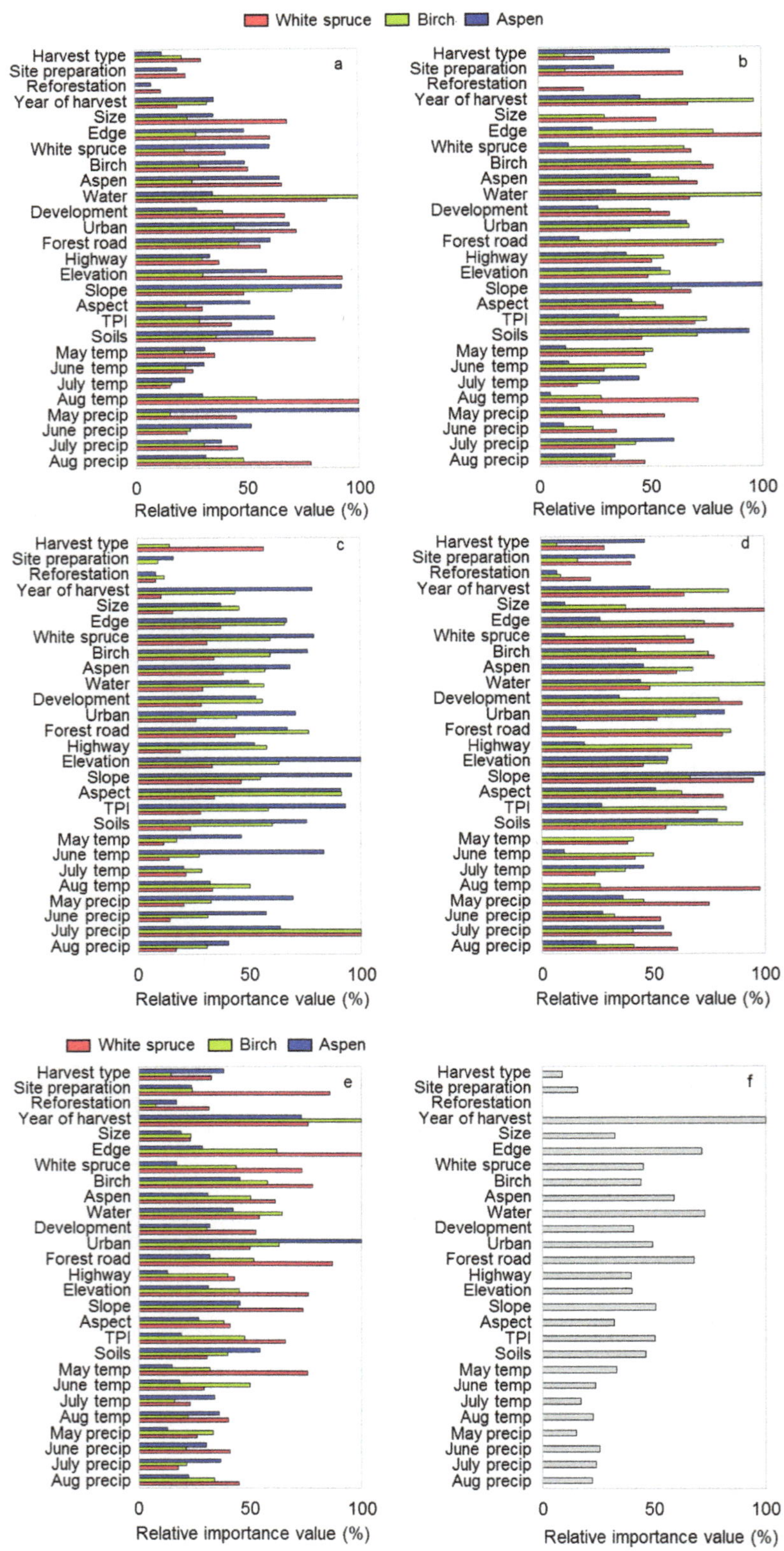

Figure 2. Relative variable importance of predictors in the predictive models of (**a**) presence/absence of "all size" group; (**b**) presence/absence of saplings; (**c**) dominance of "all size" group; (**d**) dominance of saplings; (**e**) basal area; and (**f**) biomass. The importance value for any predictor is determined by averaging the number of times it is selected as a tree node over all trees and the squared improvements in error rate resulting from these nodes [69]. A relative importance value of 100 is assigned to the most important predictor, and relatively scaled values are assigned to other predictors based on the most important predictor. Missing bars mean that the TreeNet algorithm found the predictor to have zero importance for the prediction.

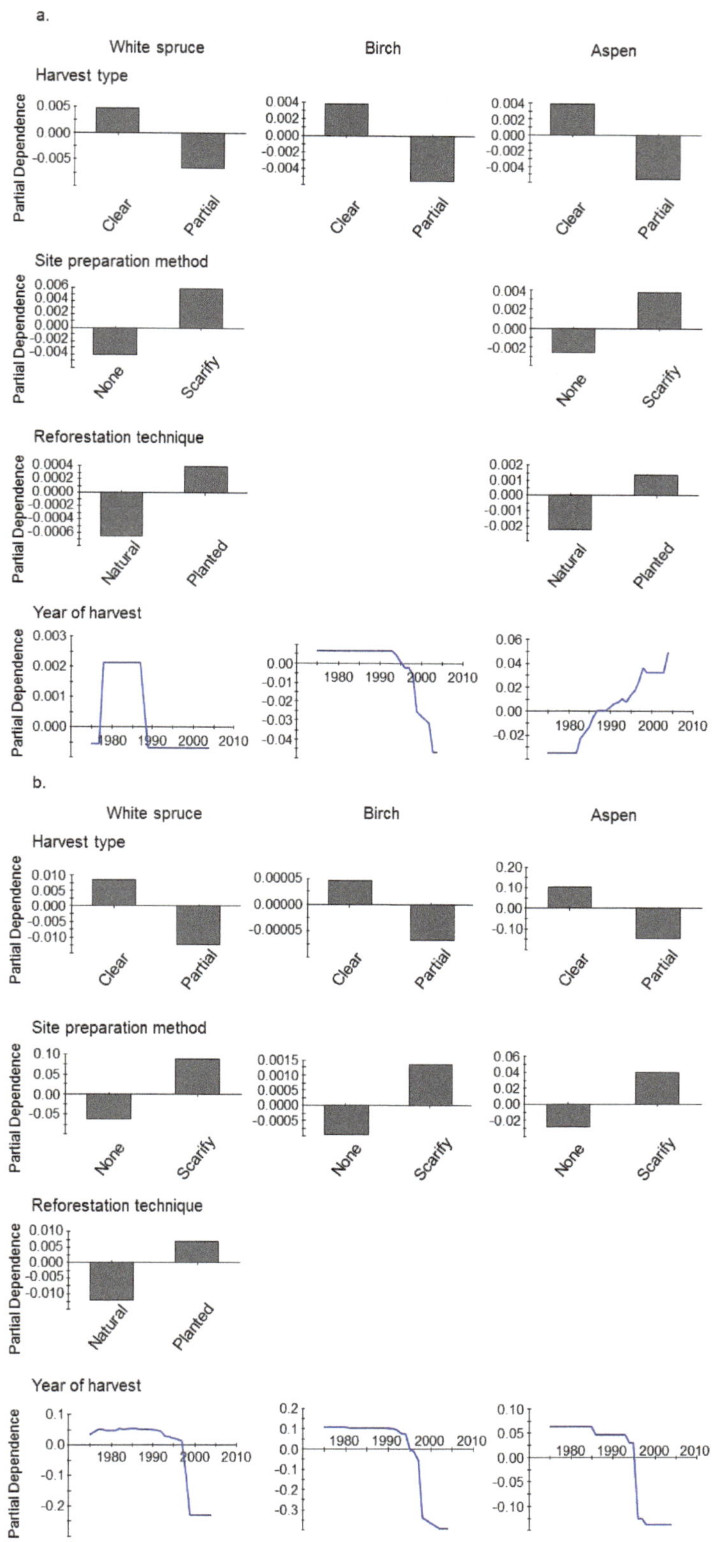

Figure 3. *Cont.*

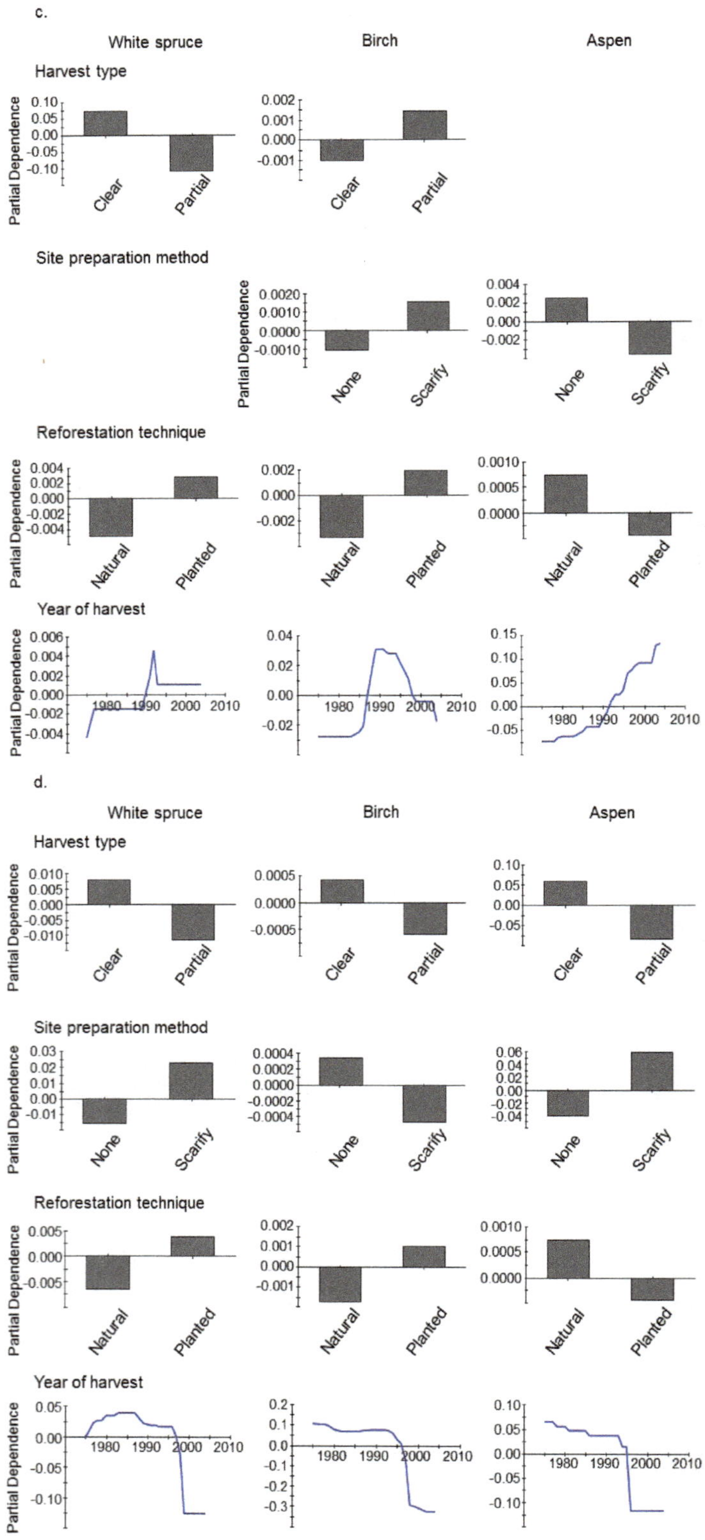

Figure 3. *Cont.*

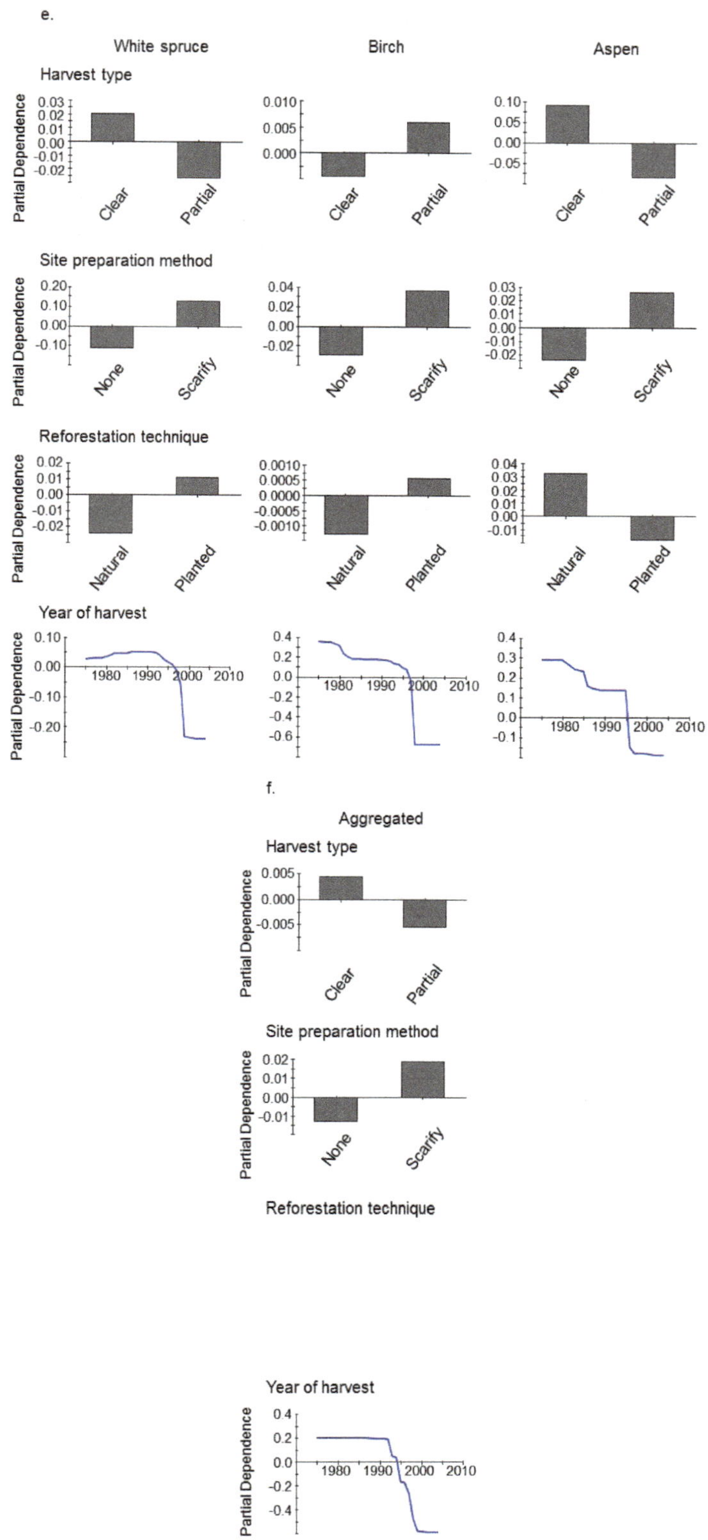

Figure 3. Partial dependence plots of harvest type, site preparation method, reforestation technique, and year of harvest for the predictive models of (**a**) presence/absence of "all size" group; (**b**) presence/absence of sapling; (**c**) dominance of "all size" group; (**d**) dominance of saplings; (**e**) basal area; and (**f**) biomass. Partial dependence plots show the relationship between the response and any given predictor by representing the dependence of the response on the predictor variable when all other variables are held at their mean [67]. *Y*-axes are the partial dependence value of a prediction being 1 (present/high). An empty space means that the TreeNet algorithm found the predictor to have zero importance for the prediction.

Aspen presence, dominance of sapling, and basal area were greatest in clearcut and scarified units (Figure 3a,b,d,e), although aspen dominance of "all size" group was greater in unscarified than in scarified units (Figure 3c). However, the effects of type of harvest and site preparation were limited on the aspen "all size" group (Figures 2 and 3a,c). Planting spruce seedlings resulted in a lower aspen dominance and basal area, but slightly greater aspen presence (Figure 3a–e). Aspen presence and dominance of the "all size" group were greater in units that were logged more recently, while aspen presence and dominance of saplings, and basal area were greater in units that were logged in earlier years (Figure 3a–e).

Biomass tended to be greater in clearcut and/or scarified units than in partial cut and/or unscarified units (Figure 3f). Reforestation technique (planted vs. natural regeneration) did not contribute to biomass prediction (Figure 2f). Year of harvest was the most important variable for the biomass prediction (Figure 2f). Biomass accumulation was low until 15 years after harvest, and became high after 20 years post-harvest (Figure 3f).

4. Discussion

The trends identified in post-harvest regeneration using a robust TreeNet algorithm provide a unique and useful basis for forest harvest management. However, the predictions need to be interpreted with a recognition of varying contributions of each variable to each prediction indicated by the relative variable importance (Figure 2). Harvest type, site preparation technique, and especially reforestation methods were not the most important among all 27 predictors, particularly for birch. This result indicates that post-harvest regeneration outcomes cannot be successfully evaluated by management practices alone. Even so, by incorporating all the predictors TreeNet analysis identified trends in post-harvest regeneration described below (Figure 3).

The effect of harvest type on post-harvest regeneration was relatively consistent with the inference that clearcutting resulted in greater presence, dominance, and basal area of white spruce and aspen of any size groups, and greater presence of birch of any size groups when compared to partial cutting. The effect of harvest type was greater on sapling presence, dominance, and basal area than the "all size" group for both white spruce and aspen (Figure 2a–d). Clearcut units supported greater biomass accumulation (up to 40 years) than partial cuts, because of the overall greater predicted presence of the three tree species and greater basal area of trees in clearcuts than partial harvest stands (Figure 3). Both white spruce and aspen experience optimal growth under full light conditions [70,71], which were created by clearcutting. Clearcutting also promotes aspen suckering [71]. Although greater growth of white spruce in clearcuts is consistent with results from individual research plots in Interior Alaska [3] and in Alberta boreal mixed wood [72], our results now demonstrate that this effect was also achieved at the operational and landscape scale.

On the other hand, although birch was more likely to appear in clearcuts than in partial cuts, birch dominance of the "all size" group and birch basal area were greater in partial cuts than in clearcuts (Figure 3). This result is somewhat inconsistent with previous studies that demonstrate greater growth of birch under greater amounts of sunlight often present in clearcuts [73–75]. There was perhaps less competition in partial cuts than in clearcuts due to the lower presence and dominance of regenerating white spruce and aspen, which could have allowed greater birch dominance and growth. Even so, the contribution of harvest type to predictions of birch in harvest regeneration outcomes was low, and so environmental factors appear to affect birch regeneration more than harvest type. This is also consistent with the ability of birch to regenerate prolifically, both asexually from sprouts and from seeds, across a wide range of disturbances [73].

Although we found that clearcutting resulted in greater subsequent white spruce presence, it should be noted that a lack of seed trees can become a limiting factor for white spruce regeneration [25,76]. During mast years, white spruce seeds are wind dispersed, with the greatest number of seeds falling within 100 to 150 m from the source tree [30]. Timoney and Peterson [25] found that in boreal Canada spruce recruitment following clearcutting was poor due to the size of the clearcut (most clearcuts exceeded 100 ha). In contrast, in Interior Alaska clearcut sites supported similar white spruce regeneration density as units that received a shelterwood harvest, a regeneration harvest technique that leaves white spruce seed trees on the harvest site [5]. The size of clearcut units was only 1.3 ha in the earlier shelterwood study [5]. Such small clearcuts provide ample opportunities for unharvested trees outside the units to disperse seeds into the units. Because most clearcut units in our study were smaller than 10 ha, it is reasonable to infer that similar seed dispersal processes took place in the operational harvests as in the shelterwood research study. Our analysis of the configuration of the sampled harvest units shows that over 90% of plots were within 100 m of the harvest unit perimeter, with the greatest distance of 150 m. Harvest units in reality rarely approached a circular configuration, which for a unit of 10 ha in size would create a maximum distance from the harvest edge of 180 m. This means that in our sampled regeneration units of 10 ha or smaller, the actual distance from the harvest perimeter was generally much less than 180 m. Therefore, we tentatively conclude that harvests smaller than 10 ha would, for the most part, not need retained seed trees within the harvest stands for white spruce regeneration.

In Interior Alaska, although profound negative effects of clearcutting have not been found, the major harvest method has been shifting from clearcutting to partial cutting primarily due to social and ecological concerns about clearcutting which have been reported in other boreal regions. In Interior Alaska, clearcutting at the current (small) scale does have some advantages compared to partial cutting, particularly because clearcutting appears to promote regeneration, and is more operationally efficient and thus more economical [77]. In addition, clearcutting in general has to be small because of the predominantly small size of pure white spruce stands that are the main target of harvest. On the other hand, clearcutting (in the literal sense of complete tree removal) removes some legacy forest structures that are important to wildlife or ecological value that could be retained in a partial cutting system. As a result, clear management objectives are necessary in order to choose the optimal harvest method.

Site preparation resulted in greater presence, dominance, and basal area of spruce, birch, and aspen in most cases (Figure 3). Site preparation has been widely demonstrated to enhance seedbed quality for tree regeneration [73,78], thus promoting more vigorous trees that can achieve higher rates of both below- and above-ground growth. Several experimental studies in Interior Alaska have reported that site preparation results in higher density and/or growth of white spruce [2,3,5,79], and in the boreal forest of Canada [22,80]. Our study now establishes that these gains are also achieved in operational practices up to 40 years following harvest. Moreover, site preparation also appears to have positive effects on birch and aspen regeneration. The increases in the presence and basal area of birch and aspen due to site preparation appear to be the result of exposure of a mineral soil substrate compared to the undisturbed forest organic layer [71,73]. One exception to this, however, appears to be the lower dominance of birch saplings and the aspen "all size" group on scarified compared to unscarified sites (Figure 3). This reversal of the general scarification effect appears to be related to the greater dominance by other species, and the relative importance of site preparation on these response variables was low, indicating the magnitude of differences were small (Figure 2). In general, site preparation following harvest promoted greater tree establishment and growth, which subsequently resulted in greater biomass accumulation (Figure 3).

While we found that site preparation typically results in greater success of regeneration, site preparation can at times result in stunted stems and roots due to intense competition during the stem exclusion phase [5] For example, unit NC-305 in our study supported very dense white spruce regeneration, 31,814 $\pm$ 12,558 stems$\cdot$ha^{-1}. This unit was logged in 1987, followed immediately by site preparation, during a year of exceptionally large white spruce seed production [27]. Although the amount of biomass in this unit was greater than the average in other harvest units, the mean diameter of regenerated stems was low, so that many years of additional growth will be required to produce harvestable material. In such a case, the optimal management approach might be to limit site preparation during the mast year. In order to do this, white spruce cone crops can be estimated by the previous year's seed production and visual inspection of bud primordia [78,81].

For the last few decades, foresters in Interior Alaska have used assisted regeneration techniques for white spruce [23] in order to sustain the species on some sites and promote large, well-positioned trees from the earliest stage of stand development. In fact, planting white spruce seedlings did result in greater white spruce presence, dominance, and basal area. However, the effects of planting on white spruce regeneration overall were very low (Figure 2a–e). In addition, dominance and basal area of aspen were lower in planted units than in naturally regenerated units (Figure 3), and reforestation did not contribute to prediction of biomass accumulation (Figure 2f). This result suggests that planted spruce seedlings suppressed natural regeneration of other species, although we cannot explicitly conclude this because of the low relative importance of the reforestation predictor term, and we do not have direct measurements of suppression. Due to the fact that planting white spruce seedlings is the most expensive post-harvest regeneration procedure, foresters should carefully evaluate the necessity of planting white spruce, and use this technique primarily in circumstances where the benefits are most likely to outweigh such undesired effects. Such circumstances include a management goal of producing large dimension white spruce in the shortest possible time, or regenerating spruce when no/low white spruce seed crop is present or predicted. Another possible management option would be delayed, refill planting that is spaced off from the established natural regeneration. On the other hand, if the goal is to produce biomass for energy generation, or to retain natural genetic diversity, then planting white spruce seedlings might have no or even potentially adverse effects.

The year of timber harvest was one of the most important predictors for many of our models, especially in determining basal area and biomass accumulation (Figure 2), because tree size clearly correlates with time since harvest. As expected, basal area and biomass tend to increase over time in the modeled prediction (Figure 3). Presence and dominance of the sapling group (DBH > 2.5 cm) also increase up to 40 years following harvest for all species as a result of tree growth (Figure 3). However, presence and dominance of trees of the "all size" group did not show a simple trend like those of saplings (Figure 3). In the early stage of regeneration, tree recruitment, growth, and mortality occur at different rates for each species, resulting in variability through time in the presence and dominance of seedlings. In addition, white spruce recruitment varies greatly by year due to its sporadic seed production cycle [27]. White spruce presence appears to increase up to 35 years following harvest, birch presence appears to increase up to about 25 years and then decline, but aspen presence shows a clear trend of decreasing for the entire study period (Figure 3). This reflects the early growth rate and the level of shade tolerance of each species [70,71,73]. White spruce has the slowest growth rates of the species measured in this study and is the most shade tolerant, while aspen has the fastest growth rates and is the least shade tolerant [70,71,73]. White spruce seedling recruitment seems to have continued for a longer period than birch and aspen, which are both shade-intolerant species that grow rapidly after disturbance [71,73]. The less shade tolerant aspen begins the self-thinning process earlier than birch (Figure 3). Birch regeneration appears to be determined largely by time since harvest and only marginally by management practices, suggesting that birch regeneration is only minimally affected by the environmental changes created by management practices.

5. Conclusions

The results of this study provide an important quantitative basis for future management planning in the western North American boreal forest. Management practices suitable or acceptable for some forms of biomass production appear to be different than practices traditionally used in the region for conventional spruce-focused management. In our study area, artificial reforestation does not appear to be superior to natural regeneration in obtaining more stems or producing greater biomass. However, clearcutting and site preparation consistently are associated with increased tree regeneration and greater basal area and biomass. As a result, clearcutting and site preparation are adequate as regeneration techniques, and planting white spruce may only be necessary in specific circumstances, such as in no/low white spruce seed crop years, or in landscapes depleted of seed trees. Finally, when biomass production of any species is the management goal, a shift from spruce harvest to birch may be possible, because birch regeneration is likely to be faster and more abundant without the additional effort required for white spruce establishment.

Acknowledgments: This research was funded by a project BAK LAP (Boreal Alaska—Learning, Adaptation, Production) supported by appropriation to the Alaska Department of Natural Resources Division of Forestry, the McIntire Stennis Cooperative Forestry Research Program (ALK13-04), and the Bonanza Creek Long-Term Ecological Research program funded by the National Science Foundation (DEB-1026415). We would like to thank the Fairbanks Area of the Alaska Department of Natural Resources Division of Forestry for providing the data, field equipment, and transportation to the study sites. We thank Salford Systems Ltd. for providing a copy of their software to the EWHALE lab by F. Huettmann, UAF for this study. Lastly, we thank R. Jess and M. Guisa for their invaluable help in the field.

Author Contributions: Miho Morimoto, Glenn Juday, and BrianYoung conceived and designed the sampling; Miho Morimoto performed the sampling and analyzed the data; Miho Morimoto and Glenn Juday wrote the paper.

Conflicts of Interest: The authors declare no conflict of interest.

References

1. Wurtz, T.L.; Gasbarro, A.F. A brief history of wood use and forest management in Alaska. *For. Chron.* **1996**, *72*, 47–50. [CrossRef]
2. Youngblood, A.; Cole, E.; Newton, M. Survival and growth response of white spruce stock types to site preparation in Alaska. *Can. J. For. Res.* **2011**, *41*, 793–809. [CrossRef]
3. Youngblood, A.P.; Zasada, J.C. White spruce artificial regeneration options on river floodplains in interior Alaska. *Can. J. For. Res.* **1991**, *21*, 423–433. [CrossRef]
4. Densmore, R.V.; Juday, G.P.; Zasada, J.C. Regeneration alternatives for upland white spruce after burning and logging in interior Alaska. *Can. J. For. Res.* **1999**, *29*, 413–423. [CrossRef]
5. Wurtz, T.L.; Zasada, J.C. An alternative to clear-cutting in the boreal forest of Alaska: A 27-year study of regeneration after shelterwood harvesting. *Can. J. For. Res.* **2001**, *31*, 999–1011. [CrossRef]
6. Chapin, F.; Fastie, C.; Viereck, L.; Ott, R.; Adams, P.; Mann, D.; Van Cleve, K.; Johnstone, J. Successional processes in the Alaskan Boreal Forest. In *Alaska's Changing Boreal Forest*; Chapin, F., Oswood, M., Van Cleve, K., Viereck, L., Verbyla, D., Eds.; Oxford University Press: New York, NY, USA, 2006; pp. 100–120.
7. Gauthier, S.; Bernier, P.; Kuuluvainen, T.; Shvidenko, A.Z.; Schepaschenko, D.G. Boreal forest health and global change. *Science* **2015**, *349*, 819–822. [CrossRef] [PubMed]
8. Foote, M.J. *Classification, Description, and Dynamics of Plant Communities after Fire in the Taiga of Interior Alaska*; U.S. Department of Agriculture, Pacific Northwest Forest and Range Experiment Station, Ed.; U.S. Department of Agriculture, Pacific Northwest Forest and Range Experiment Station: Portland, OR, USA, 1983.
9. Burton, P.J.; Parisien, M.-A.; Hicke, J.A.; Hall, R.J.; Freeburn, J.T. Large fires as agents of ecological diversity in the north american boreal forest. *Int. J. Wildland Fire* **2008**, *17*, 754–767. [CrossRef]
10. Chapin, F.S., III; Lovecraft, A.L.; Zavaleta, E.S.; Nelson, J.; Robards, M.D.; Kofinas, G.P.; Trainor, S.F.; Peterson, G.D.; Huntington, H.P.; Naylor, R.L. Policy strategies to address sustainability of alaskan boreal forests in response to a directionally changing climate. *Proc. Natl. Acad. Sci. USA* **2006**, *103*, 16637–16643. [CrossRef] [PubMed]
11. Calef, M.P.; Varvak, A.; McGuire, A.D.; Chapin, F.S., III; Reinhold, K.B. Recent changes in annual area burned in interior Alaska: The impact of fire management. *Earth Interact.* **2015**, *19*, 1–17. [CrossRef]
12. Johnstone, J.F.; Hollingsworth, T.N.; Chapin, F.S., III; Mack, M.C. Changes in fire regime break the legacy lock on successional trajectories in Alaskan boreal forest. *Glob. Chang. Biol.* **2010**, *16*, 1281–1295. [CrossRef]
13. Hollingsworth, T.N.; Johnstone, J.F.; Bernhardt, E.L.; Chapin, F.S., III. Fire severity filters regeneration traits to shape community assembly in Alaska's boreal forest. *PLoS ONE* **2013**, *8*, e56033. [CrossRef] [PubMed]
14. Shenoy, A.; Johnstone, J.F.; Kasischke, E.S.; Kielland, K. Persistent effects of fire severity on early successional forests in interior alaska. *For. Ecol. Manag.* **2011**, *261*, 381–390. [CrossRef]
15. Valentine, D.W.; Kielland, K.; Chapin, F.S., III; McCuire, A.D.; Van Cleve, K. Patterns of biogeochemistry in Alaskan boreal forests. In *Alaska's Changing Boreal Forest*; Oxford University Press: New York, NY, USA, 2006.
16. Johnstone, J.; Chapin, F. Effects of soil burn severity on post-fire tree recruitment in boreal forest. *Ecosystems* **2006**, *9*, 14–31. [CrossRef]

17. Johnstone, J.F.; Kasischke, E.S. Stand-level effects of soil burn severity on postfire regeneration in a recently burned black spruce forest. *Can. J. For. Res.* **2005**, *35*, 2151–2163. [CrossRef]
18. Johnstone, J.F.; Chapin, F.S.; Foote, J.; Kemmett, S.; Price, K.; Viereck, L. Decadal observations of tree regeneration following fire in boreal forests. *Can. J. For. Res.* **2004**, *34*, 267–273. [CrossRef]
19. Johnstone, J.F.; Rupp, T.S.; Olson, M.; Verbyla, D. Modeling impacts of fire severity on successional trajectories and future fire behavior in Alaskan boreal forests. *Landsc. Ecol.* **2011**, *26*, 487–500. [CrossRef]
20. Viereck, L.A.; Schandelmeier, L.A. *Effects of Fire in Alaska and Adjacent Canada: A Literature Review*; U.S. Department of the Interior, Bureau of Land Management, Alaska State Office: Anchorage, AK, USA, 1980.
21. Purdy, B.G.; Macdonald, S.E.; Dale, M.R.T. The regeneration niche of white spruce following fire in the mixedwood boreal forest. *Silva Fenn.* **2002**, *36*, 289–306. [CrossRef]
22. Boateng, J.O.; Heineman, J.; Bedford, L.; Harper, G.; Nemec, A.F.L. Long-term effects of site preparation and postplanting vegetation control on *picea glauca* survival, growth and predicted yield in boreal british columbia. *Scand. J. For. Res.* **2009**, *24*, 111–129. [CrossRef]
23. Alaska Division of Forestry. Forest management database. Data obtained from Alaska Division of Forestry, Fairbanks, Alaska; 2013.
24. Wurtz, T.; Ott, R.; Maishc, J. Timber harvest in interior Alaska. In *Alaska's Changing Boreal Forest*; Chapin, F., Oswood, M., Van Cleve, K., Viereck, L., Verbyla, D., Eds.; Oxford University Press: New York, NY, USA, 2006; pp. 302–308.
25. Timoney, K.P.; Peterson, G. Failure of natural regeneration after clearcut logging in wood buffalo national park, Canada. *For. Ecol. Manag.* **1996**, *87*, 89–105. [CrossRef]
26. Franklin, J.F.; Berg, D.R.; Thornburgh, D.A.; Tappeiner, J.C. Alternative silvicultural approaches to timber harvesting: Variable retention harvest systems. In *Creating a Forestry for the 21st Century: The Science of Ecosystem Management*; Kohm, K.A., Franklin, J.F., Eds.; Island Press: Washington, DC, USA, 1997; pp. 111–139.
27. Roland, C.A.; Schmidt, J.H.; Johnstone, J.F. Climate sensitivity of reproduction in a mast-seeding boreal conifer across its distributional range from lowland to treeline forests. *Oecologia* **2014**, *174*, 665–677. [CrossRef] [PubMed]
28. Juday, G.P.; Barber, V.; Rupp, S.; Zasada, J.; Wilmking, M. A 200-year perspective of climate variability and the response of white spruce in Interior Alaska. In *Climate Variability and Ecosystem Response at Long-Term Ecological RESEARCH Sites*; Greenland, D., Goodin, D.G., Smith, R.C., Eds.; Oxford University Press: New York, NY, USA, 2003.
29. Zasada, J.C.; Gregory, R.A. *Regeneration of White Spruce With Reference to Interior Alaska: A Literature Review*; Research Paper PNW-79; Institute of Northern Forestry, U.S. Forest Service: Washington, DC, USA, 1969; p. 37.
30. Youngblood, A.; Max, T.A. *Dispersal of White Spruce Seed on Willow Island in Interior Alaska*; USDA Forest Service Pacific Northwest Research Station Research Paper; USDA: Washington, DC, USA, 1992; pp. U1–U17.
31. Youngblood, A. Regenerating White Spruce in Boreal Forests of Alaska. Land and Ecosystem Management, 2012. Available online: http://www.fs.fed.us/pnw/lwm/lem/projects/youngblood_alaska.shtml (accessed on 2 May 2013).
32. Lieffers, V.I.; Macdonald, S.E.; Hogg, E.H. Ecology of and control strategies for *Calamagrostis Canadensis* in boreal forest sites. *Can. J. For. Res.* **1993**, *23*, 2070–2077. [CrossRef]
33. Alaska Division of Forestry. *Reforestation Handbook*; Alaska Department of Natural Resources Division of Forestry: Fairbanks, AK, USA, 2008.
34. Fresco, N.; Chapin, F.S., III. *Assessing the Potential for Conversion to Biomass Fuels in Interior Alaska*; U S Forest Service Pacific Northwest Research Station Research Paper PNW-RP; US Forest Service: Washington, DC, USA, 2009; pp. 1–56.
35. Alaska Energy Authority. *Alaska Wood Energy Development Task Force Project Status*; AWEDTG_Status_081315_8.5×11; AEA: Anchorage, AK, USA, 2015.
36. Janowiak, M.K.; Webster, C.R. Promoting ecological sustainability in woody biomass harvesting. *J. For.* **2010**, *108*, 16–23.
37. Burton, P.J.; Messier, C.; Adamowicz, W.L.; Kuuluvainen, T. Sustainable management of Canada's boreal forests: Progress and prospects. *Ecoscience* **2006**, *13*, 234–248. [CrossRef]

38. Larsson, S.; Danell, K. Science and the management of boreal forest biodiversity. *Scand. J. For. Res.* **2001**, *16*, 5–9. [CrossRef]
39. Roessler, J.S.; Packee, E.C. Disturbance history of the Tanana River Basin in Alaska: Management implications. In Proceedings of the Annual Tall Timbers Fire Ecology Conference. Fire and Forest Ecology: Innovative Silviculture and Vegetation Management, Tallahassee, FL, USA, 2000; pp. 46–57.
40. Youngblood, A. Regenerating white spruce in boreal forests of Alaska. Available online: http://www.fs.fed.us/pnw/lwm/lem/projects/youngblood_alaska.shtml (accessed on 14 May 2016).
41. Labau, V.J.; van Hees, W. An inventory of Alaska's boreal forests: Their extent, condition, and potential use. In Proceedings of the International Symposium on Boreal Forests: Condition, Dynamics, Anthropogenic Effects, Archangelsk, Russia, 16–26 July 1990.
42. Hanson, D. *Timber Inventory of State Forest Lands in the Tanana Valley 2013*; Alaska Department of Natural Resources Division of Forestry: Fairbanks, AK, USA, 2013.
43. Viereck, L.A.; Dyrness, C.T.; Van Cleve, K.; Foote, M.J. Vegetation, soils, and forest productivity in selected forest types in interior Alaska. *Canadian Journal of Forest Research* **1983**, *13*, 703–720. [CrossRef]
44. Ping, C.; Boone, R.; Clark, M.; Packee, E.; Swanson, D. State factor control of soil formation in interior alaska. In *Alaska's Changing Boreal Forest*; Chapin, F., Oswood, M., Van Cleve, K., Viereck, L., Verbyla, D., Eds.; Oxford University Press, Inc.: New York, NY, USA, 2006; pp. 21–38.
45. Shulski, M.; Wendler, G. *The Climate of Alaska*; University of Alaska Press: Fairbanks, AK, USA, 2007.
46. Wendler, G.; Shulski, M. A century of climate change for Fairbanks, Alaska. *Arctic* **2009**, *62*, 295–300. [CrossRef]
47. Nowacki, G.; Spencer, P.; Fleming, M.; Brock, T.; Jorgenson, T. *Ecoregions of Alaska: 2001*; U.S. Geological Survey Open-File Report 02-297 (map); U.S. Geological Survey: Reston, VA, USA, 2001.
48. The Society of American Foresters. *Dictionary of Forestry*; The Society of American Foresters: Bethesda, MD, USA, 1994.
49. Alaska Division of Forestry. *Annual Report*; Alaska Department of Natural Resources Division of Forestry: Anchorage, AK, USA, 2000.
50. Juday, G.P. Monitoring hectare-scale forest reference stands at Bonanza Creek Experimental Forest LTER. In *Long-Term Silvicultural & Ecological Studies: Results for Science and Management*; Camp, A.E., Irland, L.C., Carroll, C.J.W., Eds.; Global Institute of Sustainable Forestry, School of Forestry & Environmental Studies, Yale University: New Haven, CT, USA, 2012; pp. 31–48.
51. Bonanza Creek LTER. Available online: http://www.Lter.Uaf.Edu/ (accessed on 13 September 2016).
52. Yarie, J.; Kane, E.; Mack, M.C. Aboveground biomass equations for trees of Interior Alaska. In *Agricultural and Forestry Experiment Station Bulletin*; Univesity of Alaska Fairbanks: Fairbanks, Alaska, USA, 2007.
53. Jenness, J.; Brost, B.; Beier, P. Land Facet Corridor Designer: Extension for ArcGIS. 2013. Available online: http://www.jennessent.com/arcgis/land_facets.htm (accessed on 20 May 2014).
54. Alaska Division of Forestry. *Vegetation and Community Mapping of the Tanana Valley*; Alaska Department of Natural Resources Division of Forestry Northern Region: Fairbanks, AK, USA, 2013.
55. Web soil survey. Available online: http://websoilsurvey.sc.egov.usda.gov/App/HomePage.htm (accessed on 23 August 2016).
56. Snap data. Available online: http://ckan.snap.uaf.edu/dataset (accessed on 24 May 2016).
57. Wilmking, M.; Juday, G.P.; Barber, V.A.; Zald, H.S.J. Recent climate warming forces contrasting growth responses of white spruce at treeline in Alaska through temperature thresholds. *Glob. Chang. Biol.* **2004**, *10*, 1724–1736. [CrossRef]
58. Lloyd, A.H.; Duffy, P.A.; Mann, D.H. Nonlinear responses of white spruce growth to climate variability in interior Alaska. *Can. J. For. Res.* **2013**, *43*, 331–343. [CrossRef]
59. Beck, P.S.A.; Juday, G.P.; Alix, C.; Barber, V.A.; Winslow, S.E.; Sousa, E.E.; Heiser, P.; Herriges, J.D.; Goetz, S.J. Changes in forest productivity across alaska consistent with biome shift. *Ecol. Lett.* **2011**, *14*, 373–379. [CrossRef] [PubMed]
60. Juday, G.P.; Alix, C. Consistent negative temperature sensitivity and positive influence of precipitation on growth of floodplain Picea glauca in Interior Alaska. *Can. J. For. Res.* **2012**, *42*, 561–573. [CrossRef]
61. Van Cleve, K.; Viereck, L.A.; Dyrness, C.T. State factor control of soils and forest succession along the Tanana River in interior Alaska, USA. *Arct. Alp. Res.* **1996**, *28*, 388–400. [CrossRef]

62. Friedman, J.; Hastie, T.; Tibshirani, R. Additive logistic regression: A statistical view of boosting. *Ann. Stat.* **2000**, *28*, 337–374. [CrossRef]
63. Breiman, L. Statistical modeling: The two cultures. *Stat. Sci.* **2001**, *16*, 199–215. [CrossRef]
64. Betts, M.G.; Ganio, L.M.; Huso, M.M.P.; Som, N.A.; Huettmann, F.; Bowman, J.; Wintle, B.A. Comment on "Methods to account for spatial autocorrelation in the analysis of species distributional data: A review". *Ecography* **2009**, *32*, 374–378. [CrossRef]
65. Ohse, B.; Huettmann, F.; Ickert-Bond, S.M.; Juday, G.P. Modeling the distribution of white spruce (*Picea glauca*) for Alaska with high accuracy: An open access role-model for predicting tree species in last remaining wilderness areas. *Polar Biol.* **2009**, *32*, 1717–1729. [CrossRef]
66. Drew, C.A.; Wiersma, Y.; Huettmann, F. *Predictive Species and Habitat Modeling in Landscape Ecology: Concepts and Applications*; Springer Science & Business Media: New York, NY, USA, 2010.
67. Salford Systems. *Introducing Treenet*; Salford Systems, Ed.; Salford Systems: San Diego, CA, USA, 2013.
68. Metz, C.E. Basic principles of ROC analysis. *Sem. Nucl. Med.* **1978**, *8*, 283–298. [CrossRef]
69. Hastie, T.; Tibshirani, R.; Friedman, J. *The Elements of Statistical Learning: Data Mining, Inference and Prediction*; Springer: New York, NY, USA, 2009.
70. Nienstaedt, H.; Zasada, J.C. *Picea glauca* (moench) voss, white spruce. In *Silvics of North America: Volume 1. Conifers. Agriculture handbook 654*; Burns, R.M., Honkala, B.H., Eds.; USDA Forest Service: Washington, DC, USA, 1990; pp. 204–226.
71. Perala, D.A. *Populus tremuloides* Michx. Quaking Aspen. In *Silvics of North America*; Burns, R.M., Honkala, B.H., Eds.; USDA Forest Service: Washington, DC, USA, 1990; pp. 1082–1115.
72. Solarik, K.A.; Lieffers, V.J.; Volney, W.J.A.; Pelletier, R.; Spence, J.R. Seed tree density, variable retention, and stand composition influence recruitment of white spruce in boreal mixedwood forests. *Can. J. For. Res.* **2010**, *40*, 1821–1832. [CrossRef]
73. Safford, L.O.; Bjorkbom, J.C.; Zasada, J.C. Paper birch. In *Silvics of North America*; Burns, R.M., Honkala, B.H., Eds.; Forest Service, United States Department of Agriculture: Washington, DC, USA, 1990.
74. Marquis, D.A.; Bjorkborn, J.C.; Yelenosky, G. Effect of seedbed condition and light exposure on paper birch regeneration. *J. For.* **1964**, *62*, 876–881.
75. Perala, D.A.; Alm, A.A. Regeneration silviculture of birch: A review. *For. Ecol. Manag.* **1990**, *32*, 39–77. [CrossRef]
76. Greene, D.F.; Zasada, J.C.; Sirois, L.; Kneeshaw, D.; Morin, H.; Charron, I.; Simard, M.J. A review of the regeneration dynamics of North American boreal forest tree species. *Can. J. For. Res.* **1999**, *29*, 824–839. [CrossRef]
77. Keenan, R.J.; Kimmins, J.P. The ecological effects of clear-cutting. *Environ. Rev.* **1993**, *1*, 121–144. [CrossRef]
78. Gartner, S.M.; Lieffers, V.J.; Macdonald, S.E. Ecology and management of natural regeneration of white spruce in the boreal forest. *Environ. Rev.* **2011**, *19*, 461–478. [CrossRef]
79. Cole, E.; Youngblood, A.; Newton, M. Effects of competing vegetation on juvenile white spruce (*Picea Glauca* (Moench) Voss) growth in Alaska. *Ann. For. Sci.* **2003**, *60*, 573–583. [CrossRef]
80. Calogeropoulos, C.; Greene, D.F.; Messier, C.; Brais, S. The effects of harvest intensity and seedbed type on germination and cumulative survivorship of white spruce and balsam fir in northwestern Quebec. *Can. J. For. Res.* **2004**, *34*, 1467–1476. [CrossRef]
81. Lamontagne, J.M.; Boutin, S. Local-scale synchrony and variability in mast seed production patterns of *Picea glauca*. *J. Ecol.* **2007**, *95*, 991–1000. [CrossRef]

6

Regeneration Responses to Management for Old-Growth Characteristics in Northern Hardwood-Conifer Forests

Aviva J. Gottesman and William S. Keeton *

Rubenstein School of Environment and Natural Resources, University of Vermont, Burlington, VT 05405, USA; aviva.gottesman@uvm.edu

* Correspondence: William.Keeton@uvm.edu

Abstract: Successful tree regeneration is essential for sustainable forest management, yet it can be limited by the interaction of harvesting effects and multiple ecological drivers. In northern hardwood forests, for example, there is uncertainty whether low-intensity selection harvesting techniques will result in adequate and desirable regeneration. Our research is part of a long-term study that tests the hypothesis that a silvicultural approach called "structural complexity enhancement" (SCE) can accelerate the development of late-successional forest structure and functions. Our objective is to understand the regeneration dynamics following three uneven-aged forestry treatments with high levels of retention: single-tree selection, group selection, and SCE. Regeneration density and diversity can be limited by differing treatment effects on or interactions among light availability, competitive environment, substrate, and herbivory. To explore these relationships, manipulations and controls were replicated across 2 ha treatment units at two Vermont sites. Forest inventory data were collected pre-harvest and periodically over 13 years post-harvest. We used mixed effects models with repeated measures to evaluate the effect of treatment on seedling and sapling density and diversity (Shannon–Weiner H'). The treatments were all successful in recruiting a sapling class with significantly greater sapling densities compared to the controls. However, undesirable and prolific beech (*Fagus americana*) sprouting dominates some patches in the understory of all the treatments, creating a high degree of spatial variability in the competitive environment for regeneration. Multivariate analyses suggest that while treatment had a dominant effect, other factors were influential in driving regeneration responses. These results indicate variants of uneven-aged systems that retain or enhance elements of stand structural complexity—including old-growth characteristics—can generally foster abundant regeneration of important late successional tree species depending on site conditions, but they may require beech control where beech sprouting inhibits desired regeneration.

Keywords: old-growth; northern hardwood; regeneration; retention forestry

---

1. Introduction

Forest management for old-growth characteristics provides an important opportunity to help mitigate the effects of climate change, while providing many social, economic, and ecological benefits. The approach is to manage for high biomass stand structures such as those often found in old-growth forests [1,2]. This could entail, for example, emulating—through various retention forestry techniques—the type of natural disturbances and stand development processes leading to the development of high biomass conditions [3,4]. Prior to European settlement, forests in the northeastern U.S. were dominated by relatively frequent, gap-forming and partial disturbances that created a finely patterned mosaic of successional conditions [5,6]. These were dominated by late successional/old-growth stand structures [7], providing higher levels of some ecosystem functions, in comparison to the younger, secondary forests that predominate today.

There are a number of old-growth attributes indicative of stand structural complexity in northern hardwoods that could be promoted through retention forestry [8–11]. These include greater availability of large downed and standing woody debris (snags), large trees, tip-up mounds, horizontal variation in stand density (e.g., gaps of varying sizes and shapes), vertically complex canopies, and advanced regeneration [2,12,13]. In managing for structural complexity, silviculturists might promote a variety of ecological functions, including habitat for late-successionally-associated wildlife, hydrologic regulation [14], and increased carbon storage potential [15–17]. Structural retention systems are of great interest as a means for providing a broad array of biodiversity and ecosystem services in managed forests [18–20], but successful regeneration outcomes are required for wider adoption by the forestry profession. For example, in the northern hardwood region of eastern North America, the increased canopy cover and shade associated with high levels of retention, particularly in the context of selection harvesting systems, pose challenges for regenerating economically and ecologically desirable tree species, even though many of these are shade tolerant [21]. Our research investigates the question of whether a silvicultural practice promoting old-growth characteristics, termed "Structural Complexity Enhancement (SCE)," can regenerate desirable tree species and establish a new cohort of saplings at sufficient densities to be sustainable in comparison to conventional uneven-aged prescriptions. SCE was specifically designed to accelerate the development of old-growth forest structure and function, targeting stand structure attributes derived from previous research on old-growth forests in the U.S. Northeast [8,12,22]. Effects on stand structure [6,9], elements of biodiversity [23–25], and carbon cycling [17] have been explored previously. Research has shown that Structural Complexity Enhancement can increase herbaceous understory plant diversity [24], terrestrial amphibian populations [23], and fungal species richness [25], while enhancing stand structural complexity and promoting late-successional functions such as carbon storage [9]. However, no study has tested the effect of Structural Complexity Enhancement on the regeneration abundance and diversity in northern hardwood forests. Additionally, we are interested in whether the regeneration responses are influenced by other sources of variability, such as herbivory, substrate, light intensity, and climate, that might interact with treatment effects.

Regeneration remains a key source of uncertainty in predicting the long-term viability and efficacy of structural complexity treatments (e.g., [26–28]), including old-growth silviculture [11,18]. Challenges to successful regeneration of desirable species in selection harvesting, including gap-based silviculture, go beyond the light environment and include seed availability, desirable substrate, and competition with both herbaceous cover and advanced regeneration [6]. Herbivory by white-tailed deer (*Odocoileus virginianus*), moose (*Alces alces*), and eastern cottontail rabbit (*Sylvilagus floridanus*) can strongly influence regeneration responses and redirect compositional development [29–31]. Moreover, tree regeneration in the Northeast is increasingly limited by competition with American beech (*Fagus grandifolia*) due to its high shade tolerance and vegetative sprouting response to beech bark disease (BBD, *Nectria coccinea* var. *faginata*), causing many [21] to be skeptical of the ability of single-tree selection harvesting systems to regenerate desirable shade tolerant (e.g., sugar maple (*Acer saccharum*) and intermediate tolerant (e.g., yellow birch (*Betula alleghaniensis*)) species. The vigorous root sprouting mechanism of beech leads to the formation of dense beech thickets [32]. Beech thickets alter stand species composition through competition for light and other resources, while providing a substrate for BBD to spread from older diseased trees to the sapling class [33]. Due to the uncertainty of beech regeneration effects on the understory competitive environment, we are particularly interested in the recruitment response of common beech associates, such as sugar maple and yellow birch. Regeneration of sugar maple and yellow birch following harvesting tends to be spatially and temporally variable and dependent on interactions among multiple factors. Harvest gaps, as well as soil scarification and deadwood availability, can positively influence yellow birch establishment [34,35]. Sugar maple regenerates well on rich soils with partial canopy cover. Acid deposition can substantially deplete soil calcium on poorly buffered soils and limit sugar maple growth in some areas, another factor that could favor beech over time [36].

Ability to accurately predict regeneration responses to structural retention remains highly limited in northern hardwood-conifer systems. The first objective of this project was to explore the potential of silvicultural techniques retaining and promoting old-growth structure to successfully regenerate desirable tree species. We defined successful as the establishment of seedlings (approx. 12,300/ha minimum as suggested by Nyland [37]) and the recruitment of seedlings into the sapling class. We defined desirable tree species diversity as non-declining species diversity, where the treatments do not negatively influence species diversity from the pre-treatment baseline. The second objective was to examine varying patterns of regeneration among species, specifically American beech and sugar maple. We defined desirable regeneration as including intermediate and shade-tolerant, commercially important (i.e., for wood products and syrup production) species (sugar maple, red maple (*Acer rubrum*), and yellow birch), at densities sufficient to ensure the development of a new, merchantable cohort. Red maple was included because it is increasingly tapped for maple syrup and utilized for a variety of innovative wood products [38,39]. The third objective was to explore relationships of interest based on a priori knowledge from the regeneration literature, including those that might be modified through silvicultural practices. Relationships of interest that influence regeneration dynamics included interactions between canopy cover, substrate composition, light intensity, herbivory, climate and silvicultural treatments.

2. Materials and Methods

2.1. Study Area

The Vermont Forest Ecosystem Management Demonstration Project (FEMDP) is a study testing the long-term stand dynamics, biodiversity responses, and ecosystem service outcomes of experimental treatments designed to emulate fine-scale natural disturbance effects and promote development of late-successional/old-growth characteristics. The two study areas are located in Vermont, USA at the Mount Mansfield State Forest (MMSF) and the University of Vermont's Jericho Research Forest (JRF, Figure 1, Table 1).

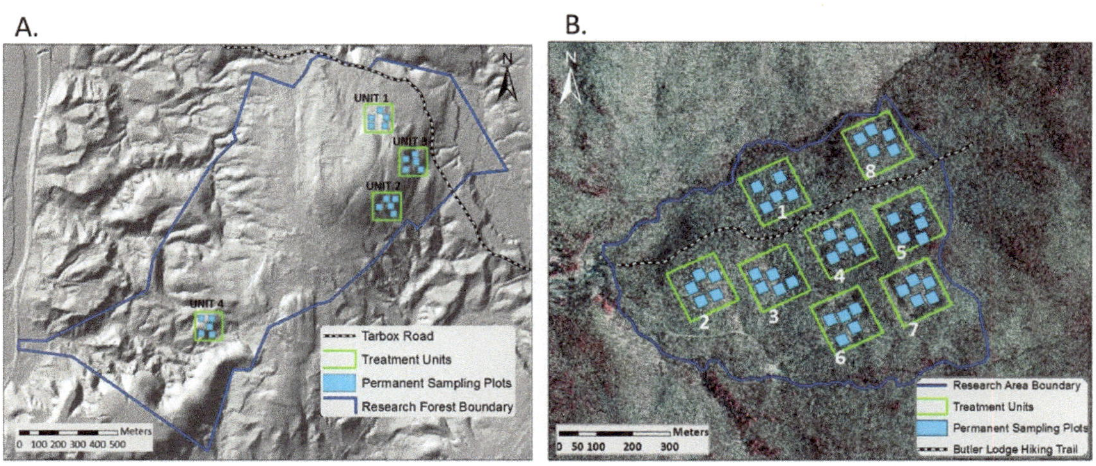

Figure 1. Regional map with locations of the two project study sites: Jericho Research Forest (**A**) and Mt. Mansfield State Forest (**B**). Also shown are treatment unit layout maps of the two study areas. Mansfield treatment manipulations: Units 1 and 8, Control; 2–3, Structural Complexity Enhancement 4–5, Single-Tree Selection; 6–7 Group Selection. Jericho manipulations: 1 and 4, Control; 2–3 Structural Complexity Enhancement.

Table 1. Site characteristics of experimental treatment units located in the Mansfield and Jericho study areas of the Forest Ecosystem Management Demonstration Project Adapted from Ford and Keeton (*in press*).

Site	Unit	Treatment	Site Index	Slope (%)	Aspect (deg)	% Hardwood	Total BA (m^2/ha)	Total Stem Density (trees/ha)	Live QMD [1] (cm)
Mansfield	1	Control	70	28.8	276	99.7	33.5	728	24.2
Mansfield	2	SCE	55	22.2	290	99.7	36.4	1044	21.1
Mansfield	3	SCE	55	13.0	260	99.7	28.5	1056	18.5
Mansfield	4	Single-Tree	60	29.6	272	95.9	33.9	750	24
Mansfield	5	Single-Tree	60	37.0	273	97.5	31.9	750	23.3
Mansfield	6	Group	60	19.4	249	98.7	30.1	1140	18.3
Mansfield	7	Group	60	26.4	250	99.4	30.8	1144	18.5
Mansfield	8	Control	55	22.3	320	98.2	27.6	1066	18.2
Jericho	1	Control	60	27.1	188	53.1	35.4	1186	19.5
Jericho	2	SCE	60	27.8	146	83.0	33.5	1040	20.2
Jericho	3	SCE	60	42.6	147	54.8	44	1034	23.3
Jericho	4	Control	60	34.2	99	74.2	30.2	940	20.2

[1] Quadratic Mean Diameter.

Both sites are mature (ca. 70–100 years), multi-aged northern hardwood-conifer forests with a documented history of timber management [9,24,40]. The Jericho Research Forest (44°26′43.70″N; 72°59′44.15″W) is located in the foothills of the Green Mountain Range, a northern extension of the Appalachian Range, at 200 to 250 m above sea level. Soils are Adams and Windsor loamy sands or sandy loams. The Mount Mansfield State Forest (44°30′23″ N; 72°50′11″ W) is on the western slopes of the Green Mountain Range. Elevations at the study area range from 470 to 660 m above sea level. Soils are primarily Peru extremely stony loams. The overstory at both sites is dominated by sugar maple, American beech, and yellow birch. At JRF, eastern hemlock (*Tsuga canadensis)* is also co-dominant, with minor components of red maple and northern red oak (*Quercus rubra).* Red spruce (*Picea rubens*) is a minor element of the canopy at MMSF [9].

2.2. Study Design

The FEMDP investigated three uneven-aged silvicultural techniques with high levels of post-harvest structure such as residual canopy cover, standing dead snags, and downed coarse woody debris. These were assigned to 2 ha treatment units in a randomized block design. At MMSF, the three treatments were all replicated twice; at JRF the SCE treatment was replicated twice. Both sites had two un-manipulated control units. To prevent soil damage and erosion, experimental manipulations (i.e., logging) were conducted in deep snow on frozen ground in the winter (January–February 2003). Marking guidelines specifically targeted retention of disease resistant beech, larger trees showing evidence of wildlife use (e.g., black bear (*Ursus americanus*) activity, cavity excavation, etc.), and tree species diversity.

Although all three treatments in the FEMDP shared structural retention as an objective, they differed in harvest approach, for example in the degree of uniform vs. variable density tree marking, silvicultural gap size and configuration, and amount of live and dead tree retention [6,9,41]. The treatments included two conventional uneven-aged prescriptions, single-tree selection (STS) and group selection (GS), which retain post-harvest structure at the high end of the range of retention parameters typical for selection harvesting in the region (Table 2). The modifications for single-tree selection set retention targets of 18.4 $m^2 \cdot ha^{-1}$ in the residual basal area, a maximum tree diameter of 60 cm at breast height (dbh, 1.37 m), and a q-factor of 1.3 to re-allocate the basal area into larger diameter classes. The group selection treatment followed the same diameter distribution-based prescription, but was applied through spatially aggregated harvesting. The group patches averaged 0.05 ha in size, a prescription intended to emulate the average size of canopy openings created by fine-scale natural disturbance in New England based on work summarized by Seymour et al. [42]. Group openings were variable in shape and some had light retention, consisting primarily of large snags and, occasionally, large diameter beech exhibiting resistance to beech bark disease (*Nectria coccinea* var. *faginata*). The treatment resulted in eight to nine groups per treatment unit, with the matrix between groups left unharvested [9].

The third treatment, SCE, employed a rotated sigmoid target (or post-harvest) diameter distribution (see [43,44]) defined by a high residual basal area (34 $m^2 \cdot ha^{-1}$), large maximum tree size (90 cm), and a variable q-factor applied to three portions of the diameter distribution (2.0 in the smallest size class, 1.1 in the medium size class, and 1.3 in the largest size class). The combined effect was a re-allocation of the basal area and growing space into larger diameter classes, while retaining abundant stems across all sizes and ages. Other structural objectives included vertically differentiated canopies, higher densities of large snags and downed logs, variable horizontal density including small (mean size of 0.02 ha) gaps, and recruitment of greater densities of large trees. The latter was achieved through partial or full crown release [8]. At one unit in each of the two study areas, trees were pulled (or pushed) over to create the tip-up mounds characteristic of old-growth northern hardwood forests.

Table 2. Silvicultural prescription details for experimental treatment manipulations at the MMSF and JRF study areas. Listed in the table is the target BDq for each treatment The BDq is the residual basal area (B), maximum target diameter (D), and q-factor (q). The q-factor is equal to the ratio of the number trees in each successively larger size class. Adapted from Keeton (2006) and Ford and Keeton *(in press)*.

Treatment	Target Residual Basal Area ($m^2 \cdot ha^{-1}$)	Max Diameter (cm)	q-Factor	Structural Objective	Silvicultural Prescription
Single-Tree Selection	18.4	60	1.3	Increased post-harvest target structural retention	Elevated target residual basal area Slash/un-merchentable bole retention
Group Selection	18.4	60	1.3	Increased post-harvest target structural retention Variable horizontal density Vertically differentiated canopy Increased horizontal complexity	Elevated target residual basal area Slash/un-merchantable bole retention Variable density marking Release advanced regeneration Spatially aggregated harvest (patches ~0.05 ha)
SCE	34	90	2.0/1.1/1.3	Re-allocation of the basal area to larger size class Vertically differentiated canopy Recruitment of greater densities of large trees Elevated coarse woody material inputs for added structure	Rotated sigmoid diameter dist. High max. D and the target basal area Retention of trees >60 cm dbh Single-tree sel. with target diameter distribution Release advanced regeneration Full (3- or 4-sided) and partial (2-sided) crown release Tree girdling/felling and leaving trees

2.3. Data Collection

Each 2 ha treatment unit contained five randomly placed 0.1 ha permanent sampling plots. The plots were buffered from edge effects through placement at least 15 m within the interior of unit boundaries and collectively represent 25% of each unit's total area. For this study, we used one year of pre-treatment (2001) data, six years of post-harvest seedling data (2003, 2004, 2006, 2007, 2013, 2015), and three years of post-harvest sapling data (2003, 2004, 2015). Tree seedlings <1 m in height were identified and counted along two 1-m wide and 31.64 m long belt transects bisecting the 0.1 ha plots. Tree saplings >1 m in height and <5 cm dbh were identified and counted within a plot (0.02 ha) nested within the 0.1 ha plots. All live and dead trees >5 cm dbh within the 0.1 ha sampling plots were tagged, measured, and identified.

Additional variables hypothesized to influence seedling establishment and survival were inventoried in 2015. These included substrate and understory composition, light intensity, herbivory, and stand structure. Herbaceous and woody shrub cover by species was measured using an ocular estimate designed to ensure precision and consistency in estimation [45]. Understory plant cover data were measured with thirteen 1-m^2 quadrats placed systematically along plot transects. The substrate data were inventoried using the same quadrats and estimation methodology as the understory data. The substrate data presented here for the purpose of assessing seed beds were categorized as fine litter, mineral soil, and coarse woody debris (>10 cm diameter). For each sapling inventoried in the 0.02 ha sapling plot, branches were assessed for the presence of deer or moose herbivory (browse). Saplings that were unequivocally browsed (i.e., torn, ragged) at any stem height (not distinguishing between moose and deer) were recorded as having the presence of browse [46].

To quantify light conditions in each treatment, hemispheric canopy photography was used to estimate the proportion of direct light (DSF) transmitted through the canopy. The camera was mounted on a tripod with a self-leveling mechanism set at 1-m height and two photographs were taken at the center of each plot. All analyses and photo selections were conducted by the same lab technician to minimize and systematize any error associated with selecting thresholds during analyses. The canopy photographs were processed and analyzed using HemiView canopy analysis software (Delta-T_Devices, Cambridge, UK).

The Palmer Drought Severity Index (PDSI) was obtained for U.S. climatological division 2 (western Vermont) from the Earth System Research Laboratory of the National Oceanic and Atmospheric Administration (NOAA; [47]). The PDSI parameters were compiled from weather stations and averaged to get a PDSI value at a spatial scale much larger than our study area, about 20,000 km^2 or half the size of Vermont. Climate data from 2001 to 2013 were used as a coarse estimate of climatic conditions including relative dryness and drought potential. PDSI has been used previously to develop predictions of seedling success under different drought scenarios [48].

2.4. Statistical Analysis

Due to uneven treatment replication among the two study areas, which limited the strength of statistical inference, analyses were performed separately by site. Multivariate analyses were used to determine if treatment had an effect on trends in the regeneration response variables over time. We used Statistical Analysis Software (SAS Institute Inc.) to build a linear mixed effects model (LME) to model the diversity response (Shannon–Weiner H') of seedling and saplings. Because seedling and sapling abundance estimates were not normally distributed, we used a generalized linear mixed model (GLIMMIX) to model total seedling and sapling densities, as well as sugar maple, red maple, yellow birch and American beech responses over time. To assess regeneration responses of those species (termed "desirable") for which there is regional concern regarding regeneration success following selection harvesting, we combined sugar maple, red maple, and yellow birch regeneration into a single response variable, termed "SM_RM_YB". GLIMMIX models use multivariate analyses to test for effects in non-normal data and are most appropriate for analyzing repeated measures [49,50]. We assumed a negative binomial distribution and a first order autoregressive covariance structure for the repeated measures. Treatment and year were modeled as fixed effects, while plots and units were modeled as random effects. Following the model design of Smith et al. [24], plots were nested within units. Using this method, the effective statistical sample size per treatment by site is 10 because plot data are pooled. The model output provided parameter estimates for the fixed effects and covariance estimates for the random effects, as well as a pairwise comparison test of the treatments at each time point to compare mean estimates between the treatments. Pre-treatment (2001) data for regeneration diversity and density were included in the model as a continuous covariate, to normalize data and allow for standardized comparisons among treatments across a range of inherent site variability. Percentage differences between pre-treatment, peak abundance, and thirteen years post-harvest were calculated following Littlefield and Keeton [51], using the Equation (1):

$$\text{Percentage difference} = ((V_H - V_B)/((V_H + V_B)/2)) \times 100 \quad (1)$$

where V_H is the post-harvest value and V_B is the pre-treatment "baseline" value. Peak abundance occurred 4-years post-treatment for the seedling class and 13-years post-harvest for the sapling class.

To test for the interaction among ecological variables having a possible influence on tree regeneration, we examined the 13-year post-harvest data set with another generalized linear mixed model. Based on the literature (for example, [52–54] and [41]), we chose variables that have been shown to influence regeneration dynamics. Predictor variables included in the model were percent leaf litter, percent browse, Direct Site Factor (DSF, obtained through analysis of hemispheric photographs), and Curtis relative-density (RD) structure index [55]. The model form for seedling density was as follows: seedling density~Percent_Browse + Percent_Fine_Litter + DSF + Curtis_RD + Site + Treatment + Site × Treatment. Collinearity diagnostics showed that covariates used in the model were not collinear, as measured by the variance inflation factor, which was 1.1 or less for all covariates. We chose DSF as a proxy for light conditions; it represents the proportion of direct solar radiation that reaches each plot, relative to that location with no sky obstructions (open canopy). The Curtis-relative density index is an aggregated stand structure index that integrates quadratic mean diameter and stem density. It is a good indicator of canopy retention as it depicts the total occupation of growing space based on tree density and size and can be used for uneven-aged management [55]. In a separate analysis, PDSI was analyzed with the Durbin-Watson statistical test which detects the presence of autocorrelation in the residuals from a time series regression analysis [56]. The Durbin-Watson test resulted in a statistical value which is significant if it falls between a given range of values. We tested the Vermont climate division 2 PDSI data against regeneration density and diversity by treatment separately.

3. Results

3.1. Objective 1: Regeneration Density and Diversity Responses

At Mount Mansfield State Forest (MMSF), the interaction between time and treatment was significant, indicating that the treatment effects changed in different patterns over time (Table 3). Analysis of total seedling density trends over time resulted in different patterns of regeneration by site (Figure 2). At Jericho Research Forest (JRF), the interaction between time and treatment was not significant ($p = 0.1283$) and SCE did not increase regeneration density significantly over background rates. At MMSF, the three harvested treatments had significantly more seedlings than the control for the first four years post-harvest ($p < 0.0001$). After 13 years, total seedling density was similar across all treatments. Seedling densities in the Control units were highly variable year to year, exhibiting pulses of recruitment and mortality, such that densities fluctuated from 2,925 seedlings/ha to 92,757 seedlings/ha. Regeneration dynamics in all the harvested treatments showed an initial pulse of recruitment in the first four years, with the percent difference of GS increasing by 103%, STS increasing by 113%, and SCE by 53% from pre-treatment densities in the seedling class. After thirteen years, GS decreased −33%, STS decreased −76%, and SCE decreased −17% from peak abundance to a similar mean seedling density.

Table 3. Type 3 Tests of Fixed Effects results from the Generalized Linear Mixed Model with total seedling (A) and sapling (B) densities at Mt. Mansfield.

A. Type III Tests of Fixed Effects MMSF Seedlings				
Effect	**Number *df***	**Density *df***	***F* Value**	**Pr > *F***
Pre-treatment	1	208	31.94	<0.0001
Treatment	3	208	24.73	<0.0001
Time	5	208	22.43	<0.0001
Treatment × Time	15	208	5.72	<0.0001
B. Type III Tests of Fixed Effects MMSF Saplings				
Effect	**Number *df***	**Density *df***	***F* Value**	**Pr > *F***
Pre-treatment	1	36.85	0.17	0.6861
Treatment	3	3.244	0.65	0.6315
Time	2	61.74	2.91	0.0618
Treatment × Time	6	64.33	6.66	<0.0001

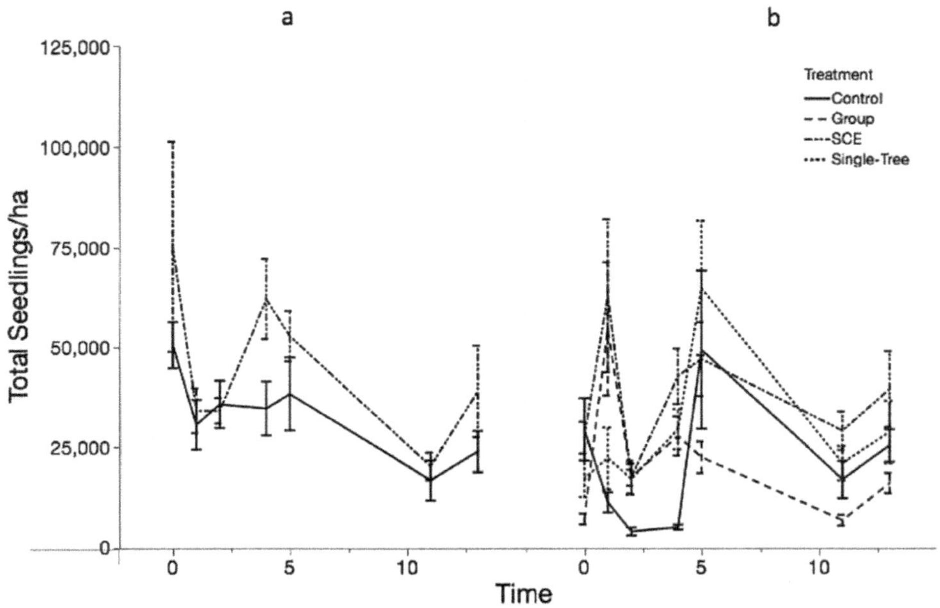

Figure 2. Total seedlings/ha by treatment over time at JRF (**a**) and MMSF (**b**). Error bars represent ±1 standard error from the mean.

While total seedling densities initially increased and then declined to the background rate detected in the Control over the 13-year period, total sapling densities display the opposite trend (Table 4). At MMSF, total sapling densities did not increase significantly over the first two years post-treatment. However, after 13 years, all treatments were successful at regenerating and recruiting a new cohort into the sapling class, at densities significantly greater than the Control (Figure 3, Table 5). Sapling densities in the Control declined over the 13-year time series, while GS, STS, and SCE increased after the initial two-year time lag. Although not significantly different, GS had the highest mean (and standard error), followed by STS and then SCE. According to the percent difference equation, all harvested treatments increased from pre-treatment to year 13, with GS increasing by 79%, STS by 60%, and SCE by 4% at MMSF and 54% at JRF

Table 4. Summary of seedling and sapling response thirteen years post-harvest at both sites. The table includes the mean of all species combined total stems/ha, minimum, and maximum values; mean beech stems/ha; SM_RM_YB = sugar maple, red maple, yellow birch combined means; and the Shannon–Weiner (H') value. All means shown with ±1 standard error. N represents the number of plots within the treatments.

Site	Treatment	N	Total Stems/ha Mean	Min	Max	Beech	SM_RM_YB	H-Index
			Seedlings					
JRF	Control	10	23,687 ± 5128	4269	56,293	3004 ± 1382	17,204 ± 4201	1.1 ± 0.14
	SCE	10	38,726 ± 11,438	12,018	122,233	1850 ± 464	27,973 ± 10,809	1.42 ± 0.11
MMSF	Control	10	25,206 ± 4089	11,701	50,127	1265 ± 294	21,743 ± 4271	1 ± 0.13
	Group	10	15,939 ± 2454	7748	34,156	3447 ± 553	11,196 ± 2554	1.12 ± 0.09
	SCE	10	39,374 ± 9506	14,231	111,006	6578 ± 1983	29,823 ± 9687	1.16 ± 0.08
	Single-Tree	9	28,586 ± 7812	4111	66,414	1125 ± 289	26,320 ± 7592	0.81 ± 0.1
			Saplings					
JRF	Control	10	1036 ± 226	160	2160	768 ± 238	92 ± 70	0.56 ± 0.12
	SCE	10	2260 ± 471	240	4480	1596 ± 501	16 ± 12	0.63 ± 0.12
MMSF	Control	10	1204 ± 208	320	2400	736 ± 144	292 ± 100	0.93 ± 0.1
	Group	10	5912 ± 1874	2200	22,400	4788 ± 1831	652 ± 206	0.67 ± 0.09
	SCE	10	3228 ± 382	1280	4760	2240 ± 433	476 ± 206	0.81 ± 0.12
	Single-Tree	9	3609 ± 674	1520	7520	964 ± 117	1662 ± 549	1.25 ± 0.06

At JRF, although the average sapling density was not significantly different between the Control and SCE at each time point (Time, p = 0.3665), the treatment × time fixed effect was significant (p = 0.0195) because the SCE treatment had an increase in sapling density while the Control had a decrease in sapling density over the 13-year time period. Pre-treatment sapling densities had a significant effect on these trends (p = 0.0079).

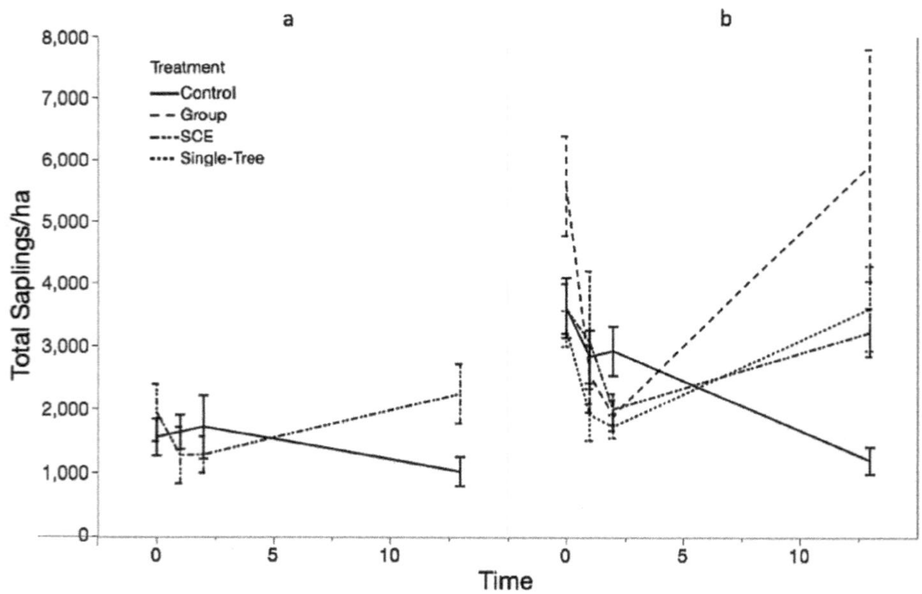

Figure 3. Total saplings/ha by treatment over time at JRF (**a**) and MMSF (**b**). Error bars represent ±1 standard error.

Table 5. Generalized Linear Mixed Model results from the pairwise comparison of Treatment × Time for total seedling (A) and sapling (B) densities at Mt. Mansfield. Estimate represents the transformed and adjusted stem counts (least squares mean). The outputs shown are immediately post-harvest (2003), and thirteen years post-harvest (2015).

			A. Seedlings				
	Treatment	**Treatment**	**Estimate**	**Standard Error**	***df***	***t* Value**	**Pr > |*t*|**
Time 1	Control	Group	−2.0489	0.3068	208	−6.68	<0.0001
	Control	SCE	−1.8638	0.2964	208	−6.29	<0.0001
	Control	STS	−0.9610	0.3079	208	−3.12	0.0021
	Group	SCE	0.1851	0.3039	208	0.61	0.5431
	Group	STS	1.0878	0.3059	208	3.56	0.0005
	SCE	STS	0.9027	0.3060	208	2.95	0.0035
Time 13	Control	Group	−0.0363	0.3065	208	−0.12	0.9060
	Control	SCE	−0.4215	0.2959	208	−1.42	0.1559
	Control	STS	−0.2814	0.3072	208	−0.92	0.3607
	Group	SCE	−0.3852	0.3045	208	−1.27	0.2072
	Group	STS	−0.2451	0.3063	208	−0.80	0.4244
	SCE	STS	0.1401	0.3060	208	0.46	
			B. Saplings				
Simple Effect Level	**Treatment**	**Treatment**	**Estimate**	**Standard Error**	***df***	***t* Value**	**Pr > |*t*|**
Time 1	Control	Group	0.1478	0.3502	10.19	0.42	0.6818
	Control	SCE	−0.079	0.3333	8.811	−0.24	0.8177
	Control	STS	0.3746	0.3398	9.447	1.10	0.2977
	Group	SCE	−0.2269	0.3504	10.2	−0.65	0.5315
	Group	STS	0.2268	0.3596	11.1	0.63	0.5411
	SCE	STS	0.4537	0.3398	9.445	1.34	0.2131
Time 13	Control	Group	−1.6100	0.3426	9.484	−4.70	0.0010
	Control	SCE	−1.0472	0.3333	8.815	−3.14	0.0122
	Control	STS	−1.1555	0.3398	9.449	−3.40	0.0073
	Group	SCE	0.5627	0.3427	9.49	1.64	0.1333
	Group	STS	0.4544	0.3518	10.35	1.29	0.2246
	SCE	STS	−0.1083	0.3398	9.443	−0.32	0.7569

We found that there were significant differences in treatment effects on species diversity. At JRF, there was no significant treatment × time effect ($p = 0.9373$). However, both the Control and SCE increased in seedling diversity (H'), with SCE having slightly higher seedling diversity than the Control at all time intervals. At MMSF, there was a significant treatment × time effect ($p = 0.0112$) on seedling diversity. The GS treatment almost doubled in seedling diversity over the 13-year time series, while STS had an increase in the first four years and then after 13 years declined to the level immediately post-harvest. The SCE treatment increased seedling diversity from year 1 to year 2 following harvest, and then leveled off, remaining constant to year 13 post-treatment. The Control varied only slightly from year to year, remaining around the same level of seedling diversity and showing no treatment × time effect. In the first year post-harvest, seedling diversity was lowest in the GS, followed by SCE, STS, and Control. After thirteen years, SCE had the highest seedling diversity, followed by the Control, then GS and finally STS (Table 4). Additionally, pre-treatment seedling diversity ($p = 0.0012$) was highly correlated with the final seedling diversity, indicating that plots that were highly diverse before the treatment maintained high diversity for the following years.

Sapling diversity at JRF declined for both the SCE and Control treatments over the 13-year time series. Although there was no treatment effect, the pre-treatment sapling diversity had a significant influence on the resulting sapling diversity. At MMSF, the pre-treatment sapling diversity also had a significant effect on the resulting sapling diversity. The Control declined in sapling diversity ($p = 0.0015$) over the 13-year time series, indicating regeneration limitations occurred regardless of the harvest. GS units showed an initial increase in sapling diversity 2 years following the harvest and then declines significantly after 13 years post-harvest ($p = 0.0042$). SCE performed comparably to

STS in regenerating and maintaining a diverse sapling cohort as demonstrated by the insignificant treatment × time effect, indicating that both treatments did not decline in sapling diversity as did the Control and GS units (see Table A1 for full species list).

3.2. *Objective 2: Competitive Interactions Among Species*

At JRF, beech seedlings had a significant treatment × time effect ($p = 0.0216$). The Control had a higher mean of beech seedlings for year 1 through year 11 post-harvest; however, the means were not significantly different between the treatments for the entire time series. At MMSF, there was a significant treatment × time effect ($p < 0.0001$). GS and SCE resulted in seedling recruitment responses that reached similar levels after 13 years, while STS and the Control had similar beech seedling densities from year to year. GS and SCE units had significantly more beech seedlings than STS and Control units, even though GS and SCE both showed decreasing beech seedling densities over the 13 years monitored.

At MMSF, there was a significant treatment × time effect ($p = 0.0040$) on beech sapling densities. Immediately post-harvest, SCE units had a greater beech sapling density than STS units ($p = 0.0257$); by year two, none of the treatment means were significantly different. However, after 13 years, the Control had significantly less beech saplings than GS ($p = 0.0020$) and SCE ($p = 0.0244$). Following the beech seedling trend at MMSF, GS and SCE displayed similar sapling densities while the Control and STS had similar sapling densities. SCE and STS treatments resulted in beech sapling densities that were not significantly different from each other 13 years post-harvest, while GS resulted in significantly more beech saplings than STS units ($p = 0.0089$).

At JRF, seedling densities for SM_RM_YB were not significantly different between the Control and SCE ($p = 0.0927$). At MMSF, there was a significant treatment × time effect ($p < 0.0001$) on SM_RM_YB, as well as a significant effect from the pre-treatment SM_RM_YB densities ($p = 0.0358$). After 4 years, the Control units had significantly fewer SM_RM_YB seedlings than the other three treatments; however, by the following year the densities had stabilized at levels that were not significantly different from one another. After 11 years, the GS treatment had significantly fewer of these species in the seedling class than SCE and STS. Following the pattern of total seedling densities, by year 13, all treatment means were not significantly different from each other (Figure 4). Although insignificant, the harvested treatments all have more SM_RM_YB seedlings than the Control. STS had a higher density of SM_RM_YB seedlings than SCE, while SCE had a higher density than GS.

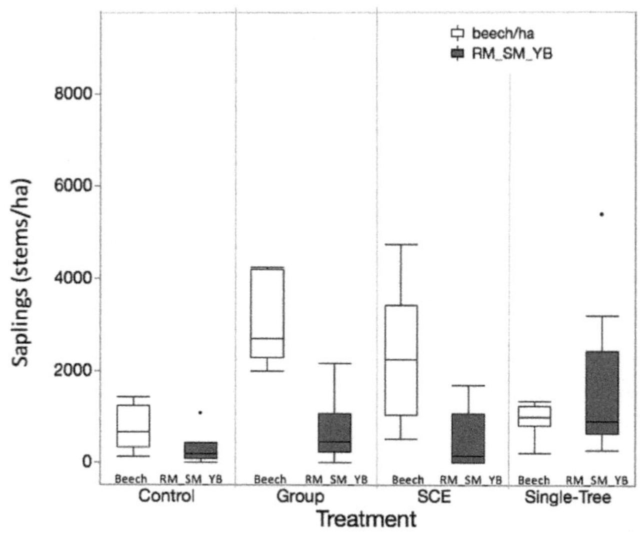

Figure 4. MMSF treatments thirteen years post-harvest showing beech saplings (white) in comparison to RM_SM_YB (red maple, sugar maple, yellow birch combined) saplings (grey). (Tukey) outlier box plot showing data distribution.

3.3. *Objective 3: Interactions with Herbivory, Substrate and Climate*

While mean densities of seedlings were similar after 13 years across all the treatments, plot level data showed large spatial variation or patchiness within the treatments (Table 4). Across both sites, the two factors found to have the most significant effect on sapling recruitment were percent browse (herbivory) and percent fine litter substrate (Table 6).

Table 6. Results from the Generalized Linear Mixed Model show that increasing browse and increasing fine litter substrate both resulted in a decline in sapling diversity (*H′*) and an increase in beech saplings. Direct Solar Fraction (DSF) and Curtis Relative-Density (RD) were not significant in the model.

Dependent Variable	Effect	Estimate	Standard Error	*df*	*t* Value	Pr > |*t*|
Sapling Diversity (*H′*)	Percent Browse	−0.00447	0.002122	48.7	−2.11	0.0403
	Percent Fine Litter Substrate	−0.01230	0.005158	46.3	−2.38	0.0213
	DSF	−0.02103	0.7502	46.4	−0.03	0.9778
	Curtis RD	−0.00542	0.03665	47.6	0.15	0.8830
Beech sapling	Percent Browse	0.01130	0.005656	49	2.00	0.0512
	Percent Fine Litter Substrate	0.02944	0.01371	46.55	2.15	0.0369
	DSF	0.4978	1.9390	46.31	0.26	0.7985
	Curtis RD	−0.06259	0.09770	48.21	−0.64	0.5248

Thirteen-years post-harvest, percent browse on saplings had a positive impact on seedling diversity ($p = 0.0286$) and a negative impact on sapling diversity ($p = 0.0403$). Increasing percent browse was positively correlated with the density of beech ($p = 0.0512$) and sugar maple saplings ($p = 0.0443$).

Percent fine litter had an influence on regeneration response similar to that of percent browse. Our data showed a positive relationship between fine litter substrate and beech sapling density ($p = 0.0369$), while there was a significantly negative relationship with sapling diversity ($p = 0.0213$).

Total seedling densities were not significantly correlated with any of the variables presenting possible non-treatment influences on regeneration dynamics. There was a significant treatment effect on total sapling densities ($p = 0.0327$), indicating that after light conditions (DSF), substrate (fine litter), herbivory, and Curtis-RD were accounted for in the model, the treatment still had the greatest effect on the amount of saplings (Table 6).

The Palmer Drought Severity Index (PDSI) values for northwestern Vermont indicated moist, non-drought conditions for most of the duration of the study. However, in the growing season (June–August) of 2001, the year the study was established, the PDSI was −1.927 which is categorized as a mild drought. This declined to a severe drought with a PDSI of −3.350 in the fall (Septeber–November). PDSI remained high with no drought until 2012 when there was a mild drought that decreased into an incipient (developing) drought in the spring of 2013.

In the Durban–Watson analysis, species diversity (*H′*) in the SCE treatment at JRF was significantly correlated with PDSI values indicative of high moisture, (DW statistic = 1.056, $p = 0.05$). None of the other treatments showed a correlation between PDSI and the diversity index. For all the treatments, the total density of seedlings was not significantly correlated with PDSI.

4. Discussion

Silvicultural techniques promoting the development of late-successional/old growth structure in northern hardwood-conifer forests have the potential to regenerate and establish a diverse new cohort of trees, including desirable species, such as sugar maple, red maple, and yellow birch, despite the relatively high levels of structural retention typical of selection harvesting systems. The effects of SCE were highly variable spatially, with successful regeneration in certain patches, while other areas were dominated by dense beech thickets. These results supported the hypothesis that SCE and other silvicultural approaches specifically intended to increase horizontal complexity in stand structure (see [57]), dependent on site conditions and other factors, have the potential to achieve regeneration levels close to or even greater than conventional uneven-aged practices, including selection systems employing larger group openings. They lend support to the efficacy of old-growth silviculture [18] for long-term sustainable management from a regeneration and recruitment perspective.

Yet, as with previous research on selection harvesting [58–60], regeneration success was clearly mixed and strongly affected by competition with beech sprouting especially. Regeneration of desirable species, in particular, ranged among patches and sites from absent or poor to very abundant.

4.1. Regeneration Response to Old-Growth Management

We found variants of uneven-aged silviculture systems designed to enhance structural complexity and old-growth characteristics adequately regenerate late-successional tree species, depending on site conditions and patch-scale controls. Although the seedling and sapling response was not as abundant as that from conventional even-aged systems, the total regeneration density was sufficient to reach full stocking according to the minimum threshold recommended by Nyland [37]. Regeneration dynamics in all the harvested treatments showed an initial pulse of recruitment in the seedling class. After thirteen years, all treatments returned to a similar mean seedling density. The decline from peak abundance (achieved in year 4 post-treatment) in seedling density can be attributed to a combination of factors, including partial canopy closure and reduced light availability, competition and density-dependent mortality, and recruitment into the sapling class [61]. The Control units at JRF had regeneration occurring in natural gaps thirteen-years post-harvest which could explain the lack of difference in regeneration between the Control and SCE treatments at that site. The treatments were all successful in recruiting a sapling class over the thirteen years post-harvest. However, 62% of the sapling layer was made up of dense beech (1843/ha on average) thickets after 13 years, a trend we found across all treatments. Beech thickets were most abundant in the GS treatment, intermediate in the SCE treatment, and least prevalent in the STS treatment.

Species diversity in the seedling class increased across all harvested treatments. Pre-treatment species diversity was as predictive as treatment type, suggesting a strong influence of initial site conditions and past stand development history on community composition. At the FEMDP sites, the spatially heterogeneous regeneration pattern for some species, such as sugar maple, red maple, and yellow birch, is clearly influenced both by light availability and competition with beech thickets, the latter being present in all treatments. Beech seedlings in all treatments showed a very large initial increase immediately post-harvest and then declined after a decade, most likely due to recruitment into the sapling class.

Previous studies have found that maple and beech tend to replace each other in stands uninfected with beach bark disease in the Upper Lake states, with saplings often occurring beneath the canopy of the other species [62,63]. Both beech and sugar maple have been found to outcompete one another following a disturbance or harvest, depending on site conditions. Sugar maple typically has faster growth rates than beech under higher light conditions in gaps (e.g., [64–66]). However, due to asexual reproduction through root sprouting and shared belowground nutritional supply, beech can often outcompete sugar maple in the understory, ultimately overtopping sexually reproduced maple [67–69]. In addition to outcompeting sugar maple, beech produces abundant beech leaf litter which contains leachate with phytotoxins that inhibit the development of sugar maple seedlings [70].

4.2. Sources of Variability in Regeneration Dynamics

Consistent with previous reports from northern hardwood forests, regeneration limitations in our study were linked to multiple factors including competition from dense beech understories and deer browse [71,72]. Adding the ecological variables into the linear mixed model showed that increasing herbivory and leaf litter were negatively associated with sapling diversity and positively associated with beech sapling densities. There was a positive association between herbivory and seedling diversity, found also in previous studies when deer populations are about 3–6 deer/km^2 [73]. The positive impact of browse on seedling diversity could be a result of disproportionate browse on common, palatable species giving rare species seedlings a competitive advantage [74]. This effect may be short-lived as seedlings move into the sapling class and there is greater competition for resources. Although beech is palatable to deer, it is less preferentially browsed in comparison to yellow birch and sugar maple. Browse and litter substrate were positively associated with beech sapling density at the FEMDP sites. Browse pressure (deer, moose, and rabbit) was highest in GS (ranging from 80% to 97%), intermediate in SCE (43%–92%), and lowest in STS (45%–91%).

In the FEMDP experiment, the SCE treatment initially increased coarse woody debris (CWD) volumes by 140% [9], though these later declined due to decay [37]. In addition to mineral soil, partially decayed softwood coarse woody debris is a preferred seedbed for yellow birch establishment and is a substrate that provides a competitive advantage for birch by altering the sapling morphology (i.e., multilayered crowns) for improved shade-tolerance [35,75]. Beech control in the treatment units could allow established yellow birch saplings to grow into merchantable size classes, such as on the CWD substrate.

4.3. Effects of Climate Variability and Drought

Although we did not see a significant treatment effect on seedling diversity at JRF, there was a relationship between climate and SCE seedling diversity. This implies that there may be an influence of drought stress, as measured by PDSI, on regeneration trends in the structural complexity enhancement treatment [48]. Mild spring drought conditions in recent years could influence early seedling establishment, potentially causing a decline in regeneration diversity. In the winter of 2012, 2013, and 2015 the PDSI indicated an incipient drought, which could relate to lower amounts of precipitations in the form of snow. A decline in snow depth can lead to decreased seedling survival due to increased browse and soil freezing stresses [76,77]. Increased sapling mortality has also been associated with root exposure due to declining snow pack [78].

Climate change models predict that increased temperatures and higher drought deficits will influence regeneration dynamics in the eastern U.S. [79–81]. Some models suggest that now dominant northern hardwood and conifer species may decline due to water deficits [82]. However, sugar maple, red maple, yellow birch, and American beech, surprisingly, showed enhanced seedling survival with increasing water deficit between 25 mm and 625 mm across their entire range [81]. These may be transient effects; if conditions become too dry or too wet seedling mortality may increase [83,84]. However, maintaining tree species diversity, as the FEMDP treatments did, at stand and landscape scales is an important strategy conferring ecosystem resilience to climate related stresses [19].

5. Management Implications

Our research suggests that management for old-growth characteristics in northern hardwood-conifer forests can promote adequate regeneration; however, site specific modifications are recommended for successful recruitment of species other than beech. Some silviculturists recommend patch- or clear-cutting to prevent prolific beech regeneration [21]. However, in our study, beech sprouting is most abundant in the group selection openings, even though advanced beech regeneration was cut (or cleaned) at the time of harvest, suggesting that more aggressive beech control would be needed for a desirable regeneration outcome. Based on our results, we suggest that low-intensity selection systems can regenerate at desirable densities [37] but may require beech control to sustain a diversity of intermediate- and shade-tolerant species. All the treatments resulted in total seedling densities above the minimum recommended by regional silvicultural guides. If regenerating commercially valuable species is the primary management objective, alternative silvicultural systems may be preferable. However, where timber is integrated with other management objectives, such as late-successional wildlife habitat, non-timber forest products, riparian functionality, or carbon storage, the regeneration densities and mixed composition resulting from selection systems that enhance structural complexity might be considered acceptable.

Removal of beech saplings alone may not be sufficient to promote sugar maple seedling establishment under selection harvesting [66]. Precutting of beech saplings and repeated cleaning entries post-harvest are potential applications of beech control [52]. Repeated cutting of beech before overstory harvest can still promote beech suckering; however, beech regeneration may remain shorter than sugar maple and yellow birch [52]. As an alternative, several studies have tested limited herbicide treatments (e.g., [85,86]) and found them to be highly effective. Broad range application of glyphosate or triclopyr in the summer months is the most efficient and most effective treatment suggested by Nyland et al. [52] based on a literature review of experimental treatments to eliminate understory interferences. Glyphosate treatments were also found to be successful at maintaining desirable species such as sugar maple and yellow birch, while minimizing American beech in a northern Maine study [87]. However, herbicide application carries risks that some managers may find unacceptable. Combined beech removal and deer exclusion, though expensive and thus often not practical, offers another means for improving regeneration outcomes [88].

Gap size can strongly influence regeneration dynamics [89]. For example, a study of northern hardwoods in Minnesota found that gaps greater than 0.02 ha had the highest regeneration densities [90]. Another long-term silvicultural study found better sugar maple establishment and survival under single-tree selection in comparison to large and medium sized group-selection openings, while yellow birch seedlings followed the reverse trend, with higher densities in larger gaps compared to smaller gaps [61]. However, these study sites were uninfected with beech bark disease. Canopy gaps increase light levels which is thought to increase species diversity [91]. However, other sources of variability may lead to unanticipated regeneration responses to gap-based silviculture [54,61,90]. These include seed availability, presence or absence of a seedbank, and seed predation [6]. As a consequence, the regeneration objective remains challenging. However, in a review of gap-based silviculture, Kern et al. [6] suggest moving towards a continuum of gap sizes, shapes, and within-gap retention, emulating the complexity associated with partial disturbances. We support this suggestion based on our findings, in which spatially complex regeneration responses likely reflected, in part, the varied light environments created by canopy openings that differed in size among the treatments. Therefore, we recommend utilizing a range of gap sizes, with smaller gaps on better quality sites and larger gaps incorporating beech control on poorer sites.

Managing forests for old-growth and late-successional characteristics is an option that some managers are considering that would provide a broader range of habitat conditions and ecosystem service co-benefits approaches [9,11,18,92]. Our research shows that SCE, like other retention-based systems, is sustainable from a regeneration perspective and thus provides a viable approach that could be integrated into holistic forest management.

Acknowledgments: This research would not have been possible without support from the Northeastern States Research Cooperative. Prior support for the FEMDP was provided by the Vermont Monitoring Cooperative, the USDA McIntire-Stennis Forest Research program, and the USDA National Research Initiative. Thanks especially to Timothy Perkins, and Anthony D'Amato for extensive input on this paper. Valuable reviews were also provided by Garrett Meigs, Sarah Ford, Andrea Urbano, and the UVM Carbon Dynamics Laboratory. Finally, we wish to thank Alan Howard for assistance with statistical modeling and analysis.

Author Contributions: The study was originally designed and implemented by W.S.K., who also coordinated early stages of data collection. Latter stages of data collection, as well as the data analysis, were conducted by A.J.G. Both authors contributed to writing and editing the paper.

Conflicts of Interest: The authors declare no conflict of interest.

Appendix A

Table A1. Seedling and sapling densities thirteen years post-harvest by species, treatment means (±1 standard error). In Group, Control, SCE the n = 10, in Single-Tree the n = 9.

	Site	JRF		MMSF			
	Treatment	Control	SCE	Control	Group	SCE	Single-Tree
Seedling							
	Striped maple	727 ± 421	1471 ± 424	885 ± 220	269 ± 82	617 ± 267	422 ± 115
	Red maple	9899 ± 3826	22,359 ± 9863	32 ± 32	95 ± 63	1439 ± 1184	53 ± 37
	Sugar maple	3052 ± 1278	4206 ± 2084	15,591 ± 4930	4096 ± 2357	18,770 ± 9065	19,889 ± 8417
	Mountain maple	0 ± 0	0 ± 0	395 ± 195	127 ± 66	300 ± 166	158 ± 83
	Yellow birch	4254 ± 2298	1407 ± 838	6120 ± 1301	7005 ± 1390	9614 ± 1953	6378 ± 1665
	Sweet birch	0 ± 0	0 ± 0	0 ± 0	47 ± 47	0 ± 0	0 ± 0
	Paper birch	0 ± 0	0 ± 0	63 ± 35	79 ± 26	79 ± 49	88 ± 60
	Bitternut hickory	0 ± 0	0 ± 0	364 ± 313	237 ± 237	1059 ± 621	35 ± 35
	American beech	3004 ± 1382	1850 ± 464	1265 ± 294	3447 ± 553	6578 ± 1983	1125 ± 289
	White ash	901 ± 329	2325 ± 720	32 ± 21	348 ± 196	680 ± 236	141 ± 81
	Hophornbeam	1344 ± 680	2625 ± 1078	16 ± 16	0 ± 0	0 ± 0	0 ± 0
	Red spruce	0 ± 0	0 ± 0	411 ± 177	174 ± 60	158 ± 91	193 ± 51
	White pine	32 ± 21	79 ± 35	0 ± 0	0 ± 0	0 ± 0	0 ± 0
	Pin cherry	0 ± 0	16 ± 16	0 ± 0	0 ± 0	0 ± 0	0 ± 0
	Black cherry	0 ± 0	0 ± 0	32 ± 21	16 ± 16	32 ± 32	105 ± 46
	Red oak	63 ± 35	1502 ± 346	0 ± 0	0 ± 0	47 ± 34	0 ± 0
	Eastern hemlock	411 ± 283	885 ± 385	0 ± 0	0 ± 0	0 ± 0	0 ± 0
Sapling							
	Striped maple	20 ± 14	296 ± 84	80 ± 36	176 ± 75	396 ± 90	720 ± 127
	Sugar maple	80 ± 58	4 ± 4	56 ± 25	4 ± 4	116 ± 58	98 ± 46
	Mountain maple	0 ± 0	0 ± 0	0 ± 0	0 ± 0	0 ± 0	18 ± 18
	Yellow birch	12 ± 12	12 ± 9	236 ± 93	648 ± 205	360 ± 160	1564 ± 559
	Sweet birch	0 ± 0	0 ± 0	0 ± 0	8 ± 5	0 ± 0	0 ± 0
	American beech	768 ± 238	1596 ± 501	736 ± 144	4788 ± 1831	2240 ± 433	964 ± 117
	White pine	8 ± 8	12 ± 12	0 ± 0	8 ± 5	16 ± 11	0 ± 0
	Hophornbeam	8 ± 8	244 ± 192	0 ± 0	0 ± 0	0 ± 0	0 ± 0
	Red spruce	0 ± 0	0 ± 0	96 ± 36	272 ± 68	100 ± 26	244 ± 74
	Black cherry	0 ± 0	0 ± 0	0 ± 0	8 ± 8	0 ± 0	0 ± 0
	Eastern hemlock	140 ± 68	96 ± 46	0 ± 0	0 ± 0	0 ± 0	0 ± 0

References

1. Nunery, J.S.; Keeton, W.S. Forest Carbon Storage in the Northeastern United States: Net Effects of Harvesting Frequency, Post-Harvest Retention, and Wood Products. *For. Ecol. Manag.* **2010**, *259*, 1363–1375. [CrossRef]
2. Burrascano, S.; Keeton, W.S.; Sabatini, F.M.; Blasi, C. Commonality and Variability in the Structural Attributes of Moist Temperate Old-Growth Forests: A Global Review. *For. Ecol. Manag.* **2013**, *291*, 458–479. [CrossRef]
3. Franklin, J.F.; Lindenmayer, D.B.; MacMahon, J.A.; McKee, A.; Magnuson, J.; Perry, D.A.; Waide, R.; Foster, D. Threads of Continuity. *Conserv. Pract.* **2000**, *1*, 8–17. [CrossRef]
4. North, M.P.; Keeton, W.S. Emulating Natural Disturbance Regimes: An Emerging Approach for Sustainable Forest Management. In *Patterns and Processes in Forest Landscapes*; Lafortezza, R., Sanesi, G., Chen, J., Crow, T.R., Eds.; Springer: Dordrecht, The Netherlands, 2008; pp. 341–372.
5. Cogbill, C.V.; Burk, J.; Motzkin, G. The Forests of Presettlement New England, USA: Spatial and Compositional Patterns Based on Town Proprietor Surveys. *J. Biogeogr.* **2002**, *29*, 1279–1304. [CrossRef]
6. Kern, C.; Burton, J.I.; Raymond, P.; D'Amato, A.W.; Keeton, W.S.; Royo, A.A.; Walters, M.B.; Webster, C.R.; Willis, J.L. Challenges Facing Gap-Based Silviculture and Possible Solutions for Mesic Northern Forests in North America. *Forestry* **2017**, *90*, 4–17. [CrossRef]
7. Lorimer, C.G.; White, A.S. Scale and Frequency of Natural Disturbances in the Northeastern US: Implications for Early Successional Forest Habitats and Regional Age Distributions. *For. Ecol. Manag.* **2003**, *185*, 41–64. [CrossRef]
8. Singer, M.T.; Lorimer, C.G. Crown Release as a Potential Old-Growth Restoration Approach in Northern Hardwoods. *Can. J. For. Res.* **1997**, *27*, 1222–1232. [CrossRef]
9. Keeton, W.S. Managing for Late-Successional/Old-Growth Characteristics in Northern Hardwood-Conifer Forests. *For. Ecol. Manag.* **2006**, *235*, 129–142. [CrossRef]
10. Hanson, J.J.; Lorimer, C.G.; Halpin, C.R.; Palik, B.J. Ecological Forestry in an Uneven-Aged, Late-Successional Forest: Simulated Effects of Contrasting Treatments on Structure and Yield. *For. Ecol. Manag.* **2012**, *270*, 94–107. [CrossRef]

11. Fassnacht, K.S.; Bronson, D.R.; Palik, B.J.; D'Amato, A.W.; Lorimer, C.G.; Martin, K.J. *Accelerating the Development of Old-Growth Characteristics in Second-Growth Northern Hardwoods*; General Technical Report; U.S. Department of Agriculture, Forest Service, Northern Research Station: Newtown Square, PA, USA, 2015.
12. McGee, G.G.; Leopold, D.J.; Nyland, R.D. Structural Characteristics of Old-growth, Maturing, and Partially Cut Northern Hardwood Forests. *Ecol. Appl.* **1999**, *9*, 1316–1329. [CrossRef]
13. Keeton, W.S.; Kraft, C.E.; Warren, D.R. Mature and Old-Growth Riparian Forests: Structure, Dynamics, and Effects on Adirondack Stream Habitats. *Ecol. Appl.* **2007**, *17*, 852–868. [CrossRef] [PubMed]
14. Wirth, C.; Messier, C.; Bergeron, Y.; Frank, D.; Fankhänel, A. Old-Growth Forest Definitions: A Pragmatic View. In *Old-Growth Forests*; Springer: Berlin/Heidelberg, Germany, 2009; pp. 11–33.
15. Keeton, W.S.; Whitman, A.A.; McGee, G.C.; Goodale, C.L. Late-Successional Biomass Development in Northern Hardwood-Conifer Forests of the Northeastern United States. *For. Sci.* **2011**, *57*, 489–505.
16. McGarvey, J.C.; Thompson, J.R.; Epstein, H.E.; Shugart, H.H. Carbon Storage in Old-growth Forests of the Mid-Atlantic: Toward Better Understanding the Eastern Forest Carbon Sink. *Ecology* **2015**, *96*, 311–317. [CrossRef] [PubMed]
17. Ford, S.E.; Keeton, W.S. Enhanced Carbon Storage through Management for Old-Growth Characteristics in Northern Hardwood-Conifer Forests. *Ecosphere* **2016**, in press.
18. Bauhus, J.; Puettmann, K.; Messier, C. Silviculture for Old-Growth Attributes. *For. Ecol. Manag.* **2009**, *258*, 525–537. [CrossRef]
19. D'Amato, A.W.; Bradford, J.B.; Fraver, S.; Palik, B.J. Forest Management for Mitigation and Adaptation to Climate Change: Insights from Long-Term Silviculture Experiments. *For. Ecol. Manag.* **2011**, *262*, 803–816. [CrossRef]
20. Lindenmayer, D.B.; Franklin, J.F.; Lõhmus, A.; Baker, S.C.; Bauhus, J.; Beese, W.; Brodie, A.; Kiehl, B.; Kouki, J.; Pastur, G.M.; et al. A Major Shift to the Retention Approach for Forestry Can Help Resolve Some Global Forest Sustainability Issues: Retention Forestry for Sustainable Forests. *Conserv. Lett.* **2012**, *5*, 421–431. [CrossRef]
21. Leak, W.B.; Yamasaki, M.; Holleran, R. *Silvicultural Guide for Northern Hardwoods in the Northeast*; USDA: Washington, DC, USA, 2014.
22. Ziegler, S.S. Disturbance Regimes of Hemlock-Dominated Old-Growth Forests in Northern New York, USA. *Can. J. For. Res.* **2002**, *32*, 2106–2115. [CrossRef]
23. McKenny, H.C.; Keeton, W.S.; Donovan, T.M. Effects of Structural Complexity Enhancement on Eastern Red-Backed Salamander (*Plethodon cinereus*) Populations in Northern Hardwood Forests. *For. Ecol. Manag.* **2006**, *230*, 186–196. [CrossRef]
24. Smith, K.J.; Keeton, W.S.; Twery, M.J.; Tobi, D.R. Understory Plant Responses to Uneven-Aged Forestry Alternatives in Northern Hardwood-Conifer Forests. *Can. J. For. Res.* **2008**, *38*, 1303–1318. [CrossRef]
25. Dove, N.C.; Keeton, W.S. Structural Complexity Enhancement Increases Fungal Species Richness in Northern Hardwood Forests. *Fungal Ecol.* **2015**, *13*, 181–192. [CrossRef]
26. Price, D.T.; Zimmermann, N.E.; van der Meer, P.J.; Lexer, M.J.; Leadley, P.; Jorritsma, I.T.M.; Schaber, J.; Clark, D.F.; Lasch, P.; McNulty, S.; et al. Regeneration in Gap Models: Priority Issues for Studying Forest Responses to Climate Change. *Clim. Chang.* **2011**, *51*, 475–508. [CrossRef]
27. Dodson, E.K.; Burton, J.I.; Puettmann, K.J. Multiscale Controls on Natural Regeneration Dynamics after Partial Overstory Removal in Douglas-Fir Forests in Western Oregon, USA. *For. Sci.* **2014**, *60*, 953–961. [CrossRef]
28. Kneeshaw, D.; Bergeron, Y. Applying Knowledge of Natural Disturbance Regimes to Develop an Ecosystem Management Approach in Forestry. In *Ecological Forest Management Handbook*; CRC Press: Boca Raton, FL, USA, 2015.
29. Augustine, D.J.; McNaughton, S.J. Ungulate Effects on the Functional Species Composition of Plant Communities: Herbivore Selectivity and Plant Tolerance. *J. Wildl. Manag.* **1998**, *62*, 1165–1183. [CrossRef]
30. Collins, B. Ground Layer Competition and Herbivory Effects on Cherrybark Oak (*Quercus pagoda* Raf.) Regeneration in Experimental Canopy Gaps. *J. Torrey Bot. Soc.* **2003**, *130*, 147–157. [CrossRef]
31. Andreozzi, H.A.; Pekins, P.J.; Langlais, M.L. Impact of Moose Browsing on Forest Regeneration in Northeast Vermont. *Alces J. Devoted Biol. Manag. Moose* **2014**, *50*, 67–79.

32. Houston, D.R. Beech Bark Disease—The Aftermath Forests Are Structured for a New Outbreak. *J. For.* **1975**, *73*, 660–663.
33. Giencke, L.M.; Dovčiak, M.; Mountrakis, G.; Cale, J.A.; Mitchell, M.J. Beech Bark Disease: Spatial Patterns of Thicket Formation and Disease Spread in an Aftermath Forest in the Northeastern United States. *Can. J. For. Res.* **2014**, *44*, 1042–1050. [CrossRef]
34. Gauthier, M.-M.; Lambert, M.-C.; Bédard, S. Effects of Harvest Gap Size, Soil Scarification, and Vegetation Control on Regeneration Dynamics in Sugar Maple-Yellow Birch Stands. *For. Sci.* **2016**, *62*, 237–246. [CrossRef]
35. Lambert, J.-B.; Ameztegui, A.; Delagrange, S.; Messier, C. Birch and Conifer Deadwood Favour Early Establishment and Shade Tolerance in Yellow Birch Juveniles Growing in Sugar Maple Dominated Stands. *Can. J. For. Res.* **2016**, *46*, 114–121. [CrossRef]
36. Huggett, B.A.; Schaberg, P.G.; Hawley, G.J.; Eagar, C. Long-Term Calcium Addition Increases Growth Release, Wound Closure, and Health of Sugar Maple (*Acer saccharum*) Trees at the Hubbard Brook Experimental Forest. *Can. J. For. Res.* **2007**, *37*, 1692–1700. [CrossRef]
37. Nyland, R.D. *Silviculture: Concepts and Applications*; McGraw-Hill Series in Forest Resources; Waveland Press: Long Grove, IL, USA, 2007.
38. Blankenhorn, P.R.; Labosky, P., Jr.; Janowiak, J.J.; Manbeck, H. The Development of Preservation Treatment Recommendations for Red Oak and Red Maple Glued-Laminated Timber Bridge Members. *For. Prod. J.* **1999**, *49*, 87.
39. Wang, X.; Ross, R.J.; Green, D.W.; Brashaw, B.; Englund, K.; Wolcott, M. Stress Wave Sorting of Red Maple Logs for Structural Quality. *Wood Sci. Technol.* **2004**, *37*, 531–537. [CrossRef]
40. Hannah, P.R. Species Composition and Dynamics in Two Hardwood Stands in Vermont: A Disturbance History. *For. Ecol. Manag.* **1999**, *120*, 105–116. [CrossRef]
41. D'Amato, A.W.; Catanzaro, P.F.; Fletcher, L.S. Early Regeneration and Structural Responses to Patch Selection and Structural Retention in Second-Growth Northern Hardwoods. *For. Sci.* **2015**, *61*, 183–189. [CrossRef]
42. Seymour, R.S.; White, A.S.; deMaynadier, P.G. Natural Disturbance Regimes in Northeastern North America—Evaluating Silvicultural Systems Using Natural Scales and Frequencies. *For. Ecol. Manag.* **2002**, *155*, 357–367. [CrossRef]
43. Goff, F.G.; West, D. Canopy-Understory Interaction Effects on Forest Population Structure. *For. Sci.* **1975**, *21*, 98–108.
44. Goodburn, J.M.; Lorimer, C.G. Population Structure in Old-Growth and Managed Northern Hardwoods: An Examination of the Balanced Diameter Distribution Concept. *For. Ecol. Manag.* **1999**, *118*, 11–29. [CrossRef]
45. Peet, R.K.; Wentworth, T.R.; White, P.S. A Flexible, Multipurpose Method for Recording Vegetation Composition and Structure. *Castanea* **1998**, *63*, 262–274.
46. Faison, E.K.; Motzkin, G.; Foster, D.R.; McDonald, J.E. Moose Foraging in the Temperate Forests of Southern New England. *Northeast. Nat.* **2010**, *17*, 1–18. [CrossRef]
47. Create a monthly or seasonal time series of climate variables. Available online: https://www.esrl.noaa.gov/psd/data/timeseries/ (accessed on 1 May 2016).
48. Zwolinski, J.; Donald, D.; Van Laar, A.; Groenewald, W. Regeneration Procedures of *Pinus radiata* in the Southern Cape Province: Part V: Post Planting Mortality and Growth of Trees in Response to the Experimental Treatments and Planting Site Environment. *S. Afr. For. J.* **1994**, *168*, 7–21.
49. Bolker, B.M.; Brooks, M.E.; Clark, C.J.; Geange, S.W.; Poulsen, J.R.; Stevens, M.H.H.; White, J.-S.S. Generalized Linear Mixed Models: A Practical Guide for Ecology and Evolution. *Trends Ecol. Evol.* **2009**, *24*, 127–135. [CrossRef] [PubMed]
50. Zuur, A.F.; Ieno, E.N.; Smith, G.M. *Analysing Ecological Data. Statistics for Biology and Health*; Springer: New York, NY, USA; London, UK, 2007.
51. Littlefield, C.E.; Keeton, W.S. Bioenergy Harvesting Impacts on Ecologically Important Stand Structure and Habitat Characteristics. *Ecol. Appl.* **2012**, *22*, 1892–1909. [CrossRef] [PubMed]
52. Nyland, R.D.; Bashant, A.L.; Bohn, K.K.; Verostek, J.M. Interference to Hardwood Regeneration in Northeastern North America: Controlling Effects of American Beech, Striped Maple, and Hobblebush. *North. J. Appl. For.* **2006**, *23*, 122–132.

53. Ward, J.; Worthley, T.E.; Smallidge, P.J.; Bennet, K.P. *Northeastern Forest Regeneration Handbook*; USDA: Newtown Square, PA, USA, 2013.
54. Kern, C.; Reich, P.B.; Montgomery, R.A.; Strong, T.F. Do Deer and Shrubs Override Canopy Gap Size Effects on Growth and Survival of Yellow Birch, Northern Red Oak, Eastern White Pine, and Eastern Hemlock Seedlings? *For. Ecol. Manag.* **2012**, *267*, 134–143. [CrossRef]
55. Curtis, R.O. Notes: A Simple Index of Stand Density for Douglas-Fir. *For. Sci.* **1982**, *28*, 92–94.
56. Durbin, J.; Watson, G. Testing for Serial Correlation in Least Squares Regression, III. *Biometrika* **1971**, *58*, 1–19. [CrossRef]
57. Franklin, J.F.; Spies, T.A.; van Pelt, R.; Carey, A.B.; Thornburgh, D.A.; Berg, D.R.; Lindenmayer, D.B.; Harmon, M.E.; Keeton, W.S.; Shaw, D.C.; et al. Disturbances and Structural Development of Natural Forest Ecosystems with Silvicultural Implications, Using Douglas-Fir Forests as an Example. *For. Ecol. Manag.* **2002**, *155*, 399–423. [CrossRef]
58. Mader, S.F.; Nyland, R.D. Six-Year Response of Northern Hardwoods to the Selection System. *North. J. Appl. For.* **1984**, *1*, 87–91.
59. Jones, R.H.; Nyland, R.D.; Raynal, D.J. Response of American Beech Regeneration to Selection Cutting of Northern Hardwoods in New York. *North. J. Appl. For.* **1989**, *6*, 34–36.
60. Donoso, P.J.; Nyland, R.D.; Zhang, L. Growth of Saplings after Selection Cutting in Northern Hardwoods. *North. J. Appl. For.* **2000**, *17*, 149–152.
61. Poznanovic, S.K.; Webster, C.R.; Bump, J.K. Maintaining Mid-Tolerant Tree Species with Uneven-Aged Forest Management: 9-Year Results from a Novel Group-Selection Experiment. *Forestry* **2013**, *86*, 555–567. [CrossRef]
62. Woods, K.D. Reciprocal Replacement and the Maintenance of Codominance in a Beech-Maple Forest. *Oikos* **1979**, *33*, 31–39. [CrossRef]
63. Whittaker, R.; Levin, S. The Role of Mosaic Phenomena in Natural Communities. *Theor. Popul. Biol.* **1977**, *12*, 117–139. [CrossRef]
64. McClure, J.W.; Lee, T.D.; Leak, W.B. Gap Capture in Northern Hardwoods: Patterns of Establishment and Height Growth in Four Species. *For. Ecol. Manag.* **2000**, *127*, 181–189. [CrossRef]
65. Ricard, J.-P.; Messier, C.; Delagrange, S.; Beaudet, M. Do Understory Sapling Respond to Both Light and below-Ground Competition? A Field Experiment in a North-Eastern American Hardwood Forest and a Literature Review. *Ann. For. Sci.* **2003**, *60*, 749–756. [CrossRef]
66. Nolet, P. *Coexistence et Sylviculture de L'érable À Sucre et Du Hêtre À Grandes Feuilles Dans Un Contexte de Changements Globaux*; Université du Québec à Montréal: Montréal, QC, Canada, 2016.
67. Beaudet, M.; Messier, C. Growth and Morphological Responses of Yellow Birch, Sugar Maple, and Beech Seedlings Growing under a Natural Light Gradient. *Can. J. For. Res.* **1998**, *28*, 1007–1015. [CrossRef]
68. Beaudet, M.; Messier, C.; Paré, D.; Brisson, J.; Bergeron, Y. Possible Mechanisms of Sugar Maple Regeneration Failure and Replacement by Beech in the Boisé-Des-Muir Old-Growth Forest, Québec. *Écoscience* **1999**, *6*, 264–271. [CrossRef]
69. Gasser, D.; Messier, C.; Beaudet, M.; Lechowicz, M.J. Sugar Maple and Yellow Birch Regeneration in Response to Canopy Opening, Liming and Vegetation Control in a Temperate Deciduous Forest of Quebec. *For. Ecol. Manag.* **2010**, *259*, 2006–2014. [CrossRef]
70. Hane, E.N.; Hamburg, S.P.; Barber, A.L.; Plaut, J.A. Phytotoxicity of American Beech Leaf Leachate to Sugar Maple Seedlings in a Greenhouse Experiment. *Can. J. For. Res.* **2003**, *33*, 814–821. [CrossRef]
71. Sage, R.W.; Porter, W.F.; Underwood, H.B. Windows of Opportunity: White-Tailed Deer and the Dynamics of Northern Hardwood Forests of the Northeastern US. *J. Nat. Conserv.* **2003**, *10*, 213–220. [CrossRef]
72. Horsley, S.B.; Stout, S.L.; DeCalesta, D.S. White-Tailed Deer Impact on the Vegetation Dynamics of a Northern Hardwood Forest. *Ecol. Appl.* **2003**, *13*, 98–118. [CrossRef]
73. Healy, W.M. Influence of Deer on the Structure and Composition of Oak Forests in Central Massachusetts. In *The Science of Overabundance: Deer Ecology and Population Management*; Smithsonian Institution Press: Washington, DC, USA, 1997; pp. 249–266.
74. Paine, C.E.T.; Beck, H. Seed Predation by Neotropical Rain Forest Mammals Increases Diversity in Seedling Recruitment. *Ecology* **2007**, *88*, 3076–3087. [CrossRef] [PubMed]
75. Marx, L.M.; Walters, M.B. Effects of Nitrogen Supply and Wood Species on *Tsuga canadensis* and *Betula alleghaniensis* Seedling Growth on Decaying Wood. *Can. J. For. Res.* **2006**, *36*, 2873–2884. [CrossRef]

76. Decker, K.L.M.; Wang, D.; Waite, C.; Scherbatskoy, T. Snow Removal and Ambient Air Temperature Effects on Forest Soil Temperatures in Northern Vermont. *Soil Sci. Soc. Am. J.* **2003**, *67*, 1234. [CrossRef]
77. Christenson, L.M.; Mitchell, M.J.; Groffman, P.M.; Lovett, G.M. Cascading Effects of Climate Change on Forest Ecosystems: Biogeochemical Links Between Trees and Moose in the Northeast USA. *Ecosystems* **2013**, *17*, 442–457. [CrossRef]
78. Drescher, M.; Thomas, S.C. Snow Cover Manipulations Alter Survival of Early Life Stages of Cold-Temperate Tree Species. *Oikos* **2013**, *122*, 541–554. [CrossRef]
79. Millar, C.I.; Stephenson, N.L.; Stephens, S.L. Climate Change and Forests of the Future: Managing in the Face of Uncertainty. *Ecol. Appl.* **2007**, *17*, 2145–2151. [CrossRef] [PubMed]
80. Rustad, L.; Campbell, J.; Dukes, J.S.; Huntington, T.; Lambert, K.F.; Mohan, J.; Rodenhouse, N. *Changing Climate, Changing Forests: The Impacts of Climate Change on Forests of the Northeastern United States and Eastern Canada*; USDA: Washington, DC, USA, 2012.
81. Canham, C.D.; Murphy, L. The Demography of Tree Species Response to Climate: Seedling Recruitment and Survival. *Ecosphere* **2016**, *7*, e01424. [CrossRef]
82. Iverson, L.; Prasad, A.; Matthews, S. Modeling Potential Climate Change Impacts on the Trees of the Northeastern United States. *Mitig. Adapt. Strateg. Glob. Chang.* **2008**, *13*, 487–516. [CrossRef]
83. Fisichelli, N.A.; Frelich, L.E.; Reich, P.B. Climate and Interrelated Tree Regeneration Drivers in Mixed Temperate–Boreal Forests. *Landsc. Ecol.* **2013**, *28*, 149–159. [CrossRef]
84. Fisichelli, N.; Wright, A.; Rice, K.; Mau, A.; Buschena, C.; Reich, P.B. First-Year Seedlings and Climate Change: Species-Specific Responses of 15 North American Tree Species. *Oikos* **2014**, *123*, 1331–1340. [CrossRef]
85. Mallett, A.L. *Management of Understory American Beech by Manual and Chemical Control Methods*; State University of New York, College of Environmental Science and Forestry: Syracuse, NY, USA, 2002.
86. Kochenderfer, J.D.; Zedaker, S.M.; Johnson, J.E.; Smith, D.W.; Miller, G.W. Herbicide Hardwood Crop Tree Release in Central West Virginia. *North. J. Appl. For.* **2001**, *18*, 46–54.
87. Nelson, A.S.; Wagner, R.G. Improving the Composition of Beech-Dominated Northern Hardwood Understories in Northern Maine. *North. J. Appl. For.* **2011**, *28*, 186–193.
88. Forrester, J.A.; Lorimer, C.G.; Dyer, J.H.; Gower, S.T.; Mladenoff, D.J. Response of Tree Regeneration to Experimental Gap Creation and Deer Herbivory in North Temperate Forests. *For. Ecol. Manag.* **2014**, *329*, 137–147. [CrossRef]
89. Arseneault, J.E.; Saunders, M.R.; Seymour, R.S.; Wagner, R.G. First Decadal Response to Treatment in a Disturbance-Based Silviculture Experiment in Maine. *For. Ecol. Manag.* **2011**, *262*, 404–412. [CrossRef]
90. Bolton, N.W.; D'Amato, A.W. Regeneration Responses to Gap Size and Coarse Woody Debris within Natural Disturbance-Based Silvicultural Systems in Northeastern Minnesota, USA. *For. Ecol. Manag.* **2011**, *262*, 1215–1222. [CrossRef]
91. Kern, C.; Montgomery, R.A.; Reich, P.B.; Strong, T.F. Harvest-Created Canopy Gaps Increase Species and Functional Trait Diversity of the Forest Ground-Layer Community. *For. Sci.* **2014**, *60*, 335–344. [CrossRef]
92. Harmon, M.E.; Ferrell, W.K.; Franklin, J.F. Effects on Carbon Storage of Conversion of Old-Growth Forests to Young Forests. *Science* **1990**, *247*, 699–703. [CrossRef] [PubMed]

7

Resources and Rules of the Game: Participation of Civil Society in REDD+ and FLEGT-VPA Processes in Lao PDR

Irmeli Mustalahti [1,*], Mathias Cramm [1], Sabaheta Ramcilovic-Suominen [1] and Yitagesu T. Tegegne [2]

1 Department of Geographical and Historical Studies, University of Eastern Finland, Yliopistokatu 7, 80100 Joensuu, Finland; m.cramm@gmail.com (M.C.); sabaheta.ramcilovik-suominen@uef.fi (S.R.-S.)
2 European Forest Institute, Yliopistokatu 6, 80100 Joensuu, Finland; yitagesu.tekle@efi.int
* Correspondence: irmeli.mustalahti@uef.fi

Abstract: Reducing Emissions from Deforestation and Forest Degradation (REDD+) aims to achieve its purpose by working across multiple sectors and involving multilevel actors in reducing deforestation and forest degradation in tropical countries. By contrast, the European Union (EU) Action Plan on Forest Law Enforcement, Governance and Trade (FLEGT) and its Voluntary Partnership Agreements (VPAs) focus on forestry and functions at a bilateral state level. The FLEGT Action Plan specifically aims to tackle illegal logging and improve forest governance in countries exporting tropical timber to the EU. Since illegal logging is just one driver of forest degradation, and legalisation of logging does not necessarily reduce deforestation and forest degradation, the two instruments differ in scope. However, by addressing the causes of forest degradation and their underlying governance issues, the FLEGT VPAs and REDD+ share many functional linkages at higher levels of forest policy and forest governance. The contribution and participation of civil society organisations (CSOs) and other actors are imperative to both processes. Our study is based on a survey of key actors (national and international) in REDD+ and FLEGT VPA processes in the Lao People's Democratic Republic (Lao PDR). Our analysis was guided by the theoretical perspectives of the policy arrangement approach and examination of two specific dimensions of this approach, namely resources and rules of the game. This paper argues that participation of CSOs in both processes is crucial because it facilitates and nurtures much needed cooperation between other national and international actors. The paper concludes that participation of CSOs could bring valuable information and knowledge into REDD+ and FLEGT VPA processes, thus contributing to increased legitimacy, justice and transparency.

Keywords: REDD+; FLEGT VPA; civil society organisations; participation; resources; rules of the game; policy arrangement approach; Lao PDR

---

1. Introduction

Climate change induced by human activities is well substantiated, for instance as demonstrated by the Intergovernmental Panel on Climate Change in the Fifth Assessment Report [1]. Deforestation and forest degradation are a major concern globally because they are one of the drivers of global warming. As much as 10% of the annual anthropogenic greenhouse gas emissions is estimated to be due to tropical deforestation and forest degradation [2]. Indeed, tropical forests showed a significant trend of increased annual forest loss between 2000 and 2012 [3]. Therefore, sustainable forest management (SFM) and conservation can play a key role in reducing global anthropogenic greenhouse gas emissions [4,5].

Scholars argue that countries in the tropics often have institutional, economic and political disadvantages in responding effectively to multifaceted and complex challenges, such as those related to society-nature interactions [6–9]. The international community has therefore taken a substantial interest in facilitating the national efforts of tropical countries to reduce deforestation and forest degradation because such interventions are of interest and benefit at a global scale. Different approaches have arisen in the wake of this interest. Two such approaches are (i) Reducing Emissions from Deforestation and Forest Degradation in Developing Countries, and the role of conservation, sustainable management of forests and enhancement of forest carbon stocks in developing countries (REDD+) and (ii) the European Union's Forest Law Enforcement, Governance and Trade (FLEGT) Action Plan and Voluntary Partnership Agreements (VPAs). REDD+ is being negotiated under the United Nations Framework Convention on Climate Change (UNFCCC), and developing countries are being supported by several donor countries, especially Norway, as well as international organisations such as the UN and the World Bank. The scope of REDD+ is to address the direct and underlying drivers of deforestation and forest degradation, and under this approach, countries are paid based on their performance in forest carbon sequestration. The latest climate agreement negotiated in Paris features an explicit stand-alone article [10] (Article 5) devoted to REDD+ that sends a clear message that REDD+ and forests will play a role in the fight against climate change. Meanwhile, the EU FLEGT Action Plan was developed in 2003 in response to imports of illegally sourced timber and vast illegal logging operations in supplying timber products [11]. The overall aim of the FLEGT VPA is to limit the import of illegal timber into the EU and reduce illegal logging, while strengthening land tenure, access and rights, and increasing the participation of different actors [11]. This should help lead the way to SFM (sustainable forest management), a principle that is adopted by most countries through the Rio Conventions [12]. FLEGT VPAs are negotiated at a bilateral level between the European Commission (EC) representing the EU member states and an individual partner country [13].

REDD+ works across multiple sectors and thus involves many different actors. Incorporating the views of all of these actors in REDD+ planning and implementation processes is important to ensure that REDD+ processes and finance will be accessible and fair for the different actors [14–16]. The FLEGT VPA process is negotiated and implemented as an agreement between the EC and each specific country individually. The VPA agreements focus solely on forestry and illegal logging, and also affect numerous actors at the national scale that are either directly or indirectly in touch with forests and involved in forestry related matters [17,18]. By addressing the causes of forest degradation and their underlying governance issues, negotiation and implementation of the FLEGT VPAs share many functional linkages with REDD+ processes [19,20]. It is argued that REDD+ should include safeguards that recognise and protect the continuity of multipurpose functions of the forest for local people and avoid dependence on external payments [15,21,22]. In a similar fashion, VPAs in principle have to include provisions for justice and sustainability matters [13]. In this context, the participation of CSOs is essential in terms of developing such safeguards for REDD+ and FLEGT.

The aim of this paper is to understand how and to what extent CSOs are engaged in REDD+ and in the emerging FLEGT VPA process in the Lao People's Democratic Republic (PDR). In addition, the paper aims to clarify the possible role which CSO participation plays in facilitating the implementation of the two processes. REDD+ in Lao PDR is at the stage of developing the national REDD+ strategy, while the VPA process is at the pre-negotiation phase. In this paper, we aim to explore these two processes through three research questions: (i) how are CSOs currently participating in REDD+ and VPA processes? (existing participation); (ii) how can the role of CSOs be strengthened? (potential participation); (iii) how can CSOs foster cooperation between the REDD+ and VPA processes?

In addressing these questions, the paper contributes to the literature on the role of non-state actors in environmental and forest governance processes and interventions in general terms as well as to the emerging literature on the same issue in the context of FLEGT and REDD+. While the literature on the non-state actors' role and participation is rather well developed in both general terms and with regard to FLEGT and REDD+ processes, such analysis in the case of Lao PDR is a novelty and is much needed. Considering the country's single party system, the involvement and role of non-state actors, such as CSOs, is an issue that has lagged behind, both in practice and in academic work. A limitation of this study is that we had difficulty in accessing information and objective opinion on the issues in focus, which in particular reflected the recent emergence of domestic CSOs and that their operations and involvement is highly regulated in Lao PDR. This situation is also clearly seen in the study sample which was dominated by international NGOs due to the difficulty of identifying domestic CSOs working on FLEGT and REDD+ in Lao PDR.

We use the policy arrangement approach (PAA) proposed by Bas Arts and colleagues to analyse the involvement of CSOs in FLEGT and REDD+ processes and their role in facilitating implementation of and interactions between the processes in Lao PDR. One of the key reasons for adopting this framework for our study is the concept of "political modernisation", which is the "building block" on which the PAA rests. The concept of political modernisation, as further explained in the section describing the theoretical framework, is about the change in forest governance represented by the emergence of new actors. Therefore, we felt that the PAA was particularly suited even though there are various other frameworks for facilitating analyses of environmental governance. We also opted for the PAA framework due to its strong emphasis on power and power relations—a dimension which is largely overlooked in the analytical frameworks emerging from the schools of institutionalism and institutional economics. Our conceptual understanding of PAA and its relevance are further explained in Section 3 which introduces the theoretical framework of the study. The last part of the article is organised in the following sections: Section 4 presents research methods; Section 5 presents and discusses the results; and Section 6 draws some conclusions and offers recommendations. However, first, in Section 2 we discuss the case study (REDD+ and FLEGT-VPA processes) in Lao PDR.

2. Case Study in Lao PDR: Forestry Governance, CSOs, REDD+ and FLEGT VPA

2.1. Forest Governance and Participatory Forestry in Lao PDR

Lao PDR is a tropical country in Southeast Asia that shares most of its border with Vietnam and Thailand. Forest cover in the country is high and is estimated at around 40%, but deforestation levels have been alarming. Lao PDR has vast natural resources that attract foreign direct investment with the aim of developing commercial agriculture and forestry production [23]. At the same time, the government of Lao PDR and donor countries have agreed to pursue forest conservation through global mechanisms such as REDD+ and FLEGT. Consequently, the various types of investments and conservation practices for managing natural resources and land are creating more pressure on rural communities but also opportunities to which these communities must adapt in order to secure their livelihoods [24,25].

Although Lao PDR is well-endowed with natural resources that can make a major contribution to the country's long-term economic development, the country has experienced an alarming deforestation rate in the last two decades. From 1982, the forest area has been decreasing dramatically, at an estimated annual average net deforestation rate of 0.7%, or forest loss of about 76,000 ha per annum [26]. The immediate and underlying factors causing deforestation and forest degradation in Lao PDR are well covered in [26,27]. Lao PDR state sources name the following activities as the principal causes of deforestation: shifting cultivation, commercial agriculture, forest fires, and mining operations in forested areas. Extraction of wood (legal and illegal) and unsustainable non-timber forest product harvesting are named as the major causes of forest degradation. Factors underlying these causes include population growth and infrastructure expansion, urbanisation, construction of roads, hydropower plants and increasing price and demand for raw materials (e.g., minerals, timber) in the regional and global markets. However, scholars increasingly question these factors and sources, and point to state-supported large-scale land concessions for fast-growing plantations and development projects as the main reasons for declining forest cover in Lao PDR, rather than shifting cultivation practiced by upland minorities [28–30].

In Lao PDR, villagers have been involved in so-called participatory forestry since the 1990's [31,32]. The first participatory forest management (PFM) intervention was the Joint Forest Management approach introduced by the Lao Swedish Forestry Programme [33]. In the 1990's, the long-term rights for use of natural forest could be allocated to individuals and organisations [34] (art.5, pp. 48–54). The law made it possible to develop another participatory forestry model with a deeper involvement of villagers, the Forest Management and Conservation Programme (FOMACOP). FOMACOP, supported by World Bank and Finnish development cooperation, operated as a pilot project of participatory sustainable production forestry. This project paid special attention to the building of villager organisations and entrepreneurial development as well as to the technical aspects associated with forest management carried out by villagers [35].

The concept of practising participatory forestry in production forests through village organisations lost political support in Lao PDR and FOMACOP ended in 2001 [36]. The Forest Sector Strategy [37] instead adopted a timber production approach under sustainable forest management (SFM) in cooperation with local villages. The stated goal was decentralisation of land and forest resources. However, from a critical point of view, rather than decentralising natural resources and providing greater rights and control to local villages, the initiative at the time also led to re-centralisation of power [38]. In more recent years, there has been renewed interest in and political acceptance of village forestry, which is, however, highly regulated. In this vein, the former FOMACOP was replaced by the Sustainable Forestry for Rural Development Scaling Up (SUFORD-SU) Project (another project supported by the Finnish development aid) that follows more closely the political decisions and regulations of Lao PDR. Production forest areas under SUFORD-SU are no longer allocated to a particular village locality. Instead, management of production forest areas must be implemented jointly by several villages, their village forest organisations and district forest management units [39–41].

Currently, the involvement and roles of villagers in SFM are embedded in the existing laws and regulations and in the National Growth and Poverty Eradication Strategy of Government of Lao PDR [42]. The most relevant laws dealing with CSO participation include: (i) the Forest Law [34]; (ii) Prime Minister's (PM) Decree No. 59 [43]; (iii) Regulation No. 0204 from the Ministry of Agriculture and Forestry (2003) and (iv) the Forestry Strategy [37]. These laws have apparently strengthened the involvement of CSOs in community-based forestry interventions in Lao PDR, and provide more freedom to develop village forestry models based on decentralised natural resources rights and control by local villages. In summary, the last two decades have seen a shift towards decentralisation, a shift back to recentralisation, and now recently again a shift towards decentralisation of natural resources in the Lao PDR.

2.2. *Civil Society Organisations (CSOs) in Lao PDR*

In 2009, the Government of Lao PDR [44] passed decree No. 115/PM which introduced national CSOs in the country, or non-profit associations (NPAs) as they are called. According to the decree, NPAs should contribute to socio-economic development and poverty eradication. In this paper, "CSOs" refers to Lao NGOs/NPAs, community-based organisations and mass organisations as well as international NGOs, which may have a role to play in the REDD+/VPA processes in the country. However, it should be pointed out that the Lao-based organisations are highly regulated by the state and their freedom of action and expression is still tightly monitored. In fact, the so-called "mass organisations" are one of the four key institutions where leading figures are party members [45]. During this study, we found that the national CSOs were not yet actively involved in the REDD+/FLEGT processes. Therefore, the data and results presented in this study are dominated by information from international NGOs.

In the context of Lao PDR's single-party and authoritarian ruling system, it is interpreted as quasi normal that the NPAs would be under governmental control and in line with government policies and goals [46,47]. For instance, the decree on NPAs stipulates that "undermining the national, collective and individual interest is forbidden" for NPAs [44]. Another decree [48] further constrains the activities of NPAs by imposing stricter rules for obtaining finance and by limiting the fields in which they may work. Besides this "new civil society" initiative, Lao PDR has four so-called mass organisations (including the Farmers' Union and Women's Union) that have close ties to the Party. These mass organisations have extensive organisational networks stretching from the top of the party hierarchy down to the village level in order to disseminate information and scale up activities [47,49]. In the case of new development activities, international NGOs, on the other hand, are comparatively more active in Lao PDR, and have recently been recognised by the government as "important contributors to national socio-economic advancements" [50]. However, their functional freedom also hangs in the balance, because the new draft decree places them under more stringent supervision by the Ministry of Foreign Affairs [51]. With the constant change in political position from centralised to decentralised, market to state regulated and an ad-hoc manner of institutional and regulation building, it remains to be seen what role CSOs will play in the country's development and in the processes of FLEGT and REDD+. However, there are hopes that the situation might change for the better with the appointment of the new prime minister in January, 2016, who apparently has adopted a more "open approach" to development and natural resource governance.

2.3. National REDD+ and FLEGT VPA Processes and Organisational Structures

The key challenges in forest and land governance in Lao PDR have been well addressed by several scholars (see for example references [26–32]) who all discuss two key challenges: the ambiguous and uncertain laws and regulations in land-use planning and allocations that further create opportunities for unsound resource use practices, and poor law enforcement performance. To address the challenges facing forest resources in the country, Lao PDR is participating in REDD+ and FLEGT VPA processes. Lao PDR was one of the first 14 countries to become a REDD+ country participant under the Forest Carbon Partnership Facility (FCPF) of the World Bank in 2008. The country's Readiness Preparation Proposal (R-PP), which is the most central national-level document that defines how a country actually wants to implement REDD+, stems from 2010. Official communications from Lao PDR requesting entry into VPA negotiations date from April 2012. By entering into a VPA with the European Commission (EC), Lao PDR, at least in theory, aims to address the problem of illegal logging activities, improve governance in the forest sector, guarantee access of timber products to the EU markets, build capacities, and increase revenue from timber exports.

2.3.1. REDD+ Process in Lao PDR

Since 2007, Lao PDR has been involved in international negotiations through a REDD+ mechanism, and the government has voiced support for an internationally binding agreement. REDD+ preparations are being supported by FCPF, the Forest Investment Programme, and UN-REDD. The Lao PDR R-PP was submitted in December 2010 and accepted in 2011 [26]. Currently, the country is gaining experience from different REDD+ pilot projects and other readiness activities and setting up REDD+ institutions. In 2011–2012, the country started an institutional reform which reorganised the responsibilities of government bodies in relation to REDD+. A new Ministry of Natural Resources and Environment (MoNRE) was established which also led to changes in the structure of the existing Ministry of Agriculture and Forestry (MAF). Prior to the establishment of MoNRE, all forestry and therefore REDD+ and FLEGT related matters fell under the MAF or, more precisely, under its Department of Forestry (DoF). Once MoNRE was established in 2012, the government merged the former Division of Forest Conservation and the Division of Forest Protection and Restoration at MAF's Department of Forestry (DoF) into the Department of Forest Resource Management (DFRM) at MoNRE [52]; the Division of Forest Production remained under the DoF at MAF.

The REDD+ Division was placed under the DFRM at MoNRE, while the REDD+ Office was under DoF at MAF. These unclear mandates between the ministries caused a lot of ambiguity and overlap in REDD+ institutional settings, and the projects related to REDD+ and production and those related to conservation and protection are managed by two different ministries (MAF and MoNRE, respectively). These overlapping roles have caused conflicts, but more importantly they have provided opportunities for the private sector to establish land investments and timber extraction under a grey area of legality and with little or no participation of local communities [49,53–55]. Furthermore, REDD+ activities were transferred back to MoNRE [56]. However, in late April 2016, the new government of Lao PDR decided to restructure some ministries and to reconsolidate the forest sector under one ministry—the Ministry of Agriculture and Forestry. The Minister of MoNRE officially transferred the DFRM to the Minister of MAF on 19 August 2016. According to the official "Handing over Note", the DFRM under the MoNRE was returned to MAF. The two national REDD+ offices will be merged into one office under DoF.

The R-PP defines an institutional setup for national and subnational levels for governing REDD+ in Lao PDR [57]. In accordance with this setup, a National REDD+ Task Force was established in 2008 under the leadership of the Director-General of Forestry, and composed at that time of 12 members from various organisations and an array of legal mandates [56]. Following the establishment of MoNRE and under the ongoing institutional reforms during 2012 and 2013, the Task Force was moved from MAF to MoNRE, and was expanded to include first 24 members, from 18 ministries and various cross-sectoral organisations [58,59], and then later 30 members. Initially, the Director-General of DFRM chaired the Task Force, then the Vice-Minister of MONRE took over the chairmanship. Now that forest management is again solely the responsibility of MAF, the leadership of the National REDD+ Task Force has returned again to MAF. Now MAF is considering who will lead the Task Force, and how to convene and make decisions. The National REDD+ Office acts as a secretariat to the National REDD+ Task Force. Similarly, provincial REDD+ Task Forces (PRTFs) are supported by Provincial REDD+ Offices (PROs). To date, PRTFs and PROs have been established in three provinces—Houaphan, Luang Prabang, and Champasack. Four provinces—Xayaboury, Luang Namtha, Bokeo, and Oudomxay—are also to establish PRTFs and PROs before the end of the 2016.

2.3.2. FLEGT-VPA Process in Lao PDR

Voluntary Partnership Agreements (VPAs) are a key instrument of the EU FLEGT Action Plan. The VPA is a voluntary bilateral trade agreement between the EU and the government of a timber producing and trading partner country. The VPA becomes legally binding upon signature and ratification by both parties [11]. In Lao PDR, MAF—specifically its Department of Forest Inspection (DoFI)—is the ministry with lead responsibility for negotiating the VPA. The FLEGT VPA informal negotiation started in April 2012, although it was not until June 2015 that the Government Office of Laos approved the start of negotiations. A FLEGT Standing Office was opened by the MAF in October 2013 [60]. In July 2014, a workshop on actor engagement involving participants from the government, private sector, CSOs and academic institutions was held in Lao PDR. At the conclusion of the workshop, participants identified the following as the most essential issues to be resolved in order to advance the VPA process: (1) establishing a national steering committee for the VPA process; (2) building the capacity of actors; (3) strengthening communication on FLEGT among actor groups [61]. During 2014, a number of workshops and events organised by the EU and German agency for international cooperation (GIZ) were held to promote EU FLEGT in Lao PDR and in Southeast Asia in general. The options for inclusive participation, building and involvement of civil society in the process and legal enforcement were reviewed and discussed. The conclusion was that more work and a willingness to foster an open and transparent approach was needed for the FLEGT VPA to be successfully implemented in Lao PDR. These options were revisited almost three years later in December 2016 when a draft Work Plan for Forest Law Enforcement and Governance (FLEG) in ASEAN (Association of South East Asian Nations) was developed.

At the present time, Lao PDR is finalising its legality definition and—as the head of the Lao FLEGT Standing Office stated at the ASEAN meeting in December 2016—the aim is to conclude the VPA negotiations with the EU in 2018. The FLEGT VPA meetings are still ongoing and according to EUFLEGT news (accessed on 25 January 2017), the recent meeting on FLEGT VPA in Lao PDR acknowledged that engagement of private sector and civil society is the main challenge.

Concerning activities on the ground, significant support from the EU is currently organised under the auspices of the GIZ and its ProFLEGT project which aims to conclude a VPA between the EU and the government of Lao PDR. The project activities have so far focussed on support and capacity building for governmental agencies, as well as policy coordination. However, activities on the ground related to testing and implementation of the system in the three pilot provinces—including Sayaboury, Khammouane and Attapeu—have not taken place as originally planned. In addition to the Pro-FLEGT project, the EU-FAO FLEGT Programme was established in 2012. This Programme aims to improve awareness and strengthen the participation of CSOs and forest communities in the FLEGT VPA countries, and has since been active in the Lao FLEGT policy process.

For the government of Lao PDR, a VPA with the EU could provide an opportunity to resolve some of these issues and privileged access to the European market. Nevertheless, a lot will depend on whether FLEGT VPA will catalyse fundamental governance and institutional reforms, and on whether such reforms will address illegal logging and deforestation outside of the forestry sector, such as that which originates from agricultural and other development sectors. However, such a major shift away from the current governance norms is not likely to happen in the near future. This argument is based for example [62]. The main VPA impacts in the country are therefore expected to be in terms of strengthening forest governance generally.

3. Theoretical and Analytical Framework: Political Modernisation and the Policy Arrangement Approach (PAA)

Our analysis for exploring the participation of CSOs in REDD+ and FLEGT VPA processes in Lao PDR is guided by the *policy arrangement approach* (PAA). The PAA is defined as "the temporary stabilisation of the content and organisation of a policy domain" [63] (p. 96). It is helpful to examine some of the theoretical and conceptual underpinnings behind the PAA framework in order to better understand the REDD+ and FLEGT-VPA processes and the roles of CSOs. The concept of *political modernisation*, which is defined as "*a comprehensive process of changes in the political domain of society*" [63] (p. 101), may serve as a useful starting point. According to this concept, political modernisation refers to the process of shifting and redefining the relationships between the state, the market and civil society. These processes are characterised by the relative increase of political power of non-state actors, such as CSOs and private-sector actors [63–65].

Political modernisation consists of two distinct phases: *early modernity* and *late modernity*. Early modernity relates to the politics revolving around a strong nation-state model and the belief in a highly manageable society and nature [65] (p. 343). Early modernity is characterised by a regulatory state that is dominant over the market and civil society and optimism over such a state's capability to solve societal problems by rational policy making and comprehensive planning [65] (p. 343), [66] (p. 29). Late modernity is characterised by a discourse of governance, interdependence and the inevitability of the need for cooperation between government, market and society where there can be "no monopoly of knowledge, problem-solving, or steering capacity", as explained in [63] (pp. 101–102).

One way in which late modernity manifests itself is through what is understood as multilevel governance, or multiactor governance. Multilevel implies that governance takes place on multiple levels from local to global, and that these levels are not 'separated' from each other; rather, their boundaries are blurred and they are in constant interaction with each other. Multiactor refers to the number of different actors besides the state that can be found at the different levels and which constitute what is understood as governance [67,68]. Some examples of these actors in forest governance include the private sector, civil society, international actors, local communities and forest users.

The PAA framework comprises four interrelated analytical dimensions: (i) actors and their coalitions; (ii) resources and power; (iii) discourses; and (iv) rules of the game [64]. Bas Arts and colleagues state in their publications on the PAA that all four dimensions should be considered together in the analysis and that change in one of them means a change in one or more of the other three dimensions. While all these dimensions link to our key research question related to participation, we decided, however, to focus on two dimensions of PAA because those were the most crucial issues based on our pre-study analysis: *rules of the game* and *resources*, which refer to actors' capacities—economic, political, cultural, knowledge and beyond—and the extent to which actors are able to exert influence over other actors (i.e., actors' power relationships). The *rules of the game* can be understood by the following essential questions [63,65]: Who decides the agenda? Who makes the decisions? Due to the interrelatedness of the dimensions of a policy arrangement, these issues of "rules of the game" are closely related to actors' resources and thus also to power [69]. By contrast, resources are defined as assets that policy actors have or can mobilise to achieve their policy goals. The division of power and influence between these actors determines the degree to which actors can influence the policy outcomes. While different types of power capacities are certainly interwoven with each other and they can all be relevant, two of these power types may be of special interest in the context of this study: the political and the knowledge-based power capacities that are a central part of the rules of the game in any governance process.

For participation and further cooperation to be realised, actors can be presupposed to have and hold power, because power is required for actions such as formulating new rules, influencing decisions or reframing discourses that contribute to making participation/cooperation happen. Further, the achieved outcomes depend on the types of resources actors possess and how they use them, and on how actors play the "policy game" and are affected by the rules. Rules of the game decide the (non) existence of participation—i.e., who may participate and who decides this; the actors with the "strongest" resources (e.g., political and knowledge-based power) are the ones who make the decisions and achieve outcomes. The results of this study (Section 5) are presented in two sections based on our interview findings on the CSOs' resources and role in the 'rules of the game': (1) Resources of CSOs and challenges to CSO participation in REDD+ and FLEGT VPA processes; and (2) Rules of the game from the perspectives of CSOs facing challenges in fostering cooperation in REDD+ and FLEGT-VPA processes. Prior to the results and discussion, the research materials and methods are presented.

4. Materials and Methods

The first step to collect the data for this study in 2014 was an extensive review of existing scientific and grey literature, including policy documents, briefing notes and policy updates. The literature review provided important contextual information that was then used for creating an online questionnaire-based survey. This enabled the collection of more specific and relevant data for the study.

The second step in 2014 was identification of key national and international policy actors (representatives of international and national organisations, bilateral and multilateral donor agencies and national level ministries and relevant departments) leading and representing their respective organisations in the REDD+ and FLEGT related processes. In total, 51 contacts were identified and contacted (named in Table 1 as pre-survey interviews) with the help of international and national CSOs and bilateral project staff active in Laos. However, only 39 could be considered to have relevant information related to more recent development in REDD+ and FLEGT processes.

In the third step in August 2014, the identified 39 key respondents were invited to take part in an online survey. These organisations and actors were first contacted by email and then were either interviewed personally by phone or in face-to-face interviews in Lao PDR. Following the interviews, eventually 17 out of 39 respondents (response rate of 44%) took a part in the on-line survey.

In the fourth and final step, key-informant interviews were carried out with selected actors to triangulate the data and generate understanding of the historical context of forest governance in Lao PDR. Three of these informants participated in the survey, but the seven additional informants did not. The face-to-face interviews took place in Vientiane in January and April 2015. In 2016, five of informants were re-contacted by email in order to up-date the information. The distribution of organisational actors in different data collection phases is presented in Table 1.

Table 1. Distribution of organisational actors in the survey and interviews.

Organisational Type	*n* (pre-Survey Interviews 2014)	*n* (Survey 2014)	*n* (Key Informant Interviews 2015)
Government	3	2	
National NGO	1	2	1
Community-based organisation		2	
International NGO	4	5	2
Intergovernmental organisation		1	2
Foreign government agency	5	3	4
Private sector		1	
Academia		1	1
Total	13	17	10

In the analysis of the survey data, distinct approaches were employed for answers to fixed-choice questions and open-ended questions. Responses to fixed-choice questions were graphically displayed in an easily understandable format as quantitative data using Microsoft Excel, and these graphs were then transferred to Microsoft Word. For open-ended questions, analysis was conducted in Microsoft Word where responses by individual respondents were examined to identify main themes or topics in their responses. The data from the interviews conducted in Lao PDR were used to support and complement the survey data so as to provide a deeper understanding and a more refined overall context of the topic.

5. Results and Discussion

5.1. Resources of CSOs and Challenges to CSO Participation in REDD+ and FLEGT VPA Processes

This section addresses the dimension of resources of PAA, interpreted and used hereafter as the capacities of CSOs (as one policy actor) to exert influence over other actors (i.e., actors' power relationships). Figure 1 summarises the types of roles that the CSOs have in the REDD+ process in Lao PDR according to respondents. It is interesting to note that none of the respondents thought that the CSOs—although participating—were actually affecting any decisions surrounding the REDD+ process. Rather, CSOs were seen to have other kinds of roles. Most of the respondents (13 out of 17) acknowledged that the CSOs' role has been to raise awareness and engage in building capacities. For instance, a respondent from a national CSO commented that CSOs had been raising awareness on both REDD+ and FLEGT VPA processes among CSO members. Figure 1 also indicates that the CSOs had a role in terms of implementing projects on the ground. This correlates with the many different REDD+ readiness activities and pilot projects in areas of climate change, deforestation and SFM. Only one respondent (of a national CSO) mentioned that CSOs had been conducting FPIC (Free Prior Informed Consent) activities. FPIC is the principle that a community has the right to give or withhold its consent to proposed REDD+ projects that may affect the lands they customarily own, occupy or otherwise use.

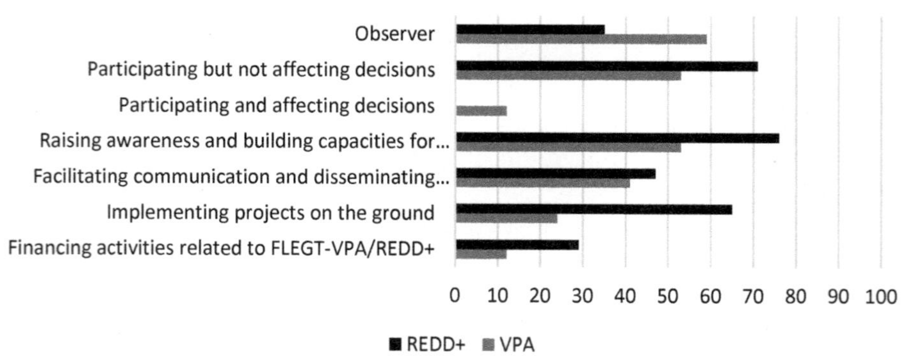

Figure 1. Role of CSOs in REDD+ and FLEGT VPA processes in Lao PDR (number of respondents = 17).

Figure 1 summarises the types of roles played by CSOs in the FLEGT VPA process in Lao PDR. The category "participating and affecting decisions" had a very low score, with only two respondents indicating this option. Even so, the result was not as drastic as in the case of REDD+ where none of the respondents indicated this option.

Nine respondents thought that the role of CSOs is to raise awareness and help build capacities for others. For instance, a respondent from a national CSO stated that they were providing information to their CSO network, which is an informal network between collaborating actors. Furthermore, as shown in Figure 1, six respondents indicated that "facilitating communication and disseminating information between different levels" was a role of CSOs in the FLEGT VPA process. Later, during face-to-face interviews, a Lao CSO respondent commented that they had been facilitating communication and disseminating information concerning FLEGT-VPA issues for local authorities and local communities and sharing lessons learnt at regional and national level. Lastly, only four of the respondents, in comparison to 11 in the case of REDD+, mentioned that CSOs were "implementing projects on the ground". The difference could partly be explained by the fact that the FLEGT VPA process was still at a very early stage, and CSOs are currently not officially involved in forest law enforcement activities. In reality, however, via their networks, social media and newspapers, they have an important role in monitoring and reporting for example illegal logging activities in Lao PDR, as commented on by an international CSO representative during pre-study interviews.

Much of the wood in Lao PDR originates from conversions of forests to other land uses, such as teak and rubber plantations, hydropower plants and mining and other infrastructure projects. The government of Laos (GoL) has been experimenting with bans on logging and/or log exports from as early as 1991 [26,49,62] when the nationwide moratorium on all logging concessions was adopted (Decree No. 67, from August 1991). However, logging—legal and illegal—has continued ever since; in the 1980s and 1990s logging was undertaken exclusively by the Lao Army and state owned enterprises, and later by international investments under the pretext of nation-wide development projects.

Several scholars have shown that private investors have so far successfully circumvented the logging ban by using the willingness of the government to grant land concessions for development projects, such as hydropower installations, plantations and mining (see for example [26,49,54]). Regrettably, as scholars and practitioners have by now concluded, many of these development projects never take place because many private investors abuse the development concessions by selling their development concessions to a third party after having logged the land and profited from timber sales. Therefore, large-scale logging has continued to a greater or lesser extent in Lao PDR despite the formal logging moratorium.

Separate questions were posed about challenges related to the participation of CSOs in REDD+ and in FLEGT VPA processes, but the data show noticeable similarities between both cases. Therefore, these results are presented jointly. Based on the responses, four main categories of challenges were identified: (i) governmental control and lack of an enabling political environment for CSOs to operate; (ii) limited understanding about REDD+ and FLEGT-VPA processes and their complexity; (iii) limited human and financial resources within the CSOs; and (iv) few national efforts in enabling REDD+ and VPA implementation in the country.

The first and main challenge raised during the survey, but also during face-to-face interviews among the respondents, is that the political system does not support or facilitate the operation of an independent civil society. Respondents also expressed a perception that the government exercises strict central control over REDD+ and related issues, as well as over the FLEGT-VPA process. One of the national key informants argued during pre-study face-to-face interviews in 2014 that it is generally difficult for non-state actors, including CSOs, to influence or affect government policy processes. These views are in line with the country's political reality, which limits the functioning of a free and independent civil society. During the survey, another national CSO respondent claimed that there is no free speech and that it will never be possible in the current political context. Moreover, the government is intending to pass new decrees that would restrict the operation of international CSOs as well as the establishment and registration of national CSOs in the country [48]. Consequently, media with a focus on the region, such as South China Morning Post (17 September 2014) and Radio Free Asia (2 October 2014), have been worried that this would curb the space for open dialogue in Lao PDR society. The new decrees would further strengthen government control over CSOs in Lao PDR, and possibly impede the funding of CSOs with foreign money.

More specifically for REDD+ and FLEGT VPA processes, during the survey, a government representative explained that there is a lack of an enabling regulatory and institutional framework for CSO participation. Progress in national REDD+ development has remained fairly limited, and this is reflected in the survey responses. REDD+ as well as FLEGT require an institutional setup that promotes and facilitates multiactor participation in order to have all relevant actors involved in the process. Additionally, a survey respondent of a foreign government agency argued that a CSO representative should be a member of the national REDD+ Task Force and participate in the decision making process on REDD+. The same respondent also noted that such a development is not likely to happen in the near future. Moreover, for the FLECT VPA process, a survey respondent of an international NGO explained that not many CSOs work on this because the topic of FLEGT VPA is a sensitive issue in Laos. The respondent further explained that forest law enforcement and monitoring of illegal logging will be difficult to implement at a local level because such actions would run counter to national and subnational income from the unlicensed logging activities. Indeed, based on several documents and articles, illegal logging and related trade are controversial issues in the country because it is argued that influential actors from the private and the public sector are involved in these trades and powerful vested interests exist [26,49,70].

The second and equally important fundamental challenge concerns the technical complexity of REDD+ and lack of knowledge and understanding about FLEGT VP. This challenge holds true for most of the actors involved in both processes. While it is important for participating actors to understand the processes and how the processes relate to each other and other factors (such as forest monitoring, forest management planning, logging licencing), there has been little information dissemination and capacity building to address the issue. Ambiguity about REDD+ will drive actors' interests down and it will be difficult for the process to gain momentum. One respondent of an international NGO stated that it is difficult to see how REDD+ relates to the communities, i.e., to their participation and benefits. In addition, on several occasions, the lack of REDD+ structures, pilot project structures and progress were pointed out as obstacles and barriers for REDD+ implementation. This implies that four years after the approval of the country's R-PP's, there is still a need for extensive capacity building among actors "outside of the country's capital". As one respondent of an international organisation put it, there is need to simplify REDD+ and make it understandable for a wider range of actors. Arguably, this is challenging when overall progress with REDD+ in Lao PDR has been slow, and institutional and regulatory developments within the process have been lagging behind. The same logic applies to the VPA process; without sufficient understanding, actors cannot properly engage in the VPA process. Moreover, awareness of the VPA process among Lao CSOs and communities is still low, as pointed out by some respondents from private sector and community-based organisations. A key informant also stated that awareness of VPA is even lower than that of REDD+.

The third significant challenge is a lack of capacities. The Lao NPAs were specifically pointed out as lacking human, financial and project management capacities in relation to REDD+ and VPA processes. According to a key informant's interviews, the most pressing issues are the lack of financial and properly trained human resources in the NPAs. A respondent of a foreign government agency also stated that NPAs do not have adequate knowledge about international policy processes, and the "big picture" surrounding those processes. Lack of these capacities and resources undoubtedly has to do with the very recent emergence of Lao civil society organisations and the limited political space for their activities and growth. This issue will need to be addressed for Lao NPAs to be able to effectively engage in REDD+ and VPA processes. However, in the current setting, this is only possible with facilitation by the national government. This is because the government, by passing laws and regulations, ultimately determines the degree to which NPAs may or may not be allowed to develop themselves.

The fourth challenge reflects the government's perceived lack of national efforts in enabling REDD+ and VPA implementation in the country. One respondent from the national level said that there is a contradiction in the national vision of development. The issues of conservation and of contradictory government policies were again mentioned by another respondent and an informant in 2016, as also evidenced by [71]. Further, a respondent of an international NGO stated that there is a lack of initiative with regard to the REDD+ process within the government. Instead of national authorities or national CBOs, the key players have been the bilateral and multilateral donor projects (such as CliPAD, and JICA's F-REDD project). The CSOs and the INGOs do not have a clear role in the REDD+ process in Lao PDR. SNV (an INGO founded in the Netherlands in 1965, with local presence in many countries in Asia, Africa and Latin America) has been active through their REDD+ pilot project, but this support was available only until October 2016.

A key informant explained that the green light from the government to involve CSOs in the FLEGT-VPA process has been pending. Understandably, it can become difficult to involve CSOs in the REDD+ or VPA process if the process lacks clear structure, direction and momentum [72], and the government is first and foremost responsible for all of these. For instance, in 2012, in the case of REDD+, the division of DoF between two ministries as a result of institutional reform was seen to cause weaknesses in its capacities; the lack of clarity in institutional mandates thus caused delays in the processing of FCPF grant financing by the World Bank [73]. However, in 2016, structural changes and integration of international experts under DoF were again taking place. This is the result of transfer of the DFRM from the Ministry of Natural Resources and Environment back to the Ministry of Agriculture and Forestry, and the REDD+ office is now being consolidated under DoF.

5.2. The "Rules of the Game" and Challenges of Realising Cooperation between REDD+ and FLEGT VPA Processes

Most respondents, in answering the question of future potential cooperation between REDD+ and FLEGT-VPA process, identified various opportunities such as matters concerning governance of forest, the law reform and enforcement. These three aspects relate to the interlinkages between the rules of the game and resources (the two analytical dimensions included in the PAA; see for example [63,65]). The "rules of the game" are closely related to actors' resources and therefore to power during the policy game: Who participates in the policy development related to REDD+ and FLEGT-VPA? Are CSOs excluded or will they participate in the future? If CSOs participate in the policy game, they need resources, which refers to actors' capacities—economic, political, cultural, knowledge and beyond – and also to the extent to which they are able to exert influence over other actors (i.e., actors' power relationships; see for example [63,65]. While different types of power capacities are certainly interwoven with access to resources, two capacities—the political and the knowledge-based power capacities—may be of special interest in the context of potential participation of CSOs in governance of forests, law reform and enforcement. These two capacities—the political and the knowledge-based power capacities—are central to the rules of the game in REDD+ and FLEGT-VPA processes.

For example, six respondents highlighted the overlap in the institutional structures of the two initiatives and in the potential participation of CSOs to foster cooperation. Foreign government agency respondents pointed out that under the process of REDD+ policy development, key actors such as CSOs should be involved so that duplications and contradictions within VPA process can be avoided. Further, the same respondents said that the work of the technical working groups of both processes should be synchronised. Misunderstanding and miscommunication were said to be the wider issues currently and the organisational setup of both MAF and MONRE were described as very complex. Due to the fact that causes of deforestation and forest degradation largely originate from outside the forest sector [74] and that REDD+ (and also VPA) requires effective cross-sectoral coordination and

CSOs participation, advanced communication mechanisms within and among ministries would play a key role.

Three respondents also acknowledged that there is potential for cooperation between the two processes in terms of addressing causes of deforestation and forest degradation. The two processes were regarded as providing opportunities for strengthened roles and rights of local communities. A foreign government agency respondent argued, for example, that REDD+ and FLEGT-VPA processes should adopt common approaches for empowering local communities to have rights and means to engage in law enforcement around forests, especially in combating poaching and illegal logging, and to receive real incentives for doing so. This view was shared by a key informant, who also noted CSOs' possible role in training local communities for these actions. A government respondent echoed this view. Beyond deforestation and forest degradation, six different respondents pointed to forest governance in general as an important topic for both processes; some areas for cooperation include access to information and greater transparency, and land law reform and enforcement. As one international expert noted: "Both (processes) are strongly linked to governance issues, so there could be scope for using both processes to improve forest governance." The respondent continued, however, that "(this discussion) would have to involve the policy sector in a broad sense—not just the forestry sector, which operates independently of for example land development sectors."

Figure 2 presents potential challenges to creating cooperation between the two processes. The two options clearly chosen most often were "insufficient capacities" (14 respondents) and 'lack of political will/interest' (13 respondents). The lack of capacities appears to be omnipresent and may undermine many efforts. These findings are linked to the concept of *resources* and refer to actors' capacities—economic, social and political—as well as to the concept of power.

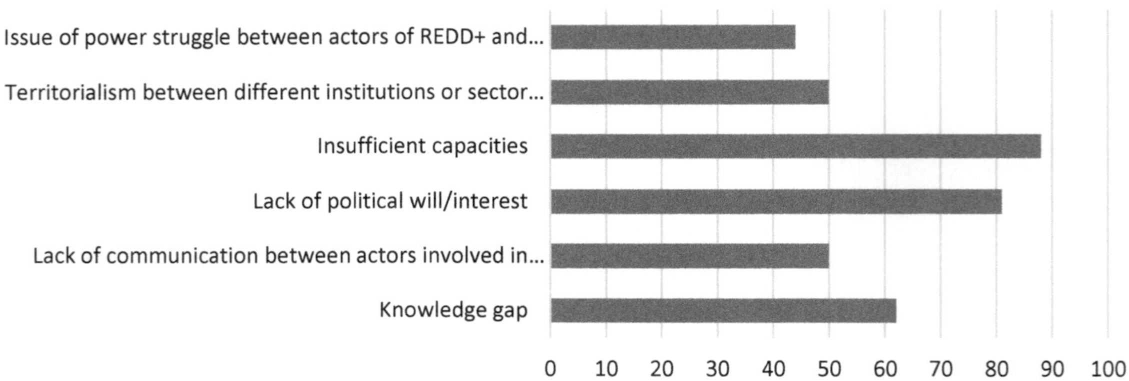

Figure 2. Challenges to cooperation between REDD+ and VPA processes (number of respondents = 16).

For example, the interviewed foreign government agency representative highlighted both processes as requiring significant investments from the Lao government, and that coordinated cooperation between REDD+ and VPA processes could lead to increased efficiency. There could consequently be a reduced need for financial and human resources, therefore such cooperation could be something to truly strive for in the country context of Lao PDR. A foreign government agency respondent explained the lack of political will or interest as related to the fact that (in the case of the VPA process) very little timber from Lao PDR goes to Europe and that benefits from REDD+ are still unclear: "So, what is the incentive for the Lao government to take them seriously?" This lack of interest could partly be explained by the third most chosen option (10 respondents), namely the "knowledge gap" about the processes and possible cooperation between them. Understanding opportunities for cooperation and its potential benefits first requires in-depth knowledge of each process, which is not yet the case for many actors in Lao PDR. An international NGO respondent argued that REDD+ and VPA concepts are new to Lao PDR and capacities of relevant government and private actors in relation to these concepts were still very low. A Lao CSO respondent explained that many people were saying that these issues were new for them and very difficult to understand, so they did not become interested in them in the first place. This indicates the need for awareness raising and capacity building.

Respondents also acknowledged that the "issue of power struggle between actors of REDD+ and VPA processes" (7 respondents), "territorialism between different institutions or sector organisations" (8 respondents) and "lack of communication between actors involved in REDD+ and FLEGT-VPA" (8 respondents) also pose challenges to bringing about cooperation between the processes. Each of these difficulties also relates to the same cross-sectoral aspect of different ministries and organisations that are involved in the same governance matters. These issues are not unique to Lao PDR (see e.g., [20] Tegegne et al., 2014 for similar findings in Cameroon and Congo Basin countries); both REDD+ and FLEGT-VPA require improved governance involving multiple actors in order to be successful. Therefore, strong leadership and coordination across sectors and institutions is instrumental to progress.

One of the respondents (a representative of a foreign government agency) explained the current lack of cooperation as being due to several factors. According to another respondent discussions and coordination between the processes cannot be constructive when the institutional setup for REDD+ is still ambiguous and the collaboration among actors generally weak. The respondent further added that the agency in charge had not shown strong leadership in coordination. Moreover, institutional and human capacity is quite limited and actors cannot understand well the requirements of both initiatives, and therefore it is quite difficult to discuss the duplication, demarcation and cooperation of the initiatives among actors.

Finally, respondents were also asked to evaluate at what organisational/administration/interest level(s) they believed that challenges to cooperation would occur (Figure 3). Here, a trend of increased "likelihood" of challenges is observed to occur from lower towards higher levels of governance. The implication is that decision-making power is concentrated at higher levels of governance rather than at lower levels, as in a 'top-down' system. This is because challenges would occur at the level where actual decisions are being made and authority is located. On the other hand, a foreign government agency respondent argued that 'once you have a successful case at village level, district government will become your ambassador to the higher levels of government'. Challenges that prevent or obstruct cooperation would still need to be tackled at all levels, with sufficient attention being paid to central and provincial levels.

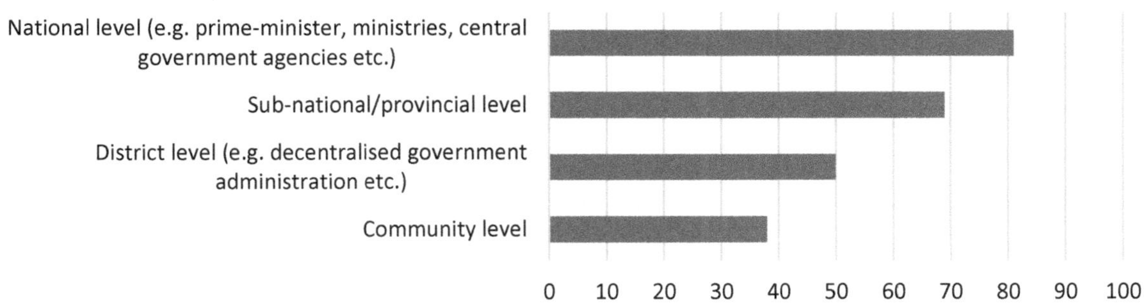

Figure 3. Perceived occurrence of challenges related to cooperation at different levels (number of respondents = 16).

The literature describes how for one process or regime to be able to "learn" from another process and apply the lessons learnt, their institutions and/or membership must overlap [19,20,75]. In the REDD+ and FLEGT-VPA processes in Lao PDR, as described in detail in Section 2.3, several albeit shifting and therefore relatively complicated institutional connections and overlaps do exist between actors and institutions involved. Based on interviews as well as national documents (such as [57]), it is possible that more and new overlaps will come into effect as both processes take shape and evolve. Currently, existing overlaps imply that the REDD+ and FLEGT-VPA processes already have interlinks in memberships and thus are able to "learn" from each other and apply lessons learnt. However, both of these processes are centrally controlled, and although CSOs could have *resources* to support the implementation of both processes, currently CSOs have limited access to set the *rules of the game*, as we further elaborate in Section 6.

6. Conclusions and Recommendations

The national CSOs in Lao PDR are composed of the newly emerging Non-Profit Associations (NPAs). However, the government does not view these organisations as independent players nor as a balancing force or counterweight to government policies, as is the case in many other countries [46]. In other words, the government does not share power with these actors (nor any other) but retains central control and power for itself [46,76]. Instead, the government recognises that CSOs should play a role in advancing development in the country, but this role should not contradict government policies and goals [48]. The political environment in Lao PDR provides the frame for the conceptual understanding of *resources*, which refers to the extent to which CSOs are able to exert influence over central control and power (actors' power relationships, as referred to in Bas Arts' conceptual frame [69]). Based on this interpretation, the *rules of the game* in REDD+ and FLEGT-VPA processes can be understood through the following crucial questions [63,65]: Who decides on the policy agenda and processes? Who may participate? Who decides the implementation?

The results of this study suggest that participation of international and local CSOs in REDD+ and FLEGT-VPA processes in Lao PDR is limited and significantly constrained. Where the CSOs are able to participate, the nature of that participation seems not to actually affect decisions that steer the processes, but serves merely to raise awareness and to a lesser degree to bring about implementation of local pilot project activities. Moreover, for the most part, the same kinds of challenges exist and hinder meaningful participation of local CSOs in both REDD+ and FLEGT-VPA; i.e., awareness and knowledge of the processes are low, and local CSOs lack capacities to a significant extent. The political system and governance structures of Lao PDR are also unsupportive of open and free participation by local as well as international CSOs. The respondents felt that local CSOs need better knowledge of REDD+ as well as of FLEGT-VPA and that experiences from other countries could be utilised to link these processes together and provide meaningful ways for CSOs to participate in them. While evidence of existing forms of cooperation between REDD+ as well as of FLEGT-VPA was limited, respondents identified a potential for bringing about cooperation in the future. For this to happen, the respondents again stressed the importance of building the capacities of all the actors involved. Three respondents also acknowledged that there is potential for cooperation between the two processes/international regimes in terms of addressing causes of deforestation and forest degradation (c.f., similar experiences related to the two international regimes in Cameroon and the Republic of Congo [20]).

Interaction of international regimes is premised on the fact that these regimes influence the development and performance of each other in several ways [77]. The interactions between policies depend on the overall political and institutional setting in which policies or regimes exist and interact [78]. This paper contributes to the literature on these interactions by investigating *rules of the game* and how CSOs could strengthen the interactions between FLEGT and REDD+ in Lao PDF country context.

Country context is important when it comes to effectiveness and pace of REDD+ readiness [79] or other natural resources governance-related processes. Indeed, the seemingly slow progress of Lao PDR in REDD+ and FLEGT-VPA can be partly explained by overall weak governance, the current political environment, and lack of *resources*. Thus, we see the multifaceted nature of power in relation to availability of resources and the capacity to achieve outcomes [69]. Participation and cooperation can be presupposed to require that specific actors exercise power if they are to perform these functions. But weak capacity or willingness of the government institutions and/or other "duty bearer" agencies will hinder the formulation of the rules etc. needed to make participation/cooperation happen. For example, in Lao PDR, the issue of weak forest governance is partly seen as a result of lack of a regulatory framework and legislation. However, while some inconsistencies and loopholes do exist, the enforcement and implementation in particular of laws and policies in Lao PDR is also acknowledged to be poor [32,49,57].

The latest in over two decades of GoL logging and/or log export bans is the new Prime Minister's Order (PM's Order 15) ban on timber exports, issued in May 2016. The moratorium does not exempt project developers or infrastructure-concession operators, and requires them not to use timber to pay for infrastructure development projects. In addition, the government is in charge of logging and selling wood directly to project developers and concession operators. In December 2016, the Laotian Times reported that the PM's Order 15 had successfully curbed harmful practices in the forestry sector and timber industry. The scientific community has therefore been led to believe the implementation of the Order has so far been taken seriously. There have as yet been no formal studies to support the merits of this Order and immediate and urgent research into these claims is required.

Thompson, Baruah and Carr [16] point out that it is problematic that states are typically treated as the main parties responsible for implementing REDD+, when they do not have sufficient capacities to successfully enforce rules and regulations. The kind of outcomes that are actually achieved depend, however, on the types of resources actors possess and how they use them, and on how actors play the whole "policy game", i.e., are influenced by which rules. Rules of the game decide the (non)existence of cooperation, who may participate and who decides this. Those who make the decisions and achieve outcomes are the actors with the greatest resources (e.g., political and knowledge-based power). A recent study [79] across REDD+ countries also found that not much attention is paid to subnational-level processes. As the REDD+ and FLEGT-VPA processes in Lao PDR evolve, state actors will need to consider how CSOs can be better involved in the processes, and in what roles, so as to utilise their capacities and for the CSOs themselves to benefit from the processes.

Currently, in Lao PDR, international and national CSOs are to a great extent concerned about and affected by their opportunities to participate in natural resources governance. In this situation, where local communities are adjusting to REDD+ and FLEGT-VPA to be able to benefit from global flows of values, CSOs are being placed in an increasingly important role, serving as the voice of the community. Traditionally, CSOs have been the weakest actors in natural resources governance in Lao PDR, but CSOs now perceive their role to be growing stronger. In the case of REDD+ and FLEGT-VPA, CSOs (both national and international) are acting as the mediators between global contexts and local conditions.

In Lao PDR, international and national CSOs should and do have a role in facilitating both REDD+ and FLEGT-VPA processes. However, it is important to remember this basic rule of the game in life in general: the actors with the greatest resources can easily overrule the decision making of those with fewer resources. There is also therefore a risk that externally funded and hence greater resourced CSOs "speak on behalf" of local communities and that local governance structures are bypassed in the process. There is further research needed on how the district and provincial level governance and village organisations, such as mass organisations (e.g., youth and women's unions) and forest management units at village level, will be integrated in the REDD+ and FLEGT-VPA processes, because they are key actors in the reduction of deforestation and forest degradation and enforcement of forest law.

This study has important implications for the countries in the Southeast Asian region involved in the FLEGT VPA and REDD+ processes, and in particular those such as Vietnam and Cambodia with similar socio-political environments. The finding that CSOs have a key role to play in strengthening cooperation and interactions between REDD+ and FELGT VPA, as well as in awareness raising and pilot project implementation, make it imperative to address further the participation challenges faced by CSOs in the region.

Limited understanding of REDD+ and FELGT VPA processes and limited human resources were identified as key challenges to CSO participation. On the other hand, according to the study, CSO understanding and resources are likely to improve in step with stronger CSOs which have a key role in raising awareness about FLEGT and REDD+. In other words, creating and promoting enabling conditions for a better involvement of CSOs is an important strategy to address barriers to CSO participation. In Lao PDR, the authoritarian regime would surely benefit from a more open and inclusive approach. Such an approach would in turn increase the legitimacy of government operations as well as acceptance of external interventions, such as FLEGT VPA and REDD+ processes. Social resistance—either hidden or more strategic and organised—continues to obstruct both acceptance of state authority and implementation of international interventions in Southeast Asia [80]. We therefore argue in favour of participation of non-state actors as a strategy for smoother implementation of environmental governance interventions in general and of FLEGT and REDD+ in particular.

Acknowledgments: We would like to acknowledge all respondents in Lao PDR who took part in the study survey. This work was part of a long-term study funded by the Academy of Finland (grant number 265,159) called "REDD+: The new regime to enhance or reduce equity in global environmental governance? A comparative study in Tanzania, Mexico and Laos". The authors closely cooperated within this comparative research project. We would especially like to acknowledge Salla Rantala and Paula Williams for all their valuable comments, and a warm thanks is extended to Yidnekachew Biza for his research assistance. The European Forest Institute is also acknowledged for institutional support. We also acknowledge the proofreading support of Nick Quist Nathaniels.

Author Contributions: Irmeli Mustalahti, main author; Mathias Cramm, collecting and analyzing data; Sabaheta Ramciloviv-Suominen, REDD+ issues and literature analysis; Yitagesu T. Tegegne, FLEGT VPA issues.

Conflicts of Interest: The authors declare no conflict of interest.

References

1. Intergovernmental Panel on Climate Change (IPCC). Summary for Policymakers. In *Climate Change 2013: The Physical Science Basis*; Contribution of Working Group I to the Fifth Assessment Report of the Intergovernmental Panel on Climate Change; Stocker, T.F., Qin, D., Plattner, G.-K., Tignor, M., Allen, S.K., Boschung, J., Nails, A., Xia, Y., Bex, V., Midgley, P.M., Eds.; Cambridge University Press: Cambridge, UK; New York, NY, USA, 2013.
2. Achard, F.; Beuchle, R.; Mayaux, P.; Stibig, H.J.; Bodart, C.; Brink, A.; Carboni, S.; Desclee, B.; Donnay, F.; Eva, H.D.; et al. Proximate causes and underlng driving forces of tropical deforestation. *Glob. Chang. Biol.* **2014**, *20*, 2540–2554. [PubMed]
3. Hansen, M.C.; Potapov, P.V.; Moore, R.; Hancher, M.; Turubanova, S.A.; Tyukavina, A.; Thau, D.; Stehman, S.V.; Goetz, S.J.; Loveland, T.R.; et al. High-resolution global maps of 21st-century forest cover change. *Science* **2013**, *342*, 850–853. [CrossRef] [PubMed]
4. Lima, M.G.B.; Braña-Varela, J.; Kleymann, H.; Carter, S. *Contribution of Forests and Land Use to Closing the Gigatonne Emissions Gap by 2020*; WWF-WUR policy brief; Wageningen University and Research: Wageningen, The Netherland, 2015.
5. United Nations Framework Convention on Climate Change (UNFCCC). Key decisions relevant for reducing emissions from deforestation and forest degradation in developing countries (REDD+). Available online: http://unfccc.int/6917.php (accessed on 15 December 2016).
6. Conca, K.; Dabelko, G.D. *Green Planet Blues: Critical Perspectives on Global Environmental Politics*; Conca, K., Dabelko, G.D., Eds.; Westview Press: Boulder, CO, USA, 2014.
7. Ebeling, J.; Yasue, M. Generating carbon finance through avoided deforestation and its potential to create climatic, conservation and human development benefits. *Philos. Trans. R. Soc. B Biol. Sci.* **2008**, *363*, 1917–1924. [CrossRef]
8. Grieg-Gran, M.; Porras, I.; Wunder, S. How can market mechanisms for forest environmental services help the poor? Preliminary lessons from Latin America. *World Dev.* **2005**, *33*, 1511–1527. [CrossRef]
9. Gupta, A. Transparency in Global Environmental Governance: A Coming of Age? *Glob. Environ. Politics* **2010**, *10*, 1–9. [CrossRef]
10. The United Nations Framework Convention on Climate Change (UNFCCC). Adoption of the Paris Agreement. Available online: https://unfccc.int/resource/docs/2015/cop21/eng/l09r01.pdf (accessed on 15 December 2016).

11. European Commission. Communication from the Commission to the Council and European Parliament: Forest Law Enforcement, Governance and Trade (FLEGT) Proposal for an EU Action Plan. Available online: http://eur-lex.europa.eu/legal-content/EN/TXT/?uri=CELEX%3A52003DC0251 (accessed on 15 December 2016).
12. United Nations. Non-Legally Binding Authoritative Statement of Principles for a Global Consensus on the Management, Conservation and Sustainable Development of All Types of Forests. A/CONF.151/26 (Vol. III). Annex III. Available online: http://www.un.org/documents/ga/conf151/aconf15126-3annex3.htm (accessed on 15 December 2016).
13. Ramcilovic-Suominen, S.; Gritten, D.; Saastamoinen, D. Concept of livelihood in the FLEGT voluntary partnership agreement and the expected impacts on the livelihood of forest communities in Ghana. *Int. For. Rev.* **2010**, *12*, 361–369. [CrossRef]
14. Rica, C.; Corbera, E.; Estrada, M.; May, P.; Navarro, G.; Pacheco, P. Rights to Land, Forests and Carbon in REDD+: Insights from Mexico, Brazil and Costa Rica. *Forests* **2011**, *2*, 301–342.
15. Mustalahti, I.; Bolin, A.; Paavola, J.; Boyd, E. REDD+ reconcile local priorities and needs with global mitigation benefits? Lessons from Angai Forest, Tanzania. *Ecol. Soc.* **2012**, *17*, 16. [CrossRef]
16. Thompson, M.C.; Baruah, M.; Carr, E.R. Seeing REDD+ as a project of environmental governance. *Environ. Sci. Policy* **2011**, *14*, 100–110. [CrossRef]
17. Beeko, C.; Arts, B. The EU-Ghana VPA: A comprehensive policy analysis of its design. *Int. For. Rev.* **2010**, *12*, 221–230. [CrossRef]
18. Dooley, K.; Ozinga, S. Building on Forest Governance Reforms through FLEGT: The Best Way of Controlling Forests' Contribution to Climate Change? *Rev. Eur. Community Int. Environ. Law R* **2011**, *20*, 163–170. [CrossRef]
19. Ochieng, R.M.; Visseren-hamakers, I.J.; Nketiah, K.S. Interaction between the FLEGT-VPA and REDD+ in Ghana: Recommendations for interaction management. *For. Policy Econ.* **2013**, *32*, 32–39. [CrossRef]
20. Tegegne, Y.; Ochieng, R.; Visseren-Hamakers, I.; Lindner, M.; Fobissie, K. Comparative analysis of the interactions between the FLEGT and REDD+ regimes in Cameroon and the Republic of Congo. *Int. For. Rev.* **2014**, *6*, 602–614. [CrossRef]
21. Mustalahti, I.; Taku, T. Forest management in REDD+: New opportunity or more risks? *Scand. J. For. Res.* **2012**, *27*, 200–209. [CrossRef]
22. Mustalahti, I.; Rakotonarivo, S. REDD+ and Empowered Deliberative Democracy: Learning from Tanzania. *World Dev.* **2014**, *59*, 199–211. [CrossRef]
23. Ministry of Planning and Investment; World Bank. *Lao People's Democratic Republic—Investment and Access to Land and Natural Resources: Challenges in Promoting Sustainable Development, a Think Piece (A Basis for Dialogue)*; World Bank: Washington, DC, USA, 2011.
24. Baird, I.G. Turning Land into Capital, Turning People into Labor: Primitive Accumulation and the Arrival of Large-Scale Economic Land Concessions in the Lao People's Democratic Republic. *New Propos. J. Marx. Interdiscip. Inq.* **2011**, *5*, 10–26.
25. Wong, G.Y.; Darachanthara, S.; Soukkhamthat, T. Economic Valuation of Land Uses in Oudomxay Province, Lao PDR: Can REDD+ be Effective in Maintaining Forests? *Land* **2014**, *3*, 1059–1074. [CrossRef]
26. Lestrelin, G.; Trockenbrodt, M.; Phanvilay, K.; Thongmanivong, S.; Vongvisoul, T.; Pham, T.T.; Castella, J.C. *The Context of REDD+ in the Lao People's Democratic Republic Drivers, Agents and Institutions*; Center for International Forestry Research (CIFOR): Bogor, Indonesia, 2013.
27. Mission Aviation Fellowship. *Annual Review of REDD+ Activities in Lao PDR 2012–2013*; Mission Aviation Fellowship: Vientiane, Laos, 2013.
28. Barney, K. China and the Production of Forestlands in Lao PDR: A political ecology of transnational enclosure. In *Taking Southeast Asia to Market: Commodities, Nature, and People in the Neoliberal Age*; Cornell University Press: Ithaca, NY, USA, 2008; pp. 91–107.
29. Lestrelin, G. Land degradation in the Lao PDR: Discourses and policy. *Land Use Policy* **2010**, *27*, 424–439. [CrossRef]
30. Vongvisouk, T.; Mertz, O.; Thongmanivong, S.; Heinimann, A.; Phanvilay, K. Shifting cultivation stability and change: Contrasting pathways of land use and livelihood change in Laos. *Appl. Geogr.* **2014**, *46*, 1–10. [CrossRef]
31. Daoroung, P. Community forests in Lao PDR: The new era of participation? *Watershed* **1997**, *3*, 1–8.

32. Fujita, Y.; Phengsopha, K. The Gap between Policy and Practice in Lao PDR. In *Lessons from Forest Decentralization: Money, Justice and the Quest for Good Governance in Asia-Pacific*; Colfer, C.J.P., Dahal, G.R., Capistrano, D., Eds.; Earthscan: London, UK, 2008; pp. 117–131.
33. Program, L.F.; Makarabhirom, P.; Raintree, J.; Lao, T.; Forestry, S. *Comparison of Village Forestry Planning Models Used in Laos*; Pearmsak Makarabhirom and John Raintree Regional Community Forestry Training Center (RECOFTC): Bangkok, Thailand, 1999.
34. Government of Lao PDR. *The Forestry Law*; Government of Lao PDR: Vientiane, Laos, 1996.
35. Samountry, X.; Bounphasaisol, T.; Leuangkhamma, T.; Phiathep, O.; Wayakone, S.; Williams, P. *Evaluation of Forest Management and Conservation Programme. FOMACOP Village Forestry Pilot Model (1995–2000). Technical Report*; FOMACOP: Vientiane, Laos, 2001.
36. Mustalahti, I. Participatory forestry in the crossroads in Laos and Vietnam. Two participatory forestry case studies. In *Contextualising Natural Resource Management in the South*; Vihemäki, H., Ed.; Institute of Development Studies, University of Helsinki: Helsinki, Finland, 2007; pp. 193–236.
37. Government of Lao PDR. *Decree on Endorsement and Declaration of the Forestry Strategy to the Year 2020 of the Lao PDR*; Government of Lao PDR: Vientiane, Laos, 2005.
38. Mustalahti, I.; Lund, J.F. Where and How Can Participatory Forest Management Succeed? Learning From Tanzania, Mozambique, and Laos. *Soc. Nat. Resour. Int. J.* **2009**, *23*, 31–44. [CrossRef]
39. Phanthanousy, B.; Sayakoummane, S. *The Lao PDR—Community Forestry in Production Forest 2005*; Community Forestry Policy Forum, RECOFTC: Bangkok, Thailand, 2005.
40. Ministry of Agriculture and Forestry. SUFORD Scaling Up. Available online: http://www.suford.org/wp-content/uploads/2015/01/sufor_scaling_up_brochure.pdf (accessed on 15 December 2016).
41. World Bank. *Project Appraisal Document. Scaling–up Participatory Sustainable Forest Management Project*; Report No: 75632-LA; World Bank: Washington, DC, USA, 2013.
42. Government of Lao PDR. *National Growth and Poverty Eradication Strategy*; Government of Lao PDR: Vientiane, Lao People's Democratic Republic, 2004.
43. Government of Lao PDR. *Decree on Sustainable Management of Production Forest Areas*; Government of Lao PDR: Vientiane, Lao People's Democratic Republic, 2002.
44. Government of Lao PDR. *Decree on Associations, No. 115/PM.*; Government of Lao PDR: Vientiane, Lao People's Democratic Republic, 2009.
45. Stuart-Fox, M. The Political Culture of Corruption in Lao PDR. *Asian Stud. Rev.* **2006**, *30*, 59–75. [CrossRef]
46. Belloni, R. Development in Practice Building civil society in Lao PDR: The Decree on Associations. *Dev. Pract.* **2014**, *24*, 353–365. [CrossRef]
47. KEPA. Reflections on Lao Civil Society. Available online: www.kepa.fi/tiedostot/lao_cs_2013.pdf (accessed on 10 November 2014).
48. Government of Lao PDR. *Draft Decree on Associations and Foundations*; Government of Lao PDR: Vientiane, Lao People's Democratic Republic, 2014.
49. Barney, K.; Canby, K. *Baseline Study 2, Lao PDR: Overview of Forest Governance, Markets and Trade*; EU FLEGT Facility: Kuala Lumpur, Malaysia, 2011.
50. LIWG. INGOs Important Contributors to Development. Land issues working group: Vientiane, Laos PDR. Available online: http://www.laolandissues.org/2014/10/29/ingos-important-contributors-to-development/ (accessed on 15 December 2016).
51. Morning Post. Laos NGO Restrictions Threaten Development, Say Non-Profit Groups. Available online: http://www.scmp.com/news/asia/article/1594490/laos-ngo-restrictions-threaten-development-say-non-profit-groups (accessed 15 December 2016).
52. Ministry of Natural Resources and Environment. *MoNRE Regarding the Organization and Activities of the Department of Forest Resource Management*; Decision No. 3121; Ministry of Natural Resources and Environment: Vientiane, Lao People's Democratic Republic, 2012.
53. Barney, B.K. Power, Progress and Impoverishment: Plantations, Hydropower, Ecological Change and Community Transformation in Hinboun District, Lao PDR. Available online: http://www.cifor.org/publications/pdf_files/books/bbarney0701.pdf (accessed on 15 December 2016).
54. Baird, I. *Quotas, Powers, Patronage and Illegal Rentseeking: The Political Economy of Logging and the Timber Trade in Southern Laos*; Forest Trends: Vientiane, Lao People's Democratic Republic, 2009.

55. Kenney-Lazar, M. Land Concession, Land Tenure, and Livelihood Change: Plantation Development in Attapeu Provinc, Southern Laos. Available online: http://www.laolandissues.org/wp-content/uploads/2012/01/Kenney-Lazar-Land-Concessions-Attapeu.pdf (accessed on 15 December 2016).
56. Vongvisouk, T.; Lestrelin, G.; Castella, J.; Mertz, O.; Broegaard, R.B.; Thongmanivong, S. REDD+ on hold: Lessons from an emerging institutional setup in Laos. *Asia Pac. Viewp.* **2016**, *57*, 393–405. [CrossRef]
57. Readiness Preparation Proposal; Lao PDR. Readiness Preparation Proposal (R-PP). Available online: https://www.forestcarbonpartnership.org/sites/forestcarbonpartnership.org/files/Documents/PDF/Sep2011/LaoR-PPFinaldraftrevised21DEC2010-CLEAN.pdf (accessed on 15 October 2014).
58. Forest Carbon Asia. New National REDD+ Task Force Appointed in Laos. Available online: http://www.forestcarbonasia.org/in-the-media/new-national-redd-task-force-appointed-laos (accessed on 15 October 2014).
59. Ministry of Natural Resources and Environment. *MoNRE on REDD+ Task Force for Implementation of Reducing Emission from Deforestation and Forest Degradation*; MoNRE Agreement No. 7176/2013; Ministry of Natural Resources and Environment: Vientiane, Lao People's Democratic Republic, 2013.
60. FERN. Forest Watch FLEGT Update November 2013. Available online: http://www.fern.org/sites/fern.org/files/FLEGT%20update_0.pdf (accessed on 2 November 2014).
61. EU FLEGT Facility. Stakeholders in Lao PDR Plan Ways to Engage in VPA Processes. Available online: http://www.euflegt.efi.int/laos-news/-/asset_publisher/FWJBfN3Zu1f6/content/stakeholders-in-lao-pdr-plan-ways-to-engage-in-vpa-processes (accessed on 1 May 2016).
62. Saunders, J. *Illegal Logging and Related Trade: The Response in Lao PDR*; Chatham House: London, UK, 2014.
63. Arts, B.; Leroy, P.; van Tatenhove, J. Political modernisation and policy arrangements: A framework for understanding environmental policy change. *Public Organ. Rev.* **2006**, *6*, 93–106. [CrossRef]
64. Arts, B.; van Tatenhove, J. Political modernisation. In *Institutional Dynamics in Environmental Governance*; Leroy, P., Arts, B., Eds.; Springer: Dordrecht, The Netherlands, 2006; pp. 21–43.
65. Arts, B.; Tatenhove, J. Van Policy and power: A conceptual framework between the "old" and "new" policy idioms. *Policy Sci.* **2004**, *37*, 339–356. [CrossRef]
66. Arts, B. "Green alliances" of business and NGOs. New styles of self-regulation or "dead-end roads"? *Corp. Soc. Responsib. Environ. Manag.* **2002**, *9*, 26–36. [CrossRef]
67. Arts, B. Assessing forest governance from a "Triple G" perspective: Government, governance, governmentality. *For. Policy Econ.* **2014**, *49*, 17–22. [CrossRef]
68. Forsyth, T. Multilevel, Multiactor Governance in REDD+: Participation, integration and coordination. In *Realising REDD+: National Strategy and Policy Options*; Angelsen, A., Ed.; CIFOR: Bogor, Indonesia, 2009; pp. 113–122.
69. Arts, B. Non-State Actors in Global Governance. Three Faces of Power. Available online: http://hdl.handle.net/11858/00-001M-0000-0028-6C87-B (accessed on 15 December 2016).
70. Vientiane Times. Logging, Tax Avoidance Top 2013 Corrupt Activities in Laos. Available online: https://www.travel-impact-newswire.com/2014/03/logging-tax-avoidance-top-2013-corrupt-activities-in-laos/ (accessed on 21 October 2014).
71. Hanssen, C. Lao land concessions, development for the people? Available online: http://www.condesan.org/mtnforum/sites/default/files/publication/files/4948.pdf (accessed on 15 December 2016).
72. Thongmanivong, S.; Phanvilay, K.; Vongvisouk, T. How Laos Is Moving Forward With REDD+ Schemes. Available online: http://epress-dev.lib.uts.edu.au/journals/index.php/ijrlp/article/view/3355 (accessed on 15 December 2016).
73. FCPF. REDD Readiness Progress Fact Sheet, June 2012. Available online: https://www.forestcarbonpartnership.org/sites/forestcarbonpartnership.org/files/Documents/PDF/June2012/Lao PDR REDD Readiness Progress Sheet_June 2012.pdf (accessed on 20 October 2014).
74. Geist, H.J.; Lambin, E.F. Proximate causes and undelying driving forces of tropical deforestation. *Bioscience* **2002**, *52*, 143–150. [CrossRef]
75. Gehring, T.; Oberthür, S. The Causal Mechanisms of Interaction between International Institutions 1. *Eur. J. Int. Relat.* **2009**, *15*, 125–156. [CrossRef]
76. Jersild, A.; Shroff, R. Feasibility of Various Responses and Interventions to Build Capacity of Local Civil Society Organizations (CSOs) in the Lao PDR. Available online: https://www.adb.org/sites/default/files/publication/28968/csb-lao.pdf (accessed on 15 December 2016).

77. Van Asselt, H.; Zelli, F. Connect the dots: Managing the fragmentation of global climate governance. *Environ. Econ. Policy Stud.* **2014**, *16*, 137–155. [CrossRef]
78. Gupta, A.; Pistorius, T.; Vijge, M.J. Managing fragmentation in global environmental governance: The REDD+ Partnership as bridge organization. *Int. Environ. Agreem. Polit. Law Econ.* **2016**, *16*, 355–374.
79. Minang, P.A.; van Noordwijk, M.; Duguma, L.A.; Alemagi, D.; Do, T.H.; Bernard, F.; Agung, P.; Robiglio, V.; Catacutan, D.; Suyanto, S. REDD+ Readiness progress across countries: Time for reconsideration. *Clim. Policy* **2014**, *14*, 685–708. [CrossRef]
80. Singh, S. *Natural Potency and Political Power: Forests and State Authority in Contemporary Laos*; Series of Southeast Asia: Politics, Meaning, and Memory; University of Hawai'i Press: Honolulu, HI, USA, 2012; p. 224.

8

Forest Islands and Castaway Communities: REDD+ and Forest Restoration in Prey Lang Forest

Courtney Work

International Institute of Social Studies, Erasmus University, 2502 LT The Hague, Netherlands; courtneykwork@gmail.com

Abstract: Climate Change policies are playing an ever-increasing role in global development strategies and their implementation gives rise to often-unforeseen social conflicts and environmental degradations. A landscape approach to analyzing forest-based Climate Change Mitigation policies (CCM) and land grabs in the Prey Lang Forest landscape, Cambodia revealed two Korea-Cambodia partnership projects designed to increase forest cover that are juxtaposed in this paper. Case study data revealed a REDD+ project with little negative impact or social conflict in the project area and an Afforestation/Reforestation (A/R) project that created both social and ecological conflicts. The study concludes that forest-based CCM policies can reduce conflict through efforts at minimal transformation of local livelihoods, maximal attention to the tenure rights, responsibilities, and authority of citizens, and by improving, not degrading, the project landscapes. The paper presents the circumstances under which these guidelines are sidestepped by the A/R project, and importantly reveals that dramatic forest and livelihood transformation had already affected the community and environment in the REDD+ project site. There are deep contradictions at the heart of climate change policies toward which attention must be given, lest we leave our future generations with nothing but forest islands and castaway communities.

Keywords: Cambodia; climate change; landscape; REDD+

---

1. Introduction

On opposite sides of Prey Lang forest in central Cambodia are two Korea-Cambodia partnership projects that are part of a larger Memorandum of Understanding between Cambodia's Forest Administration and the Korean Forest Service to invest in Cambodia's forests and provide administrative assistance toward reducing climate change (Interview "NGO Forum" 15 February 2015; Interview Think Biotech 11 January 2016). The most recent is the Korea-Cambodia Tumring REDD+ Project (T-REDD), established in July 2015 between Cambodia's Forest Administration (FA) and the Korean Forest Service (KFS). The second is an afforestation/reforestation (A/R) project in Kratie Province, established in December 2010, between the Forest Administration and the Think Biotech Cambodia, co. ltd. Phnom Penh, Cambodia (TB), a subsidiary of Korea's Hanwha Corporation. Their juxtaposition as forest-based climate change mitigation (CCM) projects enacted in the same landscape highlights some of the contradictions through which such policies are enacted.

This paper engages in the emerging global discourse in forest-based CCM initiatives that facilitate international claims to forests in Cambodia. The green grabs described in this paper are an unapologetic initiative in "securing overseas forest resources" by the Korean government [1]. Issues explored through this investigation speak to a growing body of literature concerned with the structures of governance that control exclusion and access [2,3] in the wake of land grabs in both their conservation-oriented green varieties and their more traditional economically-oriented earth toned versions [4,5]. This study situates conflict at the level of political, economic, and ecological policies and practices that privilege one type of land use over another and divest certain communities of their use rights.

A landscape approach to analyzing CCM and land grabs revealed the productive juxtaposition of two case studies enacted on opposite sides of the same vital protected area, which may have been analyzed in isolation using a project-based approach. Data were collected using a method that co-produced knowledge with citizen researchers, revealing the forest as a common pool resource that was and could continue to be sustainably managed by communities interconnected through it. Further, this research method supports actions for change, which generated alternative scenarios for project implementation discussed at the end of the paper. I argue that new tools are necessary for grappling with the profound contradictions of our coming green economy and that the three-part method of landscape perspective, local collaborations, and actions for change applied in this research can make space for fresh perspectives.

The data presented here contributes to critiques of market-based approaches to climate change policies [6,7] and discourses of sustainable forestry [8,9]. The forest restoration project is enacted as a for-profit business venture that can be green-washed because it claims to enhance forest carbon sequestration to mitigate global climate change [10]. The carbon utility of plantation forests is under debate [11], and TB has negative effects on the communities and natural resources in its area. The T-REDD project circumvents some of the pitfalls encountered by other REDD+ projects [12,13] for two reasons. The first is timing. The project area was already transformed and people already divested of their access to forest resources before the enactment of the project. The second is market orientation. T-REDD is a publicly funded project not directed toward supplying carbon credits for sale on the open market; it is designed to provide direct offsets for carbon-intensive Korean business practices inside Cambodia (Interview FA 5 September 2016).

I will demonstrate in this paper two distinct discourses engendered by climate change policies. The first involves the cynical green-washing of intensive extraction and the second shows measured attempts to grapple with the shortcomings of REDD+ governance structures in order to retain threatened forest areas. The TB project explicitly devalues traditional, sustainable livelihoods and citizen authority over land use in favor of planned reforestation and biodiversity conservation with corporate authority over land use. The discourse of the T-REDD project explicitly values citizen authority, sustainable livelihoods, and forest retention within the project boundaries. What is provocative about this juxtaposition, is that the landscape of the T-REDD project is the least conducive for meeting many of the project's objectives, it is already very degraded, is easily accessible by roads, and is under great pressure from a densely populated area. The landscape of the TB project, on the other hand, retains a great deal of biodiversity, has limited pressure for land conversion, and is in a very remote part of the forest. Both areas have a strong community, engaged in sustainable forest use, but in Kampong Thom they are pressured by incoming migrants. The evidence I will present suggests that switching project landscapes, or re-adjusting boundaries, could reach the full potential of each project's objectives with maximum benefit for the social and ecological communities in each location.

To support this claim, I first introduce the project landscape and the particular methodology deployed for this study. A brief review of the relevant literature on forest plantations and on REDD+, as carbon-reducing enterprises is followed by a discussion of those initiatives in Cambodia and descriptions of both projects from the ground, informed by the work of citizen researchers. The policy documents for each project are then evaluated considering Cambodia's national policies and the social and environmental objectives for each. The closing discussion describes the altered landscape possible through flexible planning that attempts maximum benefit for local communities.

2. The Landscape: Prey Lang

Prey Lang remains one of the largest contiguous lowland forests in Southeast Asia, ranging from 300,000 to 600,000 ha, depending on who reports the figures. The newly designated Prey Lang Wildlife Sanctuary (PLWS), is 431,683 hectares and captures most, but not all, of the forest recognized by long-term residents of the area (Figure 1).

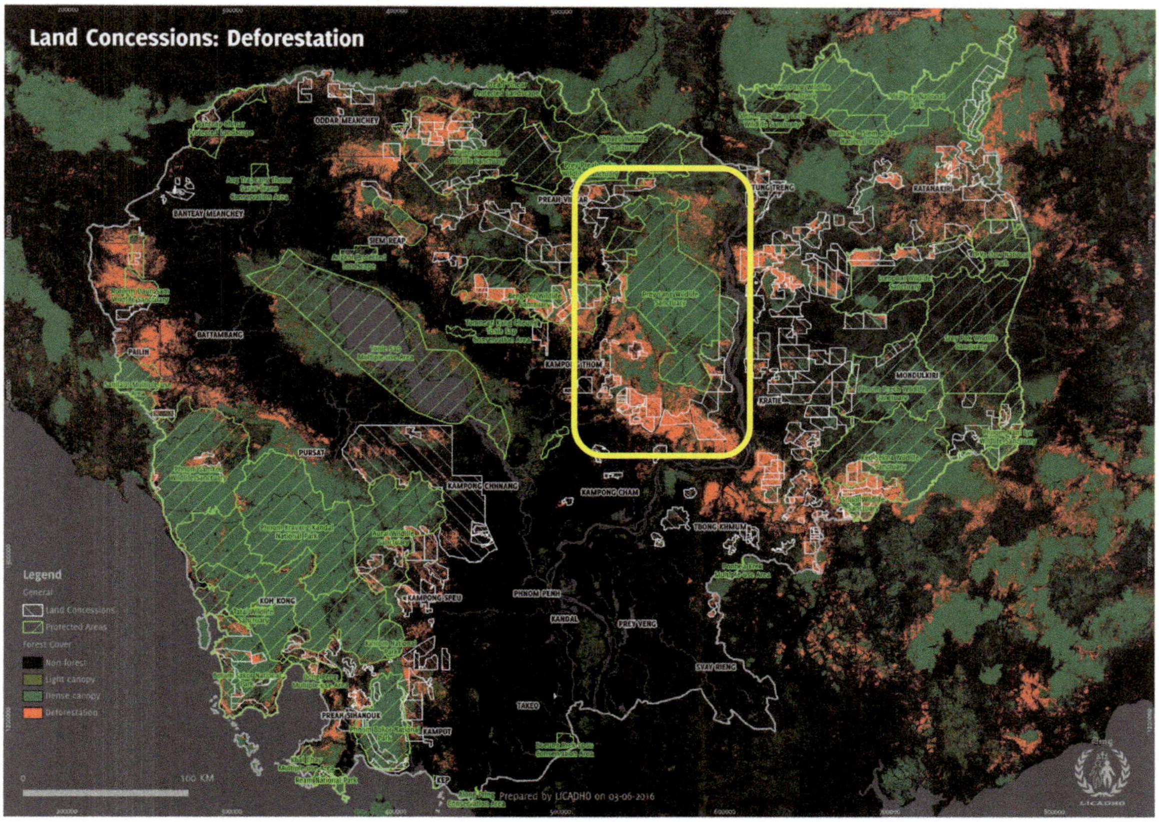

Figure 1. LICADHO (Cambodian League for the promotion and defense of human rights) forest cover map with Economic Land Concession (ELC) and protected area (PA) boundaries, 2016. Forest cover is green, deforestation is red. Green outlines are Protected Areas and white outlines are ELC [14].

The boundaries of the forest intersect four provinces: Preah Vihear, Steung Treng, Kampong Thom, and Kratie and sit between the Mekong and the Sen Rivers (Figure 2). Prey Lang is an important watershed for the Mekong and the Tonle Sap lake and the region is the site of a drawn-out struggle for forest resources that began with Cambodia's transition to a market economy in the early 1990s [15,16]. The degrading effects of this market transition on the landscape and lives of the people in Prey Lang makes the forest a target area for forest-based CCM policies [6]. The projects highlighted here, T-REDD and TB, are located in Kampong Thom and Kratie/Steung Treng, respectively (Figure 3) and each will be discussed in their affected areas and in relation to the larger landscape.

Figure 2. Prey Lang Wildlife Sanctuary (PLWS) in relation to provincial boundaries. Map from GoPhnomPenh http://gophnompenh.com/phnom-penh-overview/.

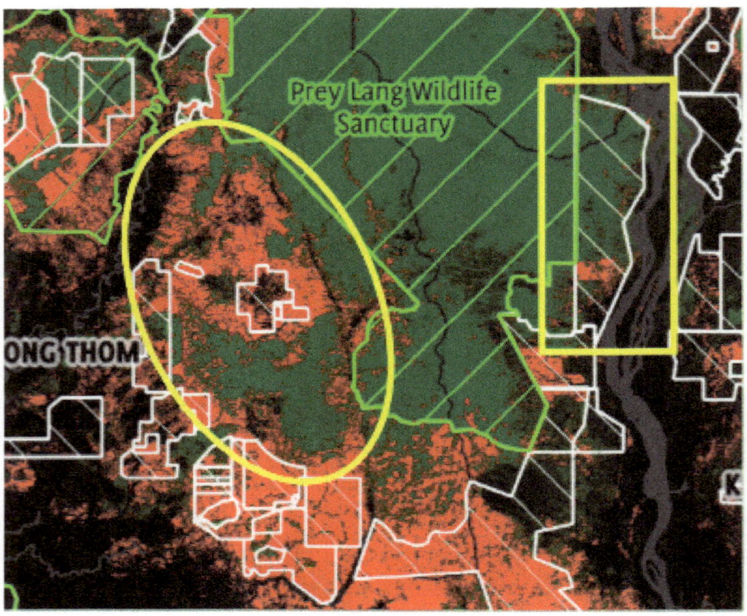

Figure 3. Tumring REDD+ Project (T-REDD) circled (project area is green); Think Biotech Cambodia, co. ltd. (TB, Phnom Penh, Cambodia) squared.

T-REDD is located at the south-western edge of Prey Lang forest in Kampong Thom province. It is an easy to access area of intense land conversion that began in the early 1990s with Forest Concessions (FCs) for industrial logging and continued through Economic Land Concessions (ELCs), a land titling campaign conducted by Prime Minister Hun Sen in 2012, and a Social Land Concession (SLC) in 2013. After Forest Concessions were cancelled in 2001 and before the land titling project in 2012, the national government awarded 16 ELCs (outlined in white) and 23 community forests (CFs) in Sandan District, Tumring, and Mien Rith communes (for more on the CF initiatives in Cambodia, please see [17–19]). Today, some stand as islands, cut off from Prey Lang forest by clear-cutting and ELCs, while others reach out from Prey Lang forest like archipelagos in a sea of cassava and rubber plantations visible inside the circle of Figure 3. They testify to the dense forest that once stood in this place and to the strength of the community that fought for them and protects them today. The T-REDD project encompasses 14 of these CFs in an 88,444 ha carbon offset initiative (Figure 4).

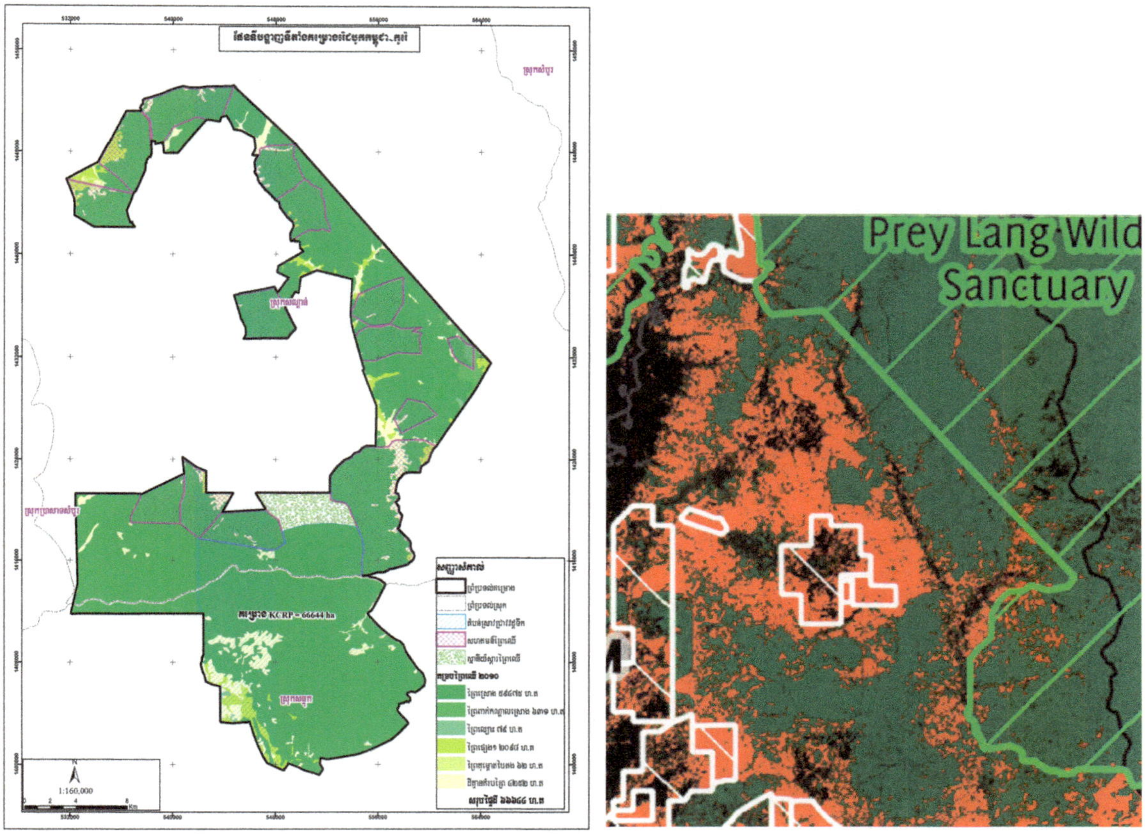

Figure 4. T-REDD project map #1 community forests (CFs) are outlined in dark green (**left**). Project area on deforestation map (**right**).

The TB project site is at the eastern boundary of Prey Lang forest, which ends at the Mekong River in Kratie and Steung Treng provinces. This is an isolated area, only accessible by boat or by crossing the forest. The area was also affected by FC, but the customary livelihood strategies of wet rice agriculture, shifting cultivation, forest product collection, resin tapping, and fishing remained strong. After FCs were curtailed in 2001, logging still continued illegally in the area, but ELCs did not immediately follow. In this case, livelihoods were altered, but customary practices remained the primary source of livelihoods. After the FCs, people also planted plantation crops and engaged in freelance logging (I use the term freelance logging to refer to for-profit logging done by local villagers. There are middle men who hire locals to work for them. It is a well-paying job and often the only job available. This is especially poignant in areas where companies impacted traditional livelihoods and left few avenues for people to buy what they used to gather from the forest. When villagers do contract logging for outsiders I argue that this is not illegal logging. It is freelance work. This is distinctly different from clearing land for market crops, which is also often an outcropping of plantation encroachment but cannot be considered freelance). Five CFs were established between 2010 and 2012 and people report an abundant lifestyle until 2012, when two companies began clearing the forest (Group Discussion 13–14 February 2016). One of these was TB, the 34,007 hectare forest restoration project that began operations in inhabited areas and encompassing four of the CFs in the areas of Kampong Cham and Beung Char commune, Sambor district, Kratie province (Figure 5), the omitted company, the Chhun Hong Rubber Better, Co. ltd. (Phnom Penh, Cambodia), will not be discussed here.

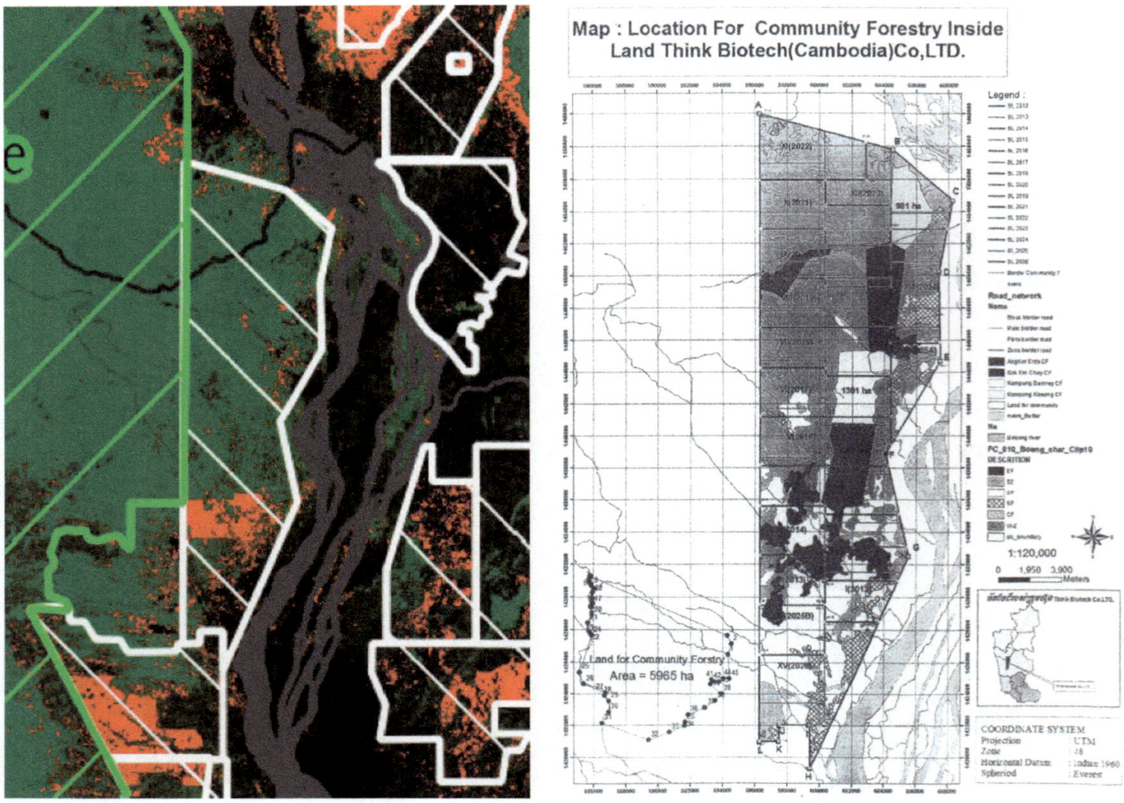

Figure 5. Think Biotech (TB) forest cover map (**left**). Think Biotech project map with four CFs labeled (**right**).

These two CCM projects involve the same governing bodies and are part of the same Memorandum of Understanding between the Korean Forest Service (KFS) and Cambodia's FA, designed to enhance forest resources, increase biodiversity, and enhance livelihoods. REDD+ does this by controlling cutting in natural forest; TB does this by removing natural forest and replacing it with, controlled plantation forest that will decrease reliance on the natural forest. The TB project is clear-cutting natural forest for a tree plantation on the most isolated edge of Prey Lang where customary cultivation practices were the norm and pressure on the forest was light. The company's dramatic land conversions have caused both environmental and social conflicts within and without its project boundaries. The T-REDD project, on the other hand, encompassed existing CFs along with remaining forested areas of state land. It was implemented on the most vulnerable boundary of Prey Lang in an area where customary livelihoods were already transformed and the forest was heavily degraded. This project has created few reported conflicts with local communities.

3. Methods

This paper is part of a broader research project, called "Mosaic: Climate change mitigation policies, land grabbing and conflict in fragile states: understanding intersections, exploring transformations in Myanmar and Cambodia". Mosaic develops a new research agenda that explicitly studies the interactions between climate change mitigation initiatives, land grabs, and resulting patterns of conflict [10]. The methodological approach has three parts. The first is to take a landscape perspective, the second is to co-produce knowledge between academic researchers and affected communities, and the third is to support actions for change. The landscape approach is outlined in the framework paper for this project [8], where the authors "conceptualize landscape as a 'place' where physical and socio-cultural elements occur in localized, spatially specific combinations and where human actors dynamically interact. Thus, a landscape is both ecologically and socially fluid and changeable, but also holds continuities .... A landscape is thus a space larger than a farm but smaller than a region, in which physical, ecological, and human dimensions co-exist as a product of socio-ecological and cultural co-evolution .... There is no single formula for determining where a landscape 'ends'." For this project, the landscape is based on the interconnected ecosystems of the large forest, and on the social relationships among the Prey Lang Community Network (PLCN), collective forest activists from all four provinces since the early days of exploitation [9].

In this research project, the co-production of knowledge emerges through three main activities: training, research, and information sharing. First, Mosaic researchers trained the PLCN in research methods and forest-based climate change mitigation policies. Second, Mosaic researchers and the PLCN conduct research related to Mosaic objectives together. The PLCN also conducts research in their areas according to their needs, for example, researching the activities of new mining operations, elite capture of forest land, or the exploitation of CFs. Data and insights from this research overlaps with Mosaic concerns and are shared and discussed with Mosaic researchers. Third, the PLCN shares data with Mosaic about forest crimes and the activities of companies, government officials, and conservation organizations. In turn, Mosaic researchers share information from their desk and urban research by translating and explaining new policy and land use initiatives that may affect the PLCN and by sharing new information learned through interviews with donors or companies. Community and academic researchers are in regular contact and alert each other to emerging concerns.

The third part of the Mosaic research initiative, supporting action for change, entails bringing what academics know about the concerns and situations of affected communities to other stakeholders informing and influencing policy and projects. It also involves helping to bridge the divide between local authorities and the PLCN.

This method brings to light the spillover effects of climate change policies like REDD+ or A/R, which include the disjuncture between project documents and project activities, the privileging of monitory benefits over biodiversity or other benefits, and the leakage of deforestation outside the project area. These policies and projects also spillover and intersect through social and ecological feedbacks that shape each other within a given landscape. When that landscape lacks strong governance and is being rapidly converted toward commercial purposes, the dynamics of land use, conflict, and cooperation are instructive for understanding the political-institutional conditions in which just and inclusive solutions might be created.

Research was conducted between January 2015 and September 2016 (with the assistance of RONG Vannrith, THUON Ratha, SONG, Danik, and SEAKCHHY Monyrath, each connected with Mosaic partner organizations in Cambodia: Equitable Cambodia and the Cambodian Peace Building Network—data was importantly supplemented with reports from the PLCN), combining participant observation methods with formal and informal interviews, group discussions, and secondary literature reviews. All interviews were conducted in Khmer or English. Grassroots-level data were collected in cooperation with local research teams. Mosaic researchers conducted research during four forest patrol activities, each consisting of five days and four nights in the forest with PLCN members and in 18 meetings with government officials, non-governmental organizations (NGOs) and international non-governmental organizations (INGOs) and community representatives related to forest governance, climate change, and conservation initiatives. These activities provide important context for the interview data collected from the four communes affected by the two projects of this paper: eight interviews with local authorities, focus group discussions with 11 of the 14 CFs in the REDD+ project (researchers were unable to reach the last three CFs and since the data received from all the others was remarkably similar and corroborated with that of other participants during forest patrol and meetings this was not pursued further), focus group discussions with seven villages affected by TB, and individual interviews with seven key informants at the community level. Additionally, interviews were conducted with eight NGOs working in the area, and six interviews with national Forest Administration officials, two interviews with TB company representatives, and one with the company that performed the Environmental Impact Assessment for TB. National authorities and company voices are underrepresented in this study, despite efforts to reach them. Especially missing are the voices of the Korean Forest Service, with whom every attempt at contact failed. Nonetheless, I feel confident that the data presented supports the claims of this paper.

The benefits of gathering data using the approach here described is that it situates two CCM projects, REDD+ and A/R, together in the same social and physical landscape where research can draw out the historical, bureaucratic, and socio-economic intersections of each project. The input from citizen researchers brings important information to bear on the land-uses and use potentials of the respective project sites, as well as understandings of community capacity for conservation initiatives. This information was instrumental in considering how changes to project implementation and governance strategies could decrease conflicts while achieving project goals. Analyzed separately, these two projects show the characteristics of conflict-light or conflict-heavy implementations of CCM policies. Thinking about them together in the same physical, social, historical, and policy landscape, however, provides space to move beyond the silos of activities within individual projects and begin to address the implications of the system as a whole.

4. Forest-Centered CCM: REDD+ and Forest Plantations

REDD+ is an international instrument designed to preserve forest resources by creating a market for the carbon that they store. The mechanisms through which carbon is captured, measured, bought, and sold vary considerably, but the use of REDD+ to offset other carbon-intensive activities continues to create inequitable outcomes by privileging forest management programs under international frameworks over more traditional methods of sustainable forest management [10]. Globally, REDD+ exhibits multiple types of implementation "running ahead of policy processes and state-driven decisions" [12] and also ahead of any internationally agreed upon process [13]. The discourse and practice is increasingly dominated by donor-driven policy narratives and technological interventions that depoliticize climate mitigation, maintaining current structures [20] and obscuring the discourses and values that underlie them. By focusing on the value of carbon and future funds from donor organizations or rich countries, the issue of deforestation and forest protection becomes "a mere footnote" [21]; diminishing the value of biodiversity [22], and the value of healthy human communities in forests [23]. Additionally, situating shifting cultivation or charcoal production as the main drivers of deforestation in REDD+ project areas, devalues the benefits of shifting cultivation, and leaves the industrial drivers of forest loss unmentioned, unquestioned, and unaffected by REDD+ [24].

As forest-based processes, the main difference between REDD+ and A/R is their approach: REDD+ is designed to keep forests standing, while A/R creates plantation forests where there are no trees. A/R projects attempt to increase industrial forest conversion through the Clean Development Mechanisms that promote managed forest cultivation and increase leafy green canopy above the bare earth for carbon capture [25,26]. In doing so, tree plantation developers that are able to invest in large-scale forestry can provide sustainable forest management and climate change mitigation [27]. In many cases, however, A/R projects are implemented in forested areas and follow the patterns of the global land grab crisis [5,28], in which small holders and indigenous people are violently dispossessed of their lands and livelihoods [29]. Case studies suggest that industrial forest plantations are neither more profitable [30], nor sequester more carbon than REDD+ or business as usual timber harvesting [31].

Both REDD+ and A/R projects settle in landscapes with existing political and economic activities [32,33] and are affected differently by various tenure and documentation issues. Communal tenure in a Cambodian REDD+ site resulted in restricted access to forest resources and resin trees [34], while this case study shows how clarifying CF tenure made protecting forests easier (*cf.* [35]), but issuing individual land titles for market crops in the same landscape put significant pressure on the forest (*cf.* [36]). REDD+ is continually confronted by its dual focus at once enhancing forest protection and increasing commercial activities [37,38]. Forest plantations tend to have less varied and more destructive effects on the social and ecological environments of their enactment [39].

The TB project documents state that the project will "stop forest clearance ... and reduce the utilization of natural forest" [40] (translation by author from original Khmer). This effectively excludes forest dependent people from using the natural forest for subsistence, and increases timber exploitation outside project boundaries because there is legal timber exported from the region (Interview community rep 15 February 2016). Additionally, it opens the region with roads and infrastructure that facilitates cutting high-value timber for elite markets [41], makes space for local elites to profit from the timber trade and also convert land to grow market crops, and it pushes marginal villagers to clear forest to replace land lost to the company and to replace subsistence economies with market-based crops. Plantation jobs are limited and the salary is incomparable to traditional livelihoods (*cf.* [42]). What is available is freelance logging and plantation crops, both of which put pressure on forest land and increase market-based economies.

REDD+, on the other hand, is not market based due to the limited market for carbon [13]. The process of offsetting was designed to function under free market principals, but after many years of attempts, REDD+ looks more like direct patronage. For example, a in transaction orchestrated by the Wildlife Conservation Society (WCS), Disney corporation purchased 14 million metric tons of carbon dioxide equivalent emissions for $2.6 million. Funds from this deal are "earmarked to help the government protect the Keo Seima Wildlife Sanctuary and will be handled through the Ministry of Environment" [43]. The influx of capital from the private sector and other governments flows into the newly created bureaucratic structures for managing climate finance in fragile host governments [44,45]. The emerging effects of this are ministries competing for environmental funding sources [46] and continued forest loss. The logics of a market economy seem to be unequal to the challenges of deforestation and of mitigating its climate changing effects.

In this environment of under-funded and donor-dependent state systems, the two cooperative projects between Cambodia and the Korean government are both framed using the discourse of climate mitigation and clean development. The next section will describe the state of carbon capture activities in which the projects under discussion were implemented.

5. Cambodia and Carbon Capture

Cambodia came early to internationally conceived climate mitigation initiatives. The first REDD+ project started in 2008. Cambodia's Forest Administration in partnership with the International Non-governmental Organization (INGO) PACT administered the Oddar Meanchey REDD project [47,48] (PACT was not the original INGO on this project and they pulled out due to complications related to the actual sale of carbon (see, [48] for more details of this case)). The United Nations Development Program (UNDP) started REDD readiness preparations in Cambodia in 2009 [49], supporting the National REDD+ Program and REDD+ readiness activities (Interview UNDP 16 March 2015). Six years later, elaborate bureaucratic structures exist and four REDD+ projects are operating with another on the way: The Oddar Meanchey project, the WCS sponsored project in the Seima Protected Area (see, [34,50]); the International Tropical Timber Association (ITTO) project "SFM management through REDD+ mechanisms in Kampong Thom province, Cambodia" (PD 740/14 Rev.2 (F)); the Tumring REDD+ project described in this paper, and a new Japanese project in Prey Lang is entering a three year planning phase. The ITTO project objective is to get experience to do REDD on the ground. It is an office-centered capacity building exercise—not designed to sell carbon (Interview FA 9 May 2016). Even with these active projects, Cambodia is still transitioning out of the "readiness" phase and preparing to move into the implementation phase of the process [51]. Tree plantations for carbon capture have not had the same uptake, but were expected to contribute to the 60% forest cover by 2015 promised in Cambodia's Millennium Development Goals (MDG) [52], and continue to play a role in achieving the 60% forest cover goal not yet reached [53].

The most successful commercial tree plantation industry in Cambodia is rubber, accounting for over 900,000 hectares in 2013–2014 [54]. This does add green leafy canopy in a manner that satisfies the Food and Agriculture Organization (FAO) guidelines for forests, but they were planted on forested land and are thus not a vehicle for carbon capture [11]. Timber plantations in Cambodia reported by FAO in 2010 covered 70,000 ha [55]. Many of these were discontinued due to the destruction of local livelihoods, alienation of local communities, negative impacts on the environment, decreased biodiversity, and reduced water quality [56,57]. Satellite data from 2013–2014 [58] shows less than 1000 ha dedicated to timber production activities in Cambodia, and TB is one of two public-private industrial tree plantation initiatives currently active in Cambodia (the Oji Paper Co., Ltd (Tokyo, Japan) will not be discussed here). These new attempts at timber plantations also have negative impacts for society and the environment, but when confronted with the destructive activities of TB by a news reporter, Cambodia's environment minister, Say Samal, conceded a loss of biodiversity, but maintained that development was more important. "We have to be realistic, we want to build our economy, we want to create jobs for our people so we have to balance that out" [54].

According to a recent study conducted for the Technical Working Group on Forest Reform in Cambodia, the implementation of industrial tree plantations lacks strategic planning, adequate forestry skills, and would benefit from more intensive and thorough study [59]. REDD+ in Cambodia suffers similar effects from rapid implementation in advance of policies or frameworks to support it, and the Disney purchase was the country's first official sale of carbon since 2008 (policies were not in place with the Ministry of Finance to effectively manage a carbon sale in Oddar Meanchey, which remains stalled at the time of this writing). In the current environment of increasing climate financing, government officials are under pressure to adopt REDD+ schemes both from the ministry's need and desire for funding, and from international discourses of climate change mitigation. The role of donor organizations and private companies pushing carbon capture schemes was critiqued by both UNDP and Forest Administration representatives in Phnom Penh. They are "approaching the government trying to push carbon capture . . . [they put] the sale of carbon before the community and before there are legal systems in place" (Interview UNDP 16 March 2015).

Not only are the legal systems to manage REDD+ still emerging, Cambodia currently has multiple systems for managing forests. This is changing, and in April 2016 almost 100,000 hectares of forest for conservation were transferred to the Ministry of Environment (MoE), while the Ministry of Agriculture, Forestry, and Fisheries (MAFF) lost control of those same areas for conversion purposes. This is an emerging and interesting conflict in the arena of climate change policies in Cambodia, as the land conflict between MAFF and MoE precedes climate funding. This environment of shifting jurisdictions amid flows of money and policy initiatives exacerbates competition among ministries for climate finance (see also, [45]), thereby obscuring the rampant deforestation.

"We have done enough damage with our money," says UNDP representative Napoleon Navarro. Working on both sides on this project, trying to facilitate government capacity building and community awareness development, Navarro contends that "you can't organize anyone around the promise of money" (Interview 16 March 2015). This sentiment was contextualized by Chhun Delux, administrator of forest finance for the Forest Administration. "Among the government and business people, there is this push for money—they are less concerned with the ecological aspects, it is a new business incentive . . . Everyone is working for their own benefit and no one looks at the big picture." He mapped out the donor tree and the breakup of the various NGO activities "...this one biodiversity, this one community forests, this one capacity building . . . see? islands. But the whole landscape [Prey Lang] is under served, which fact goes unrecognized by the officials and donors...." (Interview FA 13 July 2015).

The speculative and non-transparent nature of land conversion in Cambodia in the context of weak state institutional systems is not ideal for sustainable forest financing. This issue is exacerbated by the problem of forest financing itself, which has yet to produce a systematic framework for this endeavor. The situation in Cambodia consists of bureaucratic structures staffed by inexperienced officials, large expenditures by donor organizations, a focus on carbon over forest and community health, and continued pressure on forest resources from market-based activities like industrial plantations, selective logging, and market-crop conversion. The next section provides a closer inspection of the T-REDD project area, which has overcome some of the pitfalls related here, but remains mired in others.

6. The T-REDD Project Area

Multiple land claims, including ELC, FC, SLC, private homestead land, CF and permanent forest estate influence the T-REDD project. The history of war, forest exploitation, and marketization remain embedded in resource claims at this site. After 1993, in the post-election "transition period", abundant forests were identified by the World Bank as key export commodities [60,61]. By 1998 all forests outside of protected areas were granted as forest concessions to well-connected governors and Southeast Asian corporations [62]. This initiative provided none of the expected state-building revenue-generating effects; it was de-emphasized by 2001 [63] and replaced with ELCs.

The T-REDD site wraps around the first ELC in Prey Lang. A 6200-ha "state-led development project", the Tumring Rubber Plantation [64] was issued well inside the locally conceived boundary of the forest. People previously accustomed to shifting agriculture and collecting forest products (almost 60% of reported income, 31% and 28%, respectively) were divested of forest resources and shifting cultivation lands and asked to grow rubber [65].

In the opening ceremony for the plantation, Prime Minister Hun Sen explicitly encouraged citizens to " . . . change from collecting resin, tapping resin ... to tapping rubber" (GW 2007, 30). This was a dramatic revaluation of the land and people echoed in the TB sub-decree described below. Effects of these land-use policies are palpable for local communities, "Before the companies we rarely got sick, and if we did we could go find medicines from the forest. Now we look [for medicine] in the CF but don't find it and the people are more and more sick" (Group Discussion KT 28 February 2016). Effects are also visible in the landscape, and by 2012 the forest had given way to all the concessions outlined in white in Figure 4.

The CF initiatives started in response to the forest concessions, but due to ministerial back logs and inattention, few were in place before the implementation of ELCs. Between 2001 and 2012 a nation-wide initiative to establish community forests ensued, and communities around Prey Lang worked to establish CFs and defend the remaining unprotected forest. This was, and remains a disputed initiative in Prey Lang, as it limits community use to small designated areas. For good or for ill, by 2010, Kampong Thom had 46 CFs totaling over 35,000 ha, more than any other province [66]. Local participants say that protecting their CF "was so difficult, even the police were involved in cutting the CF" (Group Discussion KT 27 February 2016), after official ministry recognition, this task became much easier. These CFs are visible in Figure 4 as the green patches that spread from the forest, like island archipelagos.

The red deforested area surrounding the CFs and the Tumring Rubber Plantation in Figure 4 represents small-holder cassava plantations. These areas were converted in increments and sold to migrants by local elites until the roll out of Hun Sen's "Order 01 on Measures for Strengthening and Increasing the Effectiveness of the Management of Economic Land Concessions" (Order 01) in 2012. Instigated to address growing unrest over the effects of ELCs across the country, Order 01 morphed to include issuing land titles for citizens, which drove elite land capture in advance of titling (this was a terribly complicated and controversial initiative that affected people differently across the country—for a comprehensive discussion of this complex process, see [36,67,68]). Another effect of Order 01 is the replacement of ELCs with SLCs for economic endeavors. In 2013, an SLC was awarded right behind one of the T-REDD community forests. There are no landless settlers in this Social Land Concession, which is for "poor citizens who are landless or land poor" [69]. There are no landless settlers inside the SLC area, only sawmills operated by one of the same well-connected Cambodian businessmen involved in creating the Tumring Rubber Plantation [64,70].

All the land use strategies described above occurred before T-REDD and were known at the time the documents were signed; the sub-decree was not acknowledged in the first map (Figure 4), but the map was revised (Figure 6).

The once forest subsistent people now administering T-REDD were divested of their modes of production and forced into the market economy 13 years before the project. After two meetings with the Forest Administration in June and July 2015, REDD+ CF communities had low, but positive expectations. They receive patrolling equipment and 50 USD/month "if we keep the forest they will help us—if not, they won't" (Group Discussion KT 28 February 2016).

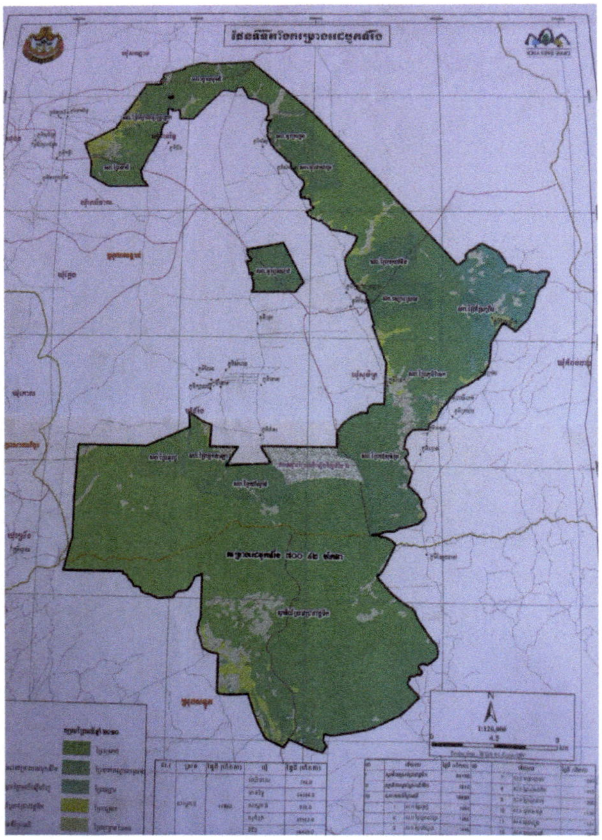

Figure 6. Photo of T-REDD Map 2 taken by author.

Community forest committees directly involved in the project reported moderate added benefits. "The FA trained us about REDD+ and now they support us with equipment and money for patrols" (Group Discussion KT 28 March 2016). "If we can get help from the government to protect [the forest] it is better for us" (Group Discussion KT 30 March 2016). In terms of the community's ability to protect the CF area, "REDD+ does not change much for us, only that we get training, equipment, and patrol money, not enough money though. It doesn't replace money we can make doing other things, so it's hard to get people to patrol. It only pays for our gas" (Group Discussion 29 February 2016). When asked directly about conflicts involving REDD+, people reported conflicts with illegal loggers, companies, elites, NGOs, or neighbors, not with REDD+, except that it did not do enough. It did not solve their problems with forest protection or getting support from the Forest Administration to arrest loggers or evict encroachers, "we call them [the FA]. Sometimes they come, but mostly not" (Group Discussion KT 29 March 2016). Neither did it capture all the CFs in the area, and the excluded CF committees wished they too would receive the benefits (Meeting, Prey Lang Working Group, 16 August 2016).

This site was purposefully chosen for REDD+ because of its high deforestation rate [71]. The project uses a simple system for measuring carbon and deforestation through a projection scenario based on historical deforestation data. Success demonstrates less deforestation than projected (Interview FA 5 September 2016). The T-REDD project documents give thoughtful attention to the challenges of REDD+ and attempt to foster community empowerment. Household uses of the forest, like shifting cultivation or charcoal production, are explicitly not considered part of the deforestation problem and community capacity to "keep the forest" is honored. T-REDD aims to, "assist community forests to scale up their forestland management area to cover the remaining permanent forest estate ... legalized as parts of the provincial and commune land-use planning framework" [72].

As the law stands at the time of this writing, however, citizens do not have the authority to patrol the larger PLWS protected area. "They give us the CF to protect while they cut the Prey Lang as they like ... even the police go too" (Group Discussion KT 28 February 2016); "they take trucks every day along the new road ... into the forest" (Group Discussion KT 29 February 2016). "If they give us authority, we can protect it [PLWS]," people say, "it has use for us—this is our resin forest" (Group Discussion KT 28 February 2016). The T-REDD project description cannot contain all the contradictions and pitfalls of REDD+ projects described above, but those it can transform suggest that that climate change policies may benefit from non-market approaches, and that effective REDD+ implementation should not degrade and at least marginally benefit local economies.

7. Think Biotech Cambodia, co. ltd.

Think Biotech was established as a public-private partnership between the Forest Administration of Cambodia and the Think Biotech Cambodia, co. ltd. in December 2010. Think Biotech is a subsidiary of the Hanwha corporation that specializes in the manufacture of explosives and military equipment. Before 2010, the Think Biotech Cambodia, co. ltd. was not listed in Hanwha's annual report and Hanwha had no experience in forestry initiatives [72].

Nonetheless, in Article 2 of the sub-decree signed by MAFF, it states that Think Biotech will establish an A/R project that will "improve soil fertility through reforestation and biodiversity conservation ... [and will be] part of Clean Development Mechanisms or other mechanisms that contribute to the reductions of greenhouse gas emissions and climate change mitigation" [40]. The conflict drivers in this case are the other objectives in Article 2 that include, "to stop slash and burn activities, and ... illegal claims to trees ... " [40]. The project deliberately alters the local economy in ways that also increase pressure on the PLWS outside the project boundary by pushing subsistence and market activities deeper into the forest.

In June 2012, the company began operations in Kampong Cham commune, Sombor District, Kratie province. Unlike in Kampong Thom, there was a period of 10 years between the slowdown in Forest Concession activity and the rise of ELCs, during which time freelance logging rose considerably. Even so, when TB came in 2012, traditional livelihoods of shifting cultivation, wet-rice agriculture, and resin collecting were the primary economic activities (Group Discussion KR 13 February 2015). As is typical of ELCs in Cambodia, TB started operations with no community consultation. At first people thought the excavators were for road development, it then became clear that they were making a business and some people took small jobs creating the tree nursery or other buildings (Interview Mr. Som No 15 February 2016).

The director of TB stated that he knew there were people living inside the concession boundaries (Interview TB 1 November 2016). The commune chief confirmed this as well, but suggested that the company cleared community land because they did not recognize it; community land for shifting cultivation looked like degraded forest (Interview KR 14 February 2015). The company signed an agreement with MAFF to begin operations in 2012 and hired CES co. ltd. (Phnom Penh, Cambodia) to conduct an environmental impact assessment (EIA). This assessment was completed in February 2013 and the project was found to have "a lot of problems" and "would be bad for the community" (Interview CES 8 November 2016). This information was shared with the MoE and with the company. According to an unpublished report by NGO Forum, the project would affect approximately 1900 families, 4412 hectares of rice fields, 3534 hectares of plantation land, 5970 hectares of community forests, 5 hectares of spirit forests, and 5 hectares of burial grounds ("NGO Forum" notes 17 November 2015).

After the negative EIA, in May of 2013, company bulldozers began clearing community lands along the old road. During this clearing, 178 households lost nearly 1000 hectares of farm and shifting cultivation land. Strong community protests kept the company from clearing more land

(Interview community rep. 14 February 2015), but there remain 400 hectares of disputed land inside the company boundaries (Interview commune chief KR 3 February 2015). In addition to community rice fields and farm lands, the company cleared nearly 5000 ha of forest land in the southern end of their project next to the affected villages (Figure 5, red deforested area). In so doing, they cleared the forest right up to the banks of the streams. The effects of this on rain-fed streams are visible (Figure 7). The sun exposure kills fish eggs in the streams and causes them to dry completely when the rains stop (Field Notes, personal communication, Mr. Som No 6 August 2016).

Figure 7. Stream *O Sro Lork* in the rainy season, 9.7.2016. Photo by Seay Monyrath.

At peak production, the company employed approximately 800 workers earning between 150–180 USD per month (Interview KR 16 February 2016), which is well below the 250–300 USD people report from tapping resin and selling market crops (Field Notes Steung Treng 19 December 2016, see also [73]). The commune chief sees the benefit of the jobs (Interview 16 February 2015), but was not pleased that the promised road was never built (Field Notes, personal communication Mr. Nak Virak 9 February 2016). The only roads built by the company were for company use and community members had to protest to gain access (Group Discussion KR 14 February 2016). Since February 2016 the company has been quiet. They laid off most of their workers and the saw mill stopped operations. In September 2016 there were about 200 workers planting saplings, many of whom are migrants (FN KR 7 September 2016).

While in the south the company seems quiet, in June of 2016 they conducted a public consultation in Steung Treng with about 100 members of the soon to be affected community. A local researcher accompanied company representatives and provincial and national level ministry officials on their mission to mark the TB boundary. At the most isolated edge of the project bordering the PLWS they placed company boundary markers and drew a map. While mapping this territory, the researcher informed the officials they will need a map of community holdings to conform to Order 01 guidelines that require companies to develop around, and not through, community lands. The Forest Administration representative told him, "we don't have to follow the rules of Order 01, this is not an ELC. It's a government partnership" (Action Research Interview ST 18 August 2016). Since the boundary mapping, community researchers have mapped 15 of the 19 resin forests in use at the present time (Figure 8).

That the company moved to map and begin developing the northernmost region of their project could be because they have cleared the area in the south right up to the boundary of the southernmost CF that is inside the concession (Figure 9). In many ways the company is frustrated with this project as well. The CFs inside project boundaries were a surprise according to the company director, and his face was visibly concerned when the author showed him the mapped boundaries of resin forests inside "his" company boundaries (Interview TB 1 November 2016).

Figure 8. Community resin forest shapefiles. Red is project boundary, white N-S line is PLWS boundary, white E-W line is provincial boundary (Steung Treng to the north, Kratie to the south). Blue are mapped forests and yellow are remaining areas to be mapped. GoogleEarth screen shot by author.

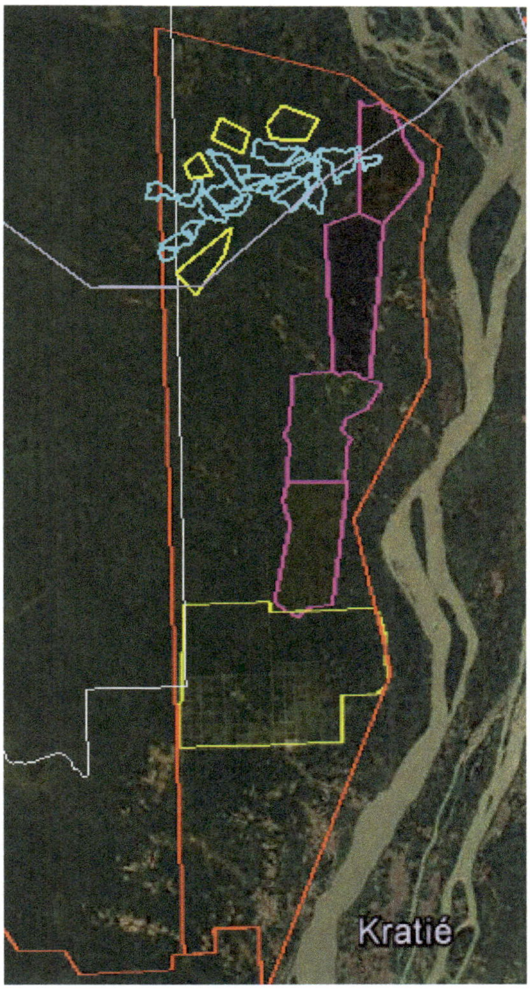

Figure 9. Google Earth screen shot with shape files. Pink are CFs and yellow is cleared company area as of September 2016. Affected area boundaries created using GPS data points collected by PLCN and author in September 2016. CF shape files are from Cambodia's Forest Administration (FA).

The situations on the ground at the TB project site differ widely from those at the T-REDD site. Most importantly, the evacuation of local subsistence practices was explicitly part of the project's goal. There are many other factors that contributed to the conflict-heavy implementation of this project, these include the lack of community consultation, ignoring EIA recommendations, clearing community land holdings, providing undesirable jobs, and restricting access to company roads. The company is actively avoiding the CF area inside their project, which is commendable, but are surprised and dismayed by the presence of community holdings inside their northern boundary. This issue is currently under negotiation.

8. Discussion

The evidence presented in this study shows two governance strategies for CCM projects as well as the social, economic, geographic, and historical circumstances of the people and the regions in which each operates. Through this analysis, focused on the potential for decreasing the incidence of conflict in CCM policies, some fundamental differences between the projects are visible. First, in the project with little conflict, both project documents and implementation practice were focused on granting tenure rights and management responsibility to participants. By contrast, in the conflict-heavy project, the language of the project documents, the implementation of project activities, and the voices of company representatives were all directed at divesting communities of their rights and responsibilities. Second, the conflict-light scenario is not focused on monitory trade or profit, a source of regular conflict in REDD+ case studies [74]. Communities know exactly what to expect from this initiative. It is not a lot and the $50/month does not compensate them for their time, but it covers the costs of patrolling. The TB project gives rise to multiple financial conflicts beyond clearing village resources for company profits: the wages are low, salaries are often late, large lay-offs have occurred, and locals accuse the company of selectively logging high-value timber from the deep forest.

These data suggest that conflict-light CCM projects can be implemented with close attention to the relationship between selective project outcomes and the physical, social, and economic landscape in which it is implemented. The evidence from these two case studies suggests that the TB restoration project is not executed in the best landscape for an industrial tree plantation and that both the social and physical landscape is better suited for a REDD+ project under the Korea-Cambodia MoU for enhancing forest resources and mitigating climate change. First, it is in an area populated with people actively protecting the forest and using the forest for subsistence. Second, it is in a remote area with very little infrastructure, which adds to transport costs for industrial production and limits encroachment, enhancing conservation. Third, it is covered with forest that in many parts of the project area is quite dense and diverse.

There is very little support for creating industrial timber plantations in already forested landscapes. Guidelines insist that A/R activities should be on degraded forest land [26], but the practice of labeling rich forest as degraded forest is widespread [75], with profoundly negative effects on social and ecological communities [28,39]. Increased deforestation beyond project boundaries is a recorded effect of forest-conversion for development. The map of the Tumring area shows clearly what this looks like (Figure 4). The TB project is in a very remote area, but is currently building roads to transport timber and equipment. Not only does this add costs to their operation, but satellite imagery shows increased forest conversion all along these roads and researchers have mapped these roads to points well inside the PLWS.

The final, and most important reason that suggests a successful REDD+ project on this site is that the community is both capable and willing to engage in forest protection activities. This most important fact, coupled with the very progressive and flexible project design of the T-REDD initiative could go a long way to protecting valuable and biodiverse forests in an isolated and naturally protected forest landscape.

Korea is innovative and is engaged in a number of development experiments; juxtaposing REDD+ and A/R is not new for them [76]. The KFS currently partners with 12 countries in forestry initiatives and is well-known for public private partnership funding arrangements [77]. The implementation of T-REDD was unique, according to the technical partner in the Korea-Cambodia T-REDD project (Meeting FA 13 July 2015), who claimed that waiting for the legal frameworks was too slow and government to government initiatives do not need them. Local sources in the community and in the Forest Administration, suggest that the government has more power to implement a REDD+ project, "the NGO has no power and local authorities have no reason to do what they ask" (Interview FA 17 December 2015); "this project is strong because it is with the government, they have to help us" (Group Discussion 29 March 2016). It is worth discussing if a Korean-led publicly engaged REDD+ development project could be instituted at the TB site and if the already cleared SLC in Tumring could be reforested for commercial timber production.

What I suggest here is a radical, perhaps impossible, undertaking of project manipulation. However, had the governments and companies involved adopted an un-cynical approach to their projects and attended to larger project landscapes for maximum project/public/forest benefit, it could have produced conflict-light CCM policy enactments.

9. Conclusions

The case studies presented here demonstrate how a particular method to analyze the relationship between CCM and conflict exposed the social and ecological impacts of two related projects. The negative impacts of the TB project hit at the intersecting agendas of a loosely accountable government and a for-profit business investment, both focused on monetary benefits T-REDD project effects cannot really be called positive, but by all accounts from participants they are not negative, and the project design makes explicit attempts to mitigate the known challenges to REDD+ implementation. These case studies also reveal a possible pathway to avoided conflict. The T-REDD and TB data suggest that forest-based CCM policies can reduce conflict through efforts at minimal transformation of local livelihoods, maximal attention to the tenure rights, responsibilities, and authority of citizens, and by improving, not degrading, the environment in project landscapes.

In the A/R study cited here, each of these conflict-reducing practices was violated, resulting in a conflict-heavy CCM enactment. TB was typical of other Cambodian economic land grabs, but was also explicitly designed to relieve citizens of their livelihoods with no considerations of tenure rights or respect for community authority. In addition, the environmental impact of large-scale clearing of forested landscapes spreads the conflict into the environment both within and beyond project boundaries. On the other hand, T-REDD did enact the conflict reducing practices. The explicit attempt to put communities at the front of managing state forest reserves marks a turning point for Cambodian forestry policy. It is important to note, however, that the project was conflict-light primarily because livelihoods had already been transformed through ELCs and other tenure schemes and its lack of reported conflict from project participants should not overshadow the "perverse logic" [78] under which it works. For $50/month the T-REDD+ project uses "if not blatantly exploits" those least responsible for causing climate change to do the "messy, time-consuming, labor-intensive and dangerous work of protecting forests, which are of global benefit" (ibid). They are put to this work to offset the carbon-intensive activities of Korean companies in Cambodia.

It is important at this historical juncture to make explicit political claims about the effects of pretending that preserving forest in one place can mitigate the large-scale emission of carbon in another place [24,37,79]. Climate change is a complex issue and each attempt to mitigate it through bureaucratic and/or market-based endeavors exposes a dangerous ambivalence. This study suggests that conflict is at the heart of CCM policies, connected as they are to the resource capture necessary for economic growth. This is a fundamental contradiction with which we must all grapple, lest we are left with nothing but forest islands and castaway communities.

Acknowledgments: Funded by the Dutch National Science Foundation (NWO), and the Department for International Development (DFID) through the CoCooN—Conflict and Cooperation in the Management of Climate Change—Integrated Project. See: http://www.iss.nl/mosaic.

Conflicts of Interest: The author declares no conflict of interest.

References

1. Jae-un, L. Korea Begins Preserving Tropical Forests in Cambodia. Korea.Net. Available online: http://www.korea.net/NewsFocus/Policies/view?articleId=127298 (accessed on 7 May 2015).
2. Ribot, J.C.; Peluso, N.L. A Theory of Access*. *Rural. Sociol.* **2009**, *68*, 153–181. [CrossRef]
3. Hall, D.; Hirsch, P.; Li, T. *Powers of Exclusion: Land Dilemmas in Southeast Asia*; University of Hawaii: Honolulu, HI, USA, 2011; in press.
4. James, F.; Leach, M.; Scoones, I. Green Grabbing: A new appropriation of nature? *J. Peasant. Stud.* **2012**, *39*, 237–261.
5. Borras, S.; Franco, J. Global Land Grabbing and Trajectories of Agrarian Change: A preliminary analysis. *J. Agrar. Chang.* **2012**, *12*, 34–59. [CrossRef]
6. Theilade, I.; Schmidt, L. *REDD + and Conservation of Prey Long Forest, Cambodia Summary of Scientific Findings 2007–2010*; No. 66; Forest and Landscape: Cornelius, OR, USA, 2011; Volume 66.
7. Touch, S.; Neef, A. Local Responses to Land Grabbing and Displacement in Rural Cambodia. In *Global Implications of Development, Climate Change and Disasters: Responses to Displacement from Asia–Pacific*; Price, S., Singer, J., Eds.; Routledge/Earthscan: London, UK; New York, NY, USA, 2015; pp. 124–141.
8. Hunsberger, C.; Corbera, E.; Borras, S.M., Jr.; Rosa, R.; de Eang, V.; Franco, J.C.; Herre, R.; Kham, S.S.; Park, C.; Pred, D.; et al. Land-Based Climate Change Mitigation, Land Grabbing, and Conflict: Towards a landscape-based and collaborative research agenda. *Can. J. Dev. Stud. Rev.* **2015**, in press.
9. Parnell, T. Story-Telling and social change: A case study of the Prey Lang Community Network. In *Conservation and Development in Cambodia: Exploring Frontiers of Change in Nature, State and Society*; Milne, S., Mahanty, S., Eds.; Routledge: London, UK; New York, NY, USA, 2015; pp. 258–279.
10. Fox, J.; Castella, J.-C.; Ziegler, A.D. Swidden, rubber and carbon: Can REDD+ work for people and the environment in Montane Mainland Southeast Asia? *Glob. Environ. Chang.* **2014**, *29*, 318–326. [CrossRef]
11. Ziegler, A.D.; Phelps, J.; Yuen, J.Q.; Webb, E.L.; Lawrence, D.; Fox, J.M.; Bruun, T.B.; Leisz, S.J.; Ryan, C.M.; Dressler, W.; et al. Carbon outcomes of major land-cover transitions in SE Asia: Great uncertainties and REDD+ policy implications. *Glob. Chang. Biol.* **2012**, *18*, 3087–3099. [CrossRef]
12. Corbera, E.; Schroeder, H. Governing and implementing REDD+. *Environ. Sci. Policy* **2011**, *14*, 89–99. [CrossRef]
13. Savaresi, A. A Glimpse into the Future of the Climate Regime: Lessons from the REDD+ Architecture. *Rev. Eur. Comp. Int. Environ. Law* **2016**, *25*, 186–196. [CrossRef]
14. Hansen, M.C.; Potapov, P.V.; Moore, R.; Hancher, M.; Turubanova, S.A.; Tyukavina, A. High-resolution global maps of forest cover change. *Science* **2013**, *342*, 850–853. [CrossRef] [PubMed]
15. Le Billon, P. The Political Ecology of Transition in Cambodia 1989–1999: War, Peace and Forest Exploitation. *Dev. Chang.* **2000**, *31*, 785–805. [CrossRef]
16. Le Billon, P.; Springer, S. Between war and peace: Violence and accommodation in the Cambodian logging sector. In *Extreme Conflict and Tropical Forests*; Springer: Dordrecht, The Netherlands, 2007; pp. 17–36.
17. Biddulph, R. In whose name and in whose interests? An actor-oriented analysis of community forestry in Bey, a Khmer village in Northeast Cambodia. In *Conservation and Development in Cambodia: Exploring Frontiers of Change in Nature, State and Society*; Milne, S., Mahanty, S., Eds.; Routledge: London, UK; New York, NY, USA, 2015; pp. 160–176.
18. Pasgaard, M.; Chea, L. Double Inequity? The Social Dimensions of Deforestation and Forest Protection in Local Communities in Northern Cambodia. *Austrian J. South East Asian Stud.* **2013**, *6*, 330–355.
19. Royal Government of Cambodia (RGC). *Community Forest Governance*; Ministry of Agriculture, Foretry, and Fisheries: Phnom Penh, Cambodia, 2003.
20. Käkönen, M.; Lebel, L.; Karhunmaa, K.; Dany, V.; Try, T. Rendering Climate Change Governable in the Least-Developed Countries: Policy Narratives and Expert Technologies in Cambodia. *Forum Dev. Stud.* **2014**, *4*, 351–376. [CrossRef]
21. Eilenberg, M. Shades of green and REDD: Local and global contestations over the value of forest versus plantation development on the Indonesian forest frontier. *Asia Pac. Viewp.* **2015**, *56*, 48–61. [CrossRef]

22. Hinsley, A.; Entwistle, A.; Pio, D.V. Does the long-term success of REDD+ also depend on biodiversity? *Oryx* **2014**, *49*, 1–6. [CrossRef]
23. Beyene, A.D.; Bluffstone, R.; Mekonnen, A. Community Controlled Forests, Carbon Sequestration and REDD+: Some Evidence from Ethiopia. *Environ. Dev. Econ.* **2016**, *21*, 249–272. [CrossRef]
24. Ingalls, M.L.; Dwyer, M.B. Missing the forest for the trees? Navigating the trade-offs between mitigation and adaptation under REDD. *Clim. Chang.* **2016**, *136*, 353–366. [CrossRef]
25. Angelsen, A.; Wertz-Kanounnikoff, S. *Moving Ahead with REDD Issues, Options and Implications*; Center for International Forestry Research: Kota Bogo, Indonesia, 2008; Volume 6.
26. United Nations Framework Convention on Climate Change (UNFCC). *Afforestation and Reforestation Projects under the Clean Development Mechanism*; UNFCC Secretariat: Bonn, Germany, 2013.
27. Cyranoski, D. Biodiversity: Logging: The new conservation. *Nature* **2007**, *446*, 608–610. [CrossRef] [PubMed]
28. Lyons, K.; Westoby, P. Carbon colonialism and the new land grab: Plantation forestry in Uganda and its livelihood impacts. *J. Rural Stud.* **2014**, *36*, 13–21. [CrossRef]
29. Neef, A.; Touch, S.; Chiengthong, J. The Politics and Ethics of Land Concessions in Rural Cambodia. *J. Agric. Environ. Eth.* **2013**, *26*, 1085–1103. [CrossRef]
30. Sasaki, N.; Yoshimoto, A. Benefits of tropical forest management under the new climate change agreement-a case study in Cambodia. *Environ. Sci. Policy* **2010**, *13*, 384–392. [CrossRef]
31. Khun, V.; Sasaki, N. Cumulative Carbon Fluxes Due to Selective Logging in Southeast Asia. *Low Carbon Econ.* **2014**, *5*, 180–191. [CrossRef]
32. Sikor, T.; Cam, H. REDD+ on the rocks? Conflict over Forest and Politics of Justice in Vietnam. *Hum. Ecol.* **2016**, *44*, 217–227. [CrossRef] [PubMed]
33. Scheidel, A.; Work, C. Large-scale forest plantations for climate change mitigation? New frontiers of deforestation and land grabbing in Cambodia. In *Global Governance/Politics, Climate Justice and Agrarian/Social Justice: Linkages and Challenges*; ICAS Colloquium: Hague, The Netherlands, 2016; p. 11.
34. Baird, I.G. "Indigenous Peoples" and land: Comparing communal land titling and its implications in Cambodia and Laos. *Asia Pac. Viewp.* **2013**, *543*, 269–281. [CrossRef]
35. Thuon, R. *REDD+ and Tenure Security of Community Forestry: A Case Study of Oddar Meanchey Community Forestry REDD+ Project*; University of Tokyo: Tokyo, Japan, 2013.
36. Milne, S. Under the leopard's skin: Land commodification and the dilemmas of Indigenous communal title in upland Cambodia. *Asia Pac. Viewp.* **2013**, *54*, 323–339. [CrossRef]
37. Vongvisouk, T.; Broegaard, R.B.; Mertz, O.; Thongmanivong, S. Rush for cash crops and forest protection: Neither land sparing nor land sharing. *Land Use Policy* **2016**, *55*, 182–192. [CrossRef]
38. Kurashima, T.; Matsuura, T.; Miyamoto, A.; Sano, M.; Tith, B.; Chann, S. Changes in Income Structure in Frontier Villages and Implications for REDD+ Benefit Sharing. *Forests* **2014**, *5*, 2865–2881. [CrossRef]
39. Olwig, M.F.; Noe, C.; Kangalawe, R.; Luoga, E. Inverting the moral economy: The case of land acquisitions for forest plantations in Tanzania Inverting the moral economy: The case of land acquisitions for forest plantations in Tanzania. *Third World Q.* **2016**, *6597*, 2316–2336.
40. Ministry of Agriculture, Forestry, and Fisheries (MAFF). *Prakas: Declare the Establishment of Bung Cha Camp for Promotion and Forest Restoration*; MAFF: Prakas, Cambodia, 2010.
41. Global Withess. *The Cost of Luxury*; Global Witness: London, UK, 2015.
42. Michaud, A. *Prey Lang Development Case: Do People Benefit from Its Development?* NGO Forum: Phnom Penh, Cambodia, 2013.
43. Kotoski, K. Disney carbon deal sets stage for more partnerships. Available online: http://www.phnompenhpost.com/business/disney-carbon-deal-sets-stage-more-partnerships (accessed on 12 August 2016).
44. Food and Agriculture Organization; United Nations Development Programme; United Nations Environment Programme. Cambodia REDD+ Roadmap. In Proceedings of UN-REDD programme 5th policy board meeting, Washington, DC, USA, 4–5 November 2010; UN-REDD: Washington, DC, USA, 2010; pp. 1–54.
45. The United Nations Programme on Reducing Emissions from Deforestation and Forest Degradation. *Myanmar REDD+ Readiness Roadmap*; The Government of the Union of Myanmar: Yangoon, Myanmar, 2013.
46. Atela, J.O.; Quinn, C.H.; Minang, P.A.; Duguma, L.A.; Houdet, J.A. Implementing REDD+ at the national level: Stakeholder engagement and policy coherences between REDD+ rules and Kenya's sectoral policies. *For. Policy Econ.* **2016**, *65*, 37–46. [CrossRef]

47. Lang, C. *Konflikte um Rohstoffe in Asien REDDheads: The People Behind REDD and the Climate Scam in Southeast Asia*; Stiftung Asienhaus: Cologne, Germany, 2016.
48. Yeang, D.; Brewster, J. *REDD+ Demonstration Activities in Cambodia: The Case of the Oddar Meanchey Community Forestry REDD+ Project*; PACT Cambodia: Phnom Penh, Cambodia, 2012.
49. The United Nations Programme on Reducing Emissions from Deforestation and Forest Degradation. UN-REDD Programme Welcomes Five New Countries. Press Release. Available online: http://www.yeangdonal.net/2009/11/un-redd-programme-welcomes-five-new.html (accessed on 7 March 2015).
50. Lynham, T.; Evans, T.; Pet, P.; Phien, S. *Monitoring Systems for Illegal Land Encroachment at the Seima Protection Forest REDD+ Demonstration Site*; Wildlife Conservation Society and the Forestry Administration of the Ministry of Agriculture, Forestry, and Fisheries: Phnom Penh, Cambodia, 2014.
51. Delux, C.; Nguon, P. *National REDD Strategy, 4th Draft*; Forest Administration of the Ministry of Agriculture, Forestry, and Fisheries: Phnom Penh, Cambodia, 2015.
52. Royal Government of Cambodia (RGC). *Cambodia Millennium Development Goals Report*; Ministry of Planning: Phnom Penh, Cambodia, 2003.
53. Royal Government of Cambodia (RGC). *National Strategic Development Plan, 2014–2018*; Ministry of Planning: Phnom Penh, Cambodia, 2014.
54. Board, J. Cambodian villagers fear for future amid forest burning dispute. Available online: http://www.channelnewsasia.com/news/asiapacific/cambodian-villagers-fear/3020786.html (accessed on 11 August 2016).
55. Forest Administration of Cambodia. *Cambodia Forestry Outlook Study*; No. Working Paper No. APFSOS II/WP/2010/32; Ministry of Agriculture, Forestry, and Fisheries: Phnom Penh, Cambodia, 2010.
56. Lang, C. The Expansion of Industrial Tree Plantations in Cambodia and Laos. *Focus Asien Schr. Asienhauses* **2006**, *29*, 24–27.
57. Middleton, C.; Sokleap, H. *Fast-Wood Plantations, Economic Concessions and Local Livelihoods in Cambodia*. 2007. Available online: http://wrm.org.uy/oldsite/countries/Cambodia/EFCT_Plantations_Report.pdf (accessed on 15 March 2015).
58. Petersen, R.; Aksenov, D.; Esipova, E.; Goldman, E.; Harris, N.; Kuksina, N.; Sargent, S.; Manisha, A.; Loboda, T.; Shevade, V. *Mapping Tree Plantations with Multispectral Imagery: Permiminary Results for Seven Tropical Countries*; Technical Note; World Resources Institute: Washington, DC, USA, 2016.
59. Ra, K.; Kimsun, T. Financial Viability of Plantations of Fast- Growing Tree Species in Cambodia. Phnom Penh, Cambodia, 2012. Available online: http://twgfr.org/download/StudyReports(2)/13-FinancialviabilityofplantationFast-growingtreespeciesinCambodia-2012.pdf (accessed on 30 January 2017).
60. World Bank. *Cambodia Agenda for Rehabilitation and Reconstruction*; World Bank: Washington, DC, USA, 1992.
61. Hughes, C. *The Political Economy of Cambodia's Transition, 1991–2001*; Routledge/Curzon: London, UK; New York, NY, USA, 2003.
62. Le Billon, P. Logging in Muddy Waters: The Politics of Forest Exploitation in Cambodia. *Crit. Asian Stud.* **2002**, *34*, 563–586. [CrossRef]
63. Milne, S. Cambodia's Unofficial Regime of Extraction: Illicit Logging in the Shadow of Transnational Governance and Investment. *Crit. Asian Stud.* **2015**, *47*, 200–228. [CrossRef]
64. Global Witness. *Cambodia 's Family Trees: Illegal Logging and the Stripping of Public Assets*; Global Witness: Phnom Penh, Cambodia, 2007.
65. Dararath, Y.; Top, N.; Lic, V. *Rubber Plantation Development in Cambodia: At What Cost?* The Economy and Environment Program for Southeast Asia (EEPSEA): Phnom Penh, Cambodia, 2011.
66. Blomley, T.; Tola, P.; Kosal, M.; Dyna, E.; Dubois, M. *Review of Community Forestry and Community Fisheries in Cambodia; Report Prepared for the Natural Resource Management and Livelihoods Programme*; Natural Resource Management and Livelihoods Programme: Phno Penh, Cambodia, 2010.
67. Grimsditch, M.; Schoenberger, L. *New Actions and Existing Policies: The Implementation and Impacts of Order 01*; NGO Forum: Phnom Penh, Cambodia, 2015.
68. Focus on the Global South. *Study on the Impacts of the Implementation of Order 01BB on Selected Communities in Rural Cambodia*; Focus on the Global South: Phnom Penh, Cambodia, 2013. Available online: http://focusweb.org/sites/www.focusweb.org/files/StudentVolunteersReport_ENG.pdf (accessed on 12 December 2012).

69. Royal Government of Cambodia. *Sub-Decree for the Transfer of 2133.99 ha of State Public Land in Samaoch Village, Mien Rith Commune, Sandan District, Kampong Thom Province, for a Social Land Concession to Give to Landless and Land Poor Citizens, Cambodia*; Royal Government of Cambodia: Phnom Penh, Cambodia, 2013.
70. Titthara, M. Deforestation Continues. Availiable online: http://www.khmertimeskh.com/news/28863/the-deforestation-continues/ (accessed on 24 August 2016).
71. Chheng, K.; Bun, R.; Williams, B. *Tumring REDD+ Project: A Joint Korea-Cambodia Project*; Ministry of Agriculture Forestry and Fisheries: Phnom Penh, Cambodia, 2015.
72. PWC, P. *Hanwha Corporation Non-Consolidated Financial Statements*; Samil Price Waterhouse Coopers: Seoul, Korea, 2010.
73. Jiao, X.; Smith-Hall, C.; Theilade, I. Rural household incomes and land grabbing in Cambodia. *Land Use Policy* **2015**, *48*, 317–328. [CrossRef]
74. Milne, S.; Adams, B. Market Masquerades: Uncovering the Politics of Community-level Payments for Environmental Services in Cambodia. *Dev. Chang.* **2012**, *43*, 133–158. [CrossRef]
75. Harms, E.; Baird, I.G. Wastelands, degraded lands and forests, and the class (ification) struggle: Three critical perspectives from mainland Southeast Asia. *Singap. J. Trop. Geogra.* **2014**, *35*, 289–294. [CrossRef]
76. REDD Desk. *Korea Indonesia Joint Project for Adaptation and Mitigation of Climate Change in Forestry*; Global Canopy Programme (GCP): Oxford, UK, 2016. Availiable online: http://theredddesk.org/countries/initiatives/korea-indonesia-joint-project-adaptation-and-mitigation-climate-change (accessed on 5 September 2016).
77. Kim, J.; Kim, J.; Shin, S.; Lee, S. *Public-Private Partnership Infrastructure Projects: Case Studies from the Republic of Korea*; Asian Development Bank: Mandaluyong, Philippines, 2011; Volume 1.
78. Frewer, T. Code REDD+ in Cambodia. Availiable online: http://asiapacific.anu.edu.au/newmandala/2015/08/11/code-redd-in-cambodia/ (accessed on 12 December 2015).
79. Käkönen, M.; Karhunmaa, K.; Bruun, O.; Kaisti, H.; Tuominen, V.; Thuon, T.; Luukkanen, J. *Climate Mitigation in the Least Carbon Emitting Countries—Dilemmas of Co-Benefits in Cambodia and Laos*; University of Turku: Turku, Finland, 2013.

9

Livelihoods and Land Uses in Environmental Policy Approaches: The Case of PES and REDD+ in the Lam Dong Province of Vietnam

Leif Tore Trædal * and Pål Olav Vedeld

Department of International Environment and Development Studies (Noragric), Faculty of Land and Society, Norwegian University of Life Sciences, Ås, N-1432, Norway; pal.vedeld@nmbu.no

* Correspondence: leif.tore.traedal@nmbu.no

Abstract: This paper explores assumptions about the drivers of forest cover change in a Payments for Environmental Services (PES) and Reduced Emissions from Deforestation and Degradation (REDD+) context in the Lam Dong Province in Vietnam. In policy discourses, deforestation is often linked to 'poor' and 'ethnic minority' households and their unsustainable practices such as the expansion of coffee production (and other agricultural activities) into forest areas. This paper applies a livelihood framework to discuss the links between livelihoods and land use amongst small-scale farmers in two communities. The findings of the livelihood survey demonstrate no clear linkages between poverty levels and unsustainable practices. In fact, the poorest segments were found to deforest the least. The ways in which current PES and REDD+ approaches are designed, do not provide appropriate solutions to address the underlying dimensions of issues at stake. The paper criticizes one-dimensional perspectives of the drivers behind deforestation and forest degradation often found in public policies and discourses. We suggest more comprehensive analyses of underlying factors encompassing the entire coffee production and land use system in this region. Addressing issues of land tenure and the scarcity of productive lands, and generating viable off-farm income alternatives seem to be crucial. Sustainable approaches for reducing deforestation and degradation could be possible through engaging with multiple stakeholders, including the business-oriented households in control of the coffee trade and of land transactions.

Keywords: Vietnam; drivers of deforestation; livelihoods; environmental policies; REDD+; PES

---

1. Introduction

Tropical deforestation has been identified as one of the main global environmental challenges, contributing to a major share of the global emissions of greenhouse gasses and to the loss of biodiversity and ecological integrity worldwide [1]. At the same time, tropical forests are key arenas for livelihoods and outcomes for indigenous people and forest communities. Over recent years, the process of establishing a global mechanism for Reduced Emissions from Deforestation and Degradation (REDD+) has increased the focus on saving tropical forests. This has prompted a number of tropical forest countries to develop policies and prepare for reducing greenhouse gas (GHG) emissions from forests. However, the lack of progress in global climate change negotiations and in establishing a global REDD+ mechanism has stalled the process and also led to a diversity of approaches and adaptations of REDD+ in many countries [2].

In this paper, we use a case study from Vietnam to explore assumed and real processes driving forest cover change, and scrutinize some of the approaches taken for REDD+. In the Central Highlands of Vietnam, small-scale coffee production has been identified as a major driver of forest cover loss, and it is also frequently associated with other negative environmental and social consequences, such as soil erosion, and economic and political marginalization of ethnic minority groups [3–5]. Various policy schemes to decrease environmental degradation and reduce poverty levels have been developed and implemented. Recently, innovative, results-based policy initiatives such as Payments for Environmental Services (PES) and REDD+ have been introduced to attempt to alleviate some of the problems. A predominant argument in forest-related policy discourses is that poverty and general low knowledge levels among ethnic groups often lead to sub-optimal and unsustainable livelihood decision making [6]. The poverty-deforestation link is not unique to Vietnam, but can be identified in readiness processes of various REDD+ countries, as documented by, for example, Dooley et al. in their review of Readiness Preparedness Proposals (R-PP) submitted to the World Bank Forest Carbon Partnership Facility (FCPF) [7].

In Vietnam, dealing with the (perceived) negative linkages between poverty levels and environmental degradation has frequently been associated with the need for awareness raising and reducing poverty rates among indigenous groups [4]. Such aims are explicitly expressed in national forest-related policy documents and discourses, such as the Lam Dong Province Provincial REDD+ Action Plan ([8], p. 8):

> *'Ethnic minority people in particular, have been carrying out deforestation and converting forest land to settlements and agricultural land to support their traditionally very large families. Awareness raising amongst both male and female members of the community is needed to reduce population growth and the deforestation associated with it.'*

This paper uses an empirical case study of the interrelationships between livelihoods and land use to investigate policy assumptions about the drivers of land use change. In Vietnam, the coffee sector, and poor people in particular, are often blamed for deforestation processes, and this sector and group of households are therefore frequently targets of environmental programs, such as PES and REDD+. The study demonstrates the shortcomings of one-dimensional analyses of drivers of land use change in policies that aim to reduce deforestation and degradation. We argue that poor people do not deforest the most, and that targeting the coffee sector will be a challenging task unless the reforms are embedded in wider structures of the coffee industry. The focus on linking the direct drivers of change to small-scale coffee production by poor households seems to mask many of the underlying factors, such as longer-term 'control' and vested interests of state-owned coffee companies and the more wealthy segments of the population in maintaining a particular mode of production. Therefore, REDD+ projects are likely to fail in achieving their goals of reducing deforestation and degradation if they are targeted only at the poor, and omit the structural determinants of the coffee sector at large.

PES has been introduced in the area to cope with smallholder expansion of coffee. Based on the findings, we reflect upon the magnitude and scale of forest conversion caused by the expansion of small-scale livelihood-driven coffee production. We also investigate the social and environmental impacts of the local PES scheme. PES is also likely to become a key component of REDD+ in Lam Dong, and the paper therefore discusses the implications of the findings in terms of practical policy implementation, particularly in view of the ongoing and planned REDD+ pilot activities in the area.

More concretely, the paper explores the following research questions: (1) How do households manage and diversify assets and resources to generate livelihood outcomes? (2) What are the impacts of PES policies on livelihoods, environmental awareness, and deforestation? (3) How do the overall production structures of the coffee economy affect livelihoods and deforestation processes? (4) What are the potential implications for emerging REDD+ policies?

The paper starts by outlining the conceptual models considered relevant for the study. This is followed-up by an introduction of the methodology and case study context. Considering the high degree of market integration and the presence of various policy schemes, analyzing the overall context and dynamics of the coffee sector and policy schemes is crucial for understanding livelihoods and land use in the study area. The final section discusses the results in view of livelihood outcomes and policy implications for REDD+ and PES schemes.

2. Conceptual Framework

This paper investigates households' livelihoods and the implications for land use and deforestation in the study area. These findings are further used to analyze policy assumptions about 'poor' and 'ethnic' households and their alleged unsustainable land-use practices that tend to be prevalent in policy documents and other discourses. Perspectives from livelihood theory, political ecology, and land-use change theory consequently inspire the conceptual framework for the study.

We applied a livelihood framework (LF) to identify differences in livelihood adaptations, and the role of small-scale coffee production in local land-use changes and deforestation. According to the LF, households combine various capitals, such as natural, physical, human, social, and financial capitals, to generate livelihood outcomes in the form of agricultural and forestry outputs, and off-farm and business oriented income [9,10]. Hence, the determinant relationship between the households' asset portfolios, livelihood strategies, and outcomes is at the core of the LF. Institutional factors, such as property regimes, markets, local values, attitudes and norms, skills, and various other social institutions and decisions taken at multiple scales, also affect access to assets and livelihood decision making in households [9,11]. The clearing of forests and expansion of agricultural areas are important components in the livelihood strategies of households [12].

Defining the relevant drivers of deforestation and degradation has been identified as one of the main challenges in developing efficient and effective REDD+ strategies and policies, and is a field that requires further research [13,14]. The land-use change literature frequently differentiates between *proximate* (direct) and *underlying* (indirect) causes of environmental change [15–17]. In our case study area, smallholder coffee production could be viewed as a main *proximate* driver of deforestation. In policy discourses, this has often been linked to livelihood and poverty as *underlying* drivers [8]. Our argument is, however, that there is a need to look beyond the livelihood and poverty dimensions in order to understand the broader issues surrounding coffee production and land-use change processes in the region.

The dynamics of the drivers of deforestation have changed over time and space in different parts of the world. In a meta-analysis of place and time-specific case studies, Rudel et al. detected a general trend from small farmers as a main agent of deforestation in the pre-1990 period, towards more agribusiness related activities in the period after 1990 [18]. In Vietnam the deforestation trends have often been characterized by large-scale deforestation processes of so-called 'slash-and-burn' and state-led deforestation in the 1970s and 1980s, towards more smallholder-oriented expansions of commercial agriculture in certain areas of the country in the 1990s and 2000s [19].

Nevertheless, making wide generalizations about the drivers of environmental degradation often masks the complexities of such processes. Tim Forsyth has termed widely accepted assumptions of the drivers of environmental change as 'environmental orthodoxies' ([20], p. 52). Such orthodoxies have been predominant in Vietnam, frequently obscuring many other underlying factors of change [21]. Notions about ethnic people's unsustainable practices in Vietnam were formerly linked to 'slash-and-burn' agriculture. However, such perceptions are also found in relation to the expansion of smallholder coffee production, as demonstrated by the quotation from the Lam Dong REDD+ Action Plan above. In this sense, the way REDD+ has been conceived in Vietnam risks ignoring important dimensions and actors involved, particularly in the coffee sector and related land-use change processes. Defining forest benefits and compensations should not only relate to the 'indigenous' and the 'poor', but acknowledge the diversity of concerns about, and benefits generated from, forests [22].

3. Methodology

In order to respond to the research objectives and questions presented above, the study adopted a case study approach. The data collection instruments included a household survey and in-depth interviews with farmers, policy makers, researchers, and government officials at national, provincial, and commune levels. The findings were triangulated against other socio-economic and qualitative studies carried out in the region, such as Hoang et al. [6], Tran [23], and Ogonowski and Enright [24].

The following sections describe the case study area and key approaches used in the collection and analyses of data.

3.1. Case Study Area

The fieldwork was carried out in the Lam Dong Province (Figure 1) which has a population of 1,259,300 (2014) [25] and covers an area of 9764.8 km^2 [26]. The province is landlocked, and characterized by upland environments. The average elevation is approximately 1500 meters above sea level. Precipitation is irregular in space and time, ranging from 1600 to about 2700 mm [26]. The variations in landscapes and precipitation contribute to exceptional bio- and ecological diversity [27].

Figure 1. Study sites of Da Nhim and Da Chais, Lam Dong Province, Vietnam, 2016. Maps generated from Google Earth and Wikipedia [28].

Historically, the economy of the province has been based on agriculture. General development trends need to be understood in the light of large-scale immigrations of people in the 1970s and 1980s, both as a result of government policies to inhabit, develop, and control the Central Highlands [29], and of people from the region seeking refuge from war and unrest. More recent immigrations have taken place through people seeking new economic opportunities, mainly related to commercial agriculture, such as the production of perennial crops (mainly coffee, rubber, and tea), vegetables, and flowers. Another factor was the resettlement of various ethnic minority groups that had traditionally practiced shifting agriculture within the region. These were situated in residential areas during the government's large-scale 'fixed agriculture' schemes of the 1970s and 1980s [21]. The combination of the displacement of people, voluntary migrations, and government land confiscations caused losses of rights to ancestral lands for various ethnic groups, and increased conflict levels over access to land and natural resources [30]. In the Lam Dong Province, land-use change and deforestation have been closely associated with the marginalization of ethnic minorities, land confiscations, forced resettlement, migrations of people, and last but not least, the integration of agriculture and local communities into a market economy [27]. Recent studies have also revealed trends of displacement of shifting cultivation for annual crops into forest margins, indirectly driven by the expansion of perennial commercial crops (mainly coffee) on agricultural land, a direct manifestation of the marginalization of ethnic groups in the Central Highlands [5].

Historically, coffee was not an important agricultural crop in Vietnam. It was initially introduced by the French colonialists as a plantation crop but during history, small-scale farmers have also widely adopted it as a cash crop [31]. During the late 1980s and 1990s, coffee production exploded, and within a decade or so, Vietnam became the world's second biggest coffee producer [32]. Key reasons for this dramatic development related to a large extent to the economic and agricultural transformation processes of the market liberalization processes (often referred to as the *doi moi*) that started in the mid-1980s [33]. The shift also related to conscious government policies of improved credit facilities, research and development, general extension activities, tenure reforms, and foreign investment [34]. The government policies have been driven by a wish to commercialize and increase the number of (agricultural) export products in order to stimulate economic growth (Figure 2). Today, coffee is grown in a variety of locations and types of farmers, ranging from small-scale household producers to larger plantations [32]. The commercial coffee production has also spurred the emergence of private businesses locally, linked mainly to coffee trading, fertilizer provision, and resale and general trade in farm supplies [35]. The great influx of people to the Central Highlands to grow coffee have, however, caused tensions and conflicts over productive land between majority Viet (Kinh) people and various ethnic minorities.

Figure 2. "Grow coffee to increase the number of products for exportation"—propaganda poster for promoting coffee production in Vietnam.

The Vietnamese state has also played a very active role in the promotion of the coffee sector in the country, and continues to have an interest in its development and continued growth. In 1995, the National Coffee Corporation (VINACAFE) was established by the government under the auspices of the Ministry of Agriculture, with the mandate to organize trade and develop the coffee sector in the country [36]. Its mandate has been very diverse, ranging from implementing government policies, research, extension services, quality control, and acting as a major commercial actor controlling a major share of the coffee export [31,37].

In the late 1990s, a collapse in global coffee affected the sector negatively, including the coffee producing households and their livelihoods, demonstrating the potential vulnerability of basing livelihoods single-mindedly on producing for a global commodity market. In recent years, fluctuating global prices, high debt levels, and bankruptcies amongst banks and key investors continue to negatively impact the sector, including the economy of VINACAFE [38]. Debt levels have reportedly also increased amongst rural households in coffee-producing zones, including the area in which this study was undertaken [6]. This indicates that the integration of rural livelihoods into a global (coffee) commodity market have caused households to become more vulnerable to commodity price fluctuations. Studies have also revealed big differences between ethnic groups in terms of abilities to adapt and cope with price fluctuations, with the Kinh demonstrating higher capacities to diversify livelihoods and income sources as compared to the minority groups [39].

The agricultural system related to coffee production in Lam Dong is characterized by the *homeland* production. ('Homeland production' here does not refer to home gardens, but rather to agricultural fields close to the house. These are normally areas of land that are denominated and certified for agricultural use of coffee in combination with persimmon and vegetables), and the *hillside* production of coffee, sometimes mixed with persimmon and maize for subsistence. The homeland areas are certified agricultural land, while the hillside coffee plots are farther away and are often the result of illegal clearing of forests. Today, coffee production is considered to be a major driver of land-use change, causing the loss and degradation of large forest areas [4,5,40]. By reviewing the production statistics of Lam Dong in the 2002–2011 period, Tan [3] demonstrated that the expansion of coffee production in Lam Dong do seem to correlate with a rise in the price of coffee.

In order to reduce environmental degradation and enhance social development in the rural areas, a number of forest policy reforms and plans have been designed and implemented since the early 1990s. These include the '327' and the '661' programs (the 661 program is also known as the Five Million Hectares Reforestation Program). These programs had the twofold objectives of conserving the remaining forests and expanding tree cover on the 'barren' lands by means of large-scale replanting of trees [41]. In the Lam Dong area, the extent of forest plantations has increased from about 27,000 ha in 1999 to almost 66,000 ha in 2011 [42]. However, official data records show that in recent years in the overall region of the Central Highlands, the negative trend of deforestation has continued.

Following the Ministry of Agriculture's forest land classification system, forests are classified as *special use*, *conservation*, or *production* forests [41]. A key approach to reducing deforestation has been to establish national parks categorized as *special use* forests and large areas categorized as *protection* forests. The latter category has the purpose of conserving key environmental functions and services, such as water provision, energy production, soil protection, and protection against extreme weather events. *Production* forests are intended to be plantations that allow the production and sale of timber and non-timber forest products, in combination with environmental protection practices [43]. In Lam Dong, the relative proportion of *protection* and *special use* forests is quite high compared to *production* forests. There are two large national parks in Lam Dong (Cat Tien and Bidoup-Nui Ba), both of which are categorized as *special use* forests. The forest classification and tenure regimes provide guidelines for what types of activities are allowed on forest land, representing a rather strict system for controlling activities and resources. This, in turn, has important implications for access to land, land use, and livelihood outcomes.

Forest tenure reforms and the distribution of forest land to individual households have been carried out in Vietnam to varying extents. In Lam Dong, state entities such as the State Forest Enterprises, Forest Management Boards, and People's Committees still control most forest land. Only 1% of forest land is managed by individual households. On the other hand, many households in the area have agricultural land certificates (*red books*) to their homeland areas [4].

3.1.1. PES in Lam Dong

Since 2008, Lam Dong has been a pilot province for the national PES program [44]. Even though the prospects for successful PES implementation were described as meagre [45], the Vietnam PES 'experiment' is today often regarded as a successful case, and frequently used as a showcase for other countries in the Mekong Region and Southeast Asia (see, for example, [46]). This is mainly due to its scale and ability to generate substantial amounts of funds.

The pilot scheme for PES was established in Lam Dong and Son La provinces already in 2008 [47], and in 2010 it was scaled up as a nation-wide policy [48]. The Vietnam Forest Sector Development Fund (VNFF) is responsible for channeling, managing, and coordinating the PES funds. During the piloting phase in Lam Dong, 9870 households were included in the scheme covering nearly 210,000 ha of forest [49].

The poverty reduction objectives of PES (and REDD+) are emphasized in various policy documents, and must be understood within the general objectives of the 2006–2020 Vietnam Forestry Development Strategy [50]. This strategy identifies poverty reduction and socio-economic development as key objectives of the forest sector. On the other hand, PES in Vietnam has been criticized for elite capture of financial resources and for failing to target the underlying causes of environmental degradation, such as contested land rights and the general lack of participation and involvement in resource management [51,52].

The critique of PES also includes a general lack of clear linkages between payments and performance [4,53]. In Lam Dong, the PES setup implies a *forest leaser model* in which households, either individually or collectively, are hired by the state forest owners to look after forest land. This work is organized in so-called 'forest protection groups' that are collectively responsible for monitoring and controlling forest land. In this sense, the contracts and duties resemble labor contracts more than anything else, and disbursements are based on participation, rather than performance. The performance component of PES was reportedly rejected in Lam Dong due to resistance and lack of understanding as to why some households might receive higher payments than others [53]. With the implementation of REDD+, however, it is planned that the performance component of forest protection will be reintroduced in Lam Dong [8].

3.1.2. REDD+ in Lam Dong

Vietnam was approved as a UN REDD Programme country in 2008, and was later granted a second phase of support. In addition, the country has received readiness support from the World Bank Forest Carbon Partnership Facility (FCPF). During the first phase of REDD+, activities were limited to testing approaches for participatory carbon monitoring, and free, prior and informed consent (FPIC). In UN REDD Phase II, five more provinces were added, all of which have developed provincial REDD+ action plans (PRAPs) that detail funding streams and more concrete activities to be tested for relieving pressure on—and increasing—forest areas in Lam Dong. In total, over the five years of implementation, the PRAP has a total budget of 1,749,275 million VND, equivalent to about 83.3 million USD [8].

In anticipation of a global results-based REDD+ mechanism, the REDD+ Action Plan for Lam Dong is meant to be a coordination mechanism for various potential funding sources that are relevant in reducing GHG emissions from forests, and increasing forest carbon stocks [8]. Potential funding sources include Official Development Assistance, PES, REDD+ projects and programs (for example, by NGOs), and state-funded budgets. REDD+ will largely build on the institutional structures that were developed through PES [54]. Support for forest protection will continue to be channeled through the VNFF and carried out through the same system of forest protection groups. The results-based component that was 'lost' in PES will be reintroduced through REDD+. At the outset, the results-based component was meant to be linked to an extensive participatory carbon monitoring component. This aspect has, however, over time been toned down, and replaced by a conventional remote sensing data approach along the United Nations Framework Convention on Climate Change (UNFCCC) guidelines. This also forms the basis for Vietnam's application for support through the FCPF Carbon

Fund [55]. Here the emissions reduction and removal potential for the six pilot provinces over the 2016–2020 period is estimated to be 20.66 $MtCO_2$, equivalent of more than 100 million USD taking a price of 5 USD per tCO_2 into account.

The PRAP identifies small-scale agriculture, in part linked to ethnic minorities 'unsustainable practices', as a main driver of deforestation in the Lam Dong province. On the other hand, national level studies of the drivers of deforestation and degradation tend to put more emphasis on the links between coffee expansion and the forces and elasticities of the global coffee commodity market [19]. The apparent void here between the national broad sweeping analyses of the drivers of change, and the provincial and local level approaches of livelihoods improvement and diversification is striking. The PRAP's REDD+ activities include components of training of farmers in agricultural techniques (with a focus on intensifying production in the form of more coffee per unit of land), and establishing village development funds to provide farmers with favorable loans in order to stimulate the establishment of alternative livelihoods [8]. In addition, it includes components of increasing the economic value of agriculture, cultivating multipurpose tree species that can contribute to diversifying income, planting more trees (forest rehabilitation) (similar to the 661 program approaches, except that tree species should be indigenous), and promoting livestock as a way of diversifying household economies. [8]. Nevertheless, how—and to what degree—REDD+ aims at targeting the underlying structures of the coffee sector at large remains unclear in current plans and approaches.

3.2. Study Sample

For the survey, we selected the two communes Da Nhim and Da Chais in the Lac Duong District of Lam Dong, with 2009 populations of 3347 and 1339, respectively [23]. A total of four villages were randomly selected from within the communes, and 25 households were interviewed in each village. This constituted a total sample of 100 households of the 915 households in the two communes. The ethnic composition of households is similar in the two communes (about 85% *K'ho*), and thus the *K'ho* constituted 85% of our total sample. The remaining households were *Kinh* (14%) and *Tay* (1%). With regard to socio-economic status and land use (in particular agricultural expansion), the Lac Duong District contains some of the more inaccessible areas in the province, as well as a high percentage of ethnic minorities [27]. In this sense, the communes are representative of the economically more marginal—but forest rich—areas of the province. In both of the communes, forest areas are predominantly of *protection* type (according to the Ministry of Agriculture forest classification system). In addition, in Da Nhim there are certain areas taken out of the official forest classification system that are categorized as *unclassified forests* that potentially could be used for 'planned deforestation' activities (including for hydropower, mining, ecotourism, and agriculture) [56].

The two communes were considered to be representative of the PES household population in the province, as they form key areas for PES and REDD+ policy implementation within the Da Nhim watershed. In this context, the Da Nhim hydropower station is the customer, buying environmental 'services' (water for energy production) from the forest owners (the Da Nhim Forest Management Board (DNFMB) and the Bidoup-Nui Ba National Park (BNBNP)) for managing the upstream forests. The forest owners in turn contract households to conserve and manage forests sustainably in order to provide water to the hydropower station. Households are trained and paid to carry out community patrolling of the forests. The approach and implications for the households in terms of training, duties, and remunerations are similar for the contractual arrangements with the DNFMB and the BNBNP. According to the information we received at the commune and province levels, households are selected based on ethnicity (minorities are prioritized) and income (the poor are prioritized). Seventy-nine of the households in the sample participated in PES, constituting 18% of the 450 households targeted through PES schemes in the area [6].

The REDD+ pilot activities of the Lam Dong Provincial REDD+ Action Plan started in the area only after we had conducted our survey. Our reflections around REDD+ impacts and implications are therefore based on reviews of policy documents and in-depth interviews that we carried out before the REDD+ activities were implemented in the communes and after the survey.

3.3. Data Analyses

3.3.1. Investigating Livelihoods

We investigated livelihood assets, activities, and different sources of income (including agriculture and various off-farm activities) from different households. This provided information about differences in land use. In order to further investigate the factors that determine income, we ran a multiple regression model of total income against various socio-economic assets [10]. We assumed that income is dependent on financial, physical, social, human, and cultural capitals. Households were also categorized into three equal-sized income groups ('poor', 'medium', and 'better off') based on yearly income in order to investigate differences in livelihood income sources, priorities, and decision making in relation to land use. Here we used pre-PES income levels (that is, total net income without PES) as the basis for the income categorization. This was done in order to capture the poverty dimensions predicted in policy documents regarding the manner in which households in the area were selected for PES (categorizing the households based on total net income, including PES, gave a different distribution of households as compared to without PES. When running the total income model, 16 households were categorized differently as compared to without PES). The mapping of livelihoods and land-use practices feeds into the study of impacts of PES and REDD+ policies in the area.

3.3.2. Measuring the Drivers of Changes in Forest Cover

Micro-studies of land use and decision making at the household level tend to be site specific, and it is often challenging to extrapolate results in space and over time [16,57]. However, we maintain that context-specific studies can introduce nuances in perceptions and orthodoxies about the drivers of land-use change that are widely accepted and often taken for granted, particularly in policies and measures that deal with environment and development issues.

In order to avoid potential biases of data, self-reporting on land use and land use change was consistently avoided in the survey. We did not ask people directly about whether they had cleared land or not. We calculated land-use changes on the respective households' land by combining in-field measures and observations. In total, 181 agricultural plots were walked and measured with GPS in the two communes. The data were stored as gpx-files and analyzed through various online tools, including Google Earth Pro and Landsat images developed by Hansen et al. [58] at a resolution of 30 m. This gave us an overview of changes in each measured land plot over the 2000 to 2014 period. In addition, feeding these data into GIS ArcMap base maps provided by the Lowering Emissions in Asia's Forests (LEAF) project gave us information about the types of forest categories in the various locations.

There are, of course, methodological challenges in combining time series data of this kind, with cross-sectional household income data, since household socio-economic factors change over time. We proceeded to use these data, with some caution, but we believe it is valuable in developing an indicative estimate of the average rate of forest-cover loss due to small-scale livelihood activities related to coffee production. Combining it with in-depth interview data on the coffee production system helps to contextualize policy discourses around the drivers of deforestation and environmental degradation.

3.3.3. Investigating Policy Effects and Implications

The study investigated the effects of PES payments on livelihood and land use within various household categories. We assumed that the poorest households had been prioritized for PES and that this would have contributed to reducing income inequality in the communes. We used the PES proportion of total income as a measure of its importance in terms of poverty reduction. In order to consider to what degree different land use and investment rationales might be attributed to the social and cultural status of households, we measured areas cultivated, input investments made, and yields per unit of land, and how these variables relate to income levels. The different ways in which various households have access to and manage land provided insights into the degree to which links between income levels and pressure on forests may be substantiated. This helped us to reflect upon the relevance and prospects of success for ongoing and planned REDD+ activities in the area.

Key strategies of PES and REDD+ are also to improve citizen environmental 'knowledge' and 'awareness'. Measures to accomplish this include information campaigns, stakeholder meetings, and the use of media (television, radio, newspapers) [23]. For measuring changes in environmental attitudes, we developed a composite indicator. This indicator combines six different questions regarding awareness and perceptions of the value of forests and the conservation of forests. The responses were valued along a Likert type scale, (1–5 scale, 1 implying low awareness, 5 high). The responses were merged into an attitude awareness indicator where the potential score is between 6 and 30.

3.3.4. Misrepresentations of Coffee Income—A Methodological Challenge

A particular challenge we faced was that about 25% of the households reported net negative agricultural income. This may be or lead to a source of misrepresentation in the wealth ranking of households when the estimate of annual income is based on one particular year. In many cases, households will incur high initial, investment costs in establishing new coffee plantations which do not yield any crop income for three to five years [59]. The survey results show that many households established new plantations in 2010/2011, implying high input costs in terms of fertilizers and pesticides, combined with very low income (if any), resulting in negative agricultural income in the survey year (2014). The new plantation efforts in 2010/2011 may be explained by favorable *robusta* coffee market price trends, and expansive population trends, especially in Da Chais. Similar results have been found in other studies carried out in the area, such as the one by Tran [23]. We therefore believe that the negative income can be attributed to the "decoupled" or "disjointed" nature of costs and income distribution that accrued differently for various households over the time leading up to the survey.

In order to adjust this misrepresentation to a 'smoother', more longer-term income measure, we calculated a mean income value per ha, based on the mean gross value of coffee production of the households that produce coffee on their plots. We found that the mean gross coffee production was 3664 kg per ha (on a total area of 82 ha across the sample). We also found mean net investment to be just over 17.0 million VND per year, which conforms well with the national cost figures presented by Thang et al. [59] of an average of 16.9 million VND per year (approximately 846 USD). Based on this, we calculated the average net income per ha to be about 4.8 million VND per year.

We used these figures as an indicator of the net coffee income by household, multiplied by the land available per individual household. In this paper, we refer to this measure as the 'coffee index-adjusted net income'. We then used this measure to categorize households into three, equal-sized income groups. The income adjustment caused 25 households to change income groups, in most cases moving one level either upwards or downwards. In order to maintain transparency in the analyses, the descriptive data (in Table 1) contain both the mean 'indexed' and 'actual' income values. We retain, however, the 'index adjusted income' in the statistical analyses (involving income groups), because we believe that it gives a more realistic picture of wealth and income levels amongst households in the communes in the study area.

Table 1. Socio-economic assets and income sources, Da Nhim and Da Chais communes, Lam Dong Province, Vietnam, 2014 (N = 100).

Variable	Mean	Std Dev	Min	Max
Total net income (1000 VND)	48,480.72	92,810.94	−24000	774,000.00
Index-adjusted net income (1000 VND)	47,684.75	91,823.96	2403.3	789,004.27
Off-farm income (1000 VND)	30,737.50	94,218.92	0	780,000.00
Indexed-adjusted agricultural income (1000 VND)	6594.40	7232.19	0	50,388.32
Livestock income (1000 VND)	221.60	2649.39	−1000.00	26,400.00
Net income: fish (1000 VND)	129.00	831.37	0	6500.00
Net income: forests (1000 VND)	1952.97	3603.71	0	25,948.00
PES income (1000 VND)	7432.00	4966.21	0	16,400.00
Other environmental schemes (1000 VND)	617.28	2746.17	0	15,600.00
Household size	5.45	2.75	2	23
Age of household head	43.66	12.71	23	76
People available to engage in adult labor	3.23	1.98	1	13
Years of education of household head	4.27	4.10	0	16
Total value of assets (1000 VND)	144,606.30	339,807.40	5000.00	3,144,000.00
Debt (1000 VND)	34,170.00	111,399.00	0	1,000,000.00
Debt:income ratio	1.23	1.25	0	15.26
Total cultivated area (ha)	1.14	0.87	0	4.51
Total area of homeland (ha)	0.35	0.43	0	1.97
Area of coffee production (ha)	0.83	0.74	0	4.22
Gross coffee production (kg)	2887.17	3911.84	0	300,000.00
Input investments per ha (1000 VND)	15,761.77	14,734.69	0	65,502.18
Coffee production (kg) per ha	3423.30	3110.30	0	12,455.52
Average distance to land plots (km)	1.68	2.17	0	10.43
Area of land cleared (ha) (2000–2014)	0.35	0.43	0	2.04
Area of uncertified land cleared (2000–2014)	0.28	0.42	0	2.04

USD1 = VND 21,000. PES, Payments for Environmental Services.

4. Results and Discussion

The objectives of this paper are threefold. First, we present general household characteristics and income levels, followed by an analysis of livelihood activities and outcomes per income group. The results mainly reflect the findings from the household survey, but in-depth interviews and secondary sources also inform the discussion—particularly in relation to understanding the production structure of the local coffee economy. The results section feeds into a discussion about how households manage resources differently in order to generate their livelihoods. Second, we measured the relative importance of PES income in household livelihoods, and the degree to which PES participation can be linked to differences between income groups, in terms of income and land use. Third, we explored the coffee economy and land use by investigating the relative importance of coffee production in household economies, and the links between household affluence and the expansion of coffee production into forest land. This discussion also reports on the general organization of the coffee economy in the area. Finally, this section ends with a discussion of future prospects for REDD+ in view of the findings of the study.

4.1. Household Characteristics and Income Levels

The average household size was 5.5 members. The average age of the household head was 43.7 years, with an average level of education of 4.3 years (see Table 1). The level of poverty in the area is high. The mean net income for all households was found to be approximately 48 million VND per annum (Table 1), equaling a mean net income per capita of 8.7 million VND per annum. This equals about 1.1 USD per person per day, i.e. a very low income level as compared with global poverty level income standards of 1 USD per person per day. Nevertheless, seen in a national context, this income level is well above the national poverty rate for rural areas of 4.8 million VND per person per year [60]. Looking at the defined income groups for the study, the income levels again appear to be very low. Both the 'poor' (2.7 million VND per person per year) and the 'medium' (4.4 million VND per person per year) are below the national poverty rate for rural areas. Nevertheless, the registered low income levels correspond with other income-related studies from the area, such as Tran [23]. There are also substantial income inequalities within the communes. The variations in net income levels demonstrate a range from about 1 to 798 million VND per household per year. The estimated Gini coefficient of the communes (0.56) shows that this is well above the national average of 0.39.

The main income source in the study area is off-farm activities (64.4%), followed by PES activities (15.6%), and agriculture (13.8%). Forestry (beyond PES) and livestock-related income are small (4.1% and 0.4%, respectively). Forest related income sources are subsistence oriented and predominantly involve collecting fuelwood. The main limiting factor for agricultural production in the area is the extent of available land [23], which partly explains the high degree of off-farm reliance amongst the sampled households. The households reported an average of 1.14 ha of cultivable land each, of which about 0.82 ha (73%) is used for coffee.

We ran an ordinary least squares (OLS) analysis to test the causal relationship between access to assets and total income (R^2 = 0.4932; F = 4.8998; $p < 0.0001$) (Table 2). We found income to be positively correlated with the level of education of the household head, distance to land plots in general (regardless of whether it was 'homeland' or 'hillside'), and the total area of certified agricultural land (homeland). We found a negative correlation between net income and the total cultivated area, indicating that the higher net income the less land households cultivate. On the other hand, and contrary to policy assumptions about the 'poor' and 'ethnic', the OLS indicates a positive correlation between income levels and clearing of land, namely that the households with higher income had cleared more land in the 2000–2014 period. Household participation in PES also correlated negatively with income. This is as expected, considering the expressed priority to include poorer households in PES.

Table 2. Total income by socio-economic assets, Dha Nhim and Da Chais communes, Lam Dong Province, Vietnam, 2014 (N = 100).

Term	Correlation Estimate	Std Error	*t* Ratio	Prob > |*t*|
Intercept	3221.43	40,867.85	0.08	0.9374
Village 1	−24,876.99	17,209.08	−1.45	0.1521
Village 2	−31,809.43	14,681.07	−2.17	0.0332 **
Village 3	3572.64	13,956.88	0.26	0.7986
Ethnicity (dummy = *K'ho* ethnicity)	8629.39	16,967.99	0.51	0.6124
Size of household	5657.58	5030.02	1.12	0.2640
Sex of household head (dummy = male)	−39,948.57	14,859.90	−2.69	0.0087 ***
Age of household head	469.95	677.98	0.69	0.4902
Adult labor	−5471.05	7347.30	−0.74	0.4586
Years of education of household head	6738.28	2456.36	2.74	0.0075 ***
Collective work Y/N (dummy = yes)	−6407.06	8595.69	−0.75	0.4582
Total value of assets (1000 VND)	0.01	0.03	0.44	0.6631
Debt (1000 VND)	−0.05	0.08	−0.64	0.5210
Total cultivated area (ha)	−39,122.85	13,779.20	−2.84	0.0057 ***
Total area of homeland (ha)	72,656.29	24,393.56	2.98	0.0038 ***
Average distance to land plots (km)	9231.01	3211.01	2.87	0.0051 ***
PES participation (Y/N) (dummy = yes)	−28,299.79	14,632.62	−1.93	0.0566 *
Area of land cleared (2000–2014)	91,889.36	27,096.12	3.39	0.0011 ***

R^2 = 0.4979; F = 4.8098; $p < 0.0001$; p-values estimate significance for differences between household assets: *** is significant at $p < 0.01$; ** is significant at $p < 0.05$; * is significant at $p < 0.1$.

Surprisingly, the OLS indicates that female-headed households generate significantly more income than male-headed ones (Table 2). However, this result is due to an outlier observation in the sample. The richest household in the sample was female headed, with about 798 million VND in total net income per annum. Omitting the outlier and re-running the test eliminated the correlation between gender and income. This also eliminated the correlation between income and location (village) (namely that location influences income levels). Since the outlier household was also ethnic minority (*K'ho*), the omission yielded a positive correlation between ethnic affiliation and income (the *Kinh* earning on average more than the *K'ho*).

In summary, the descriptive data indicate that income levels are low, with substantial differences between the wealthiest and the poorest households. The agricultural and forest-related income levels are also low. According to official statistics from 2014, the provincial mean income from agriculture, forestry, and fishery was 982,000 VND per capita per month [61]. By comparison, the survey data demonstrate an average per capita income from the agriculture and forestry of about 130,000 VND per household.

Breaking the income data down according to income levels—'poor', 'medium', and 'better off'—provides more detailed insights into livelihoods and land-use dynamics for the various income groups. The results are presented in Table 3. The 'better off' households reported significantly more assets and resources than did the 'poor' and 'medium' ones. The level of education of the household head is also on average higher for the 'better off' than for the other two income groups. The differences in average household size and the average age of the household heads were statistically insignificant.

Table 3. Socio-economic characteristics, livelihood activities, and outcomes by income groups (N = 100), Da Nhim and Da Chais Communes, Lam Dong Province, Vietnam, 2014.

Variable	Poor	Medium	Better off	p-Value
Household size	4.88	5.64	5.85	0.1276
Ethnic minorities (%)	91.20	100.00	63.60	0.0004
Age of household head	44.00	42.18	44.79	0.8030
Years of education of household head	3.53 ^{b}	3.15 ^{b}	6.15 ^{a}	0.0066
Total value of assets (1000 VND)	80,080.88 ^{b}	77,465.76 ^{b}	278,227.70 ^{a}	<0.0001 ***
Debt (1000 VND)	21,617.65	19,212.12	62,060.61	0.0866 *
Net income (coffee index-adjusted) (1000 VND)	13,081.92 ^{b}	24,803.05 ^{b}	106,217.86 ^{a}	<0.0001 ***
- Off-farm income (1000 VND)	333.82 ^{b}	4155.76 ^{b}	88,644.24 ^{a}	<0.0001 ***
- Paid work (1000 VND)	108.82 ^{b}	3422.42 ^{b}	78,352.73 ^{a}	<0.0001 ***
- Other business (1000 VND)	11.76 ^{b}	0 ^{b}	10,012.12 ^{a}	<0.0001 ***
- Income transfers (1000 VND)	213.24	733.33	279.39	0.8954
- Indexed-adjusted agricultural income (1000 VND)	2925.35 ^{b}	8611.93 ^{a}	8357.09 ^{a}	0.0007 ***
- Indexed-adjusted coffee income (1000 VND)	2294.47 ^{b}	5090.87 ^{a}	4501.64 ^{a}	0.0072 ***
- Agricultural subsistence income (1000 VND)	630.88 ^{b}	3521.06 ^{a}	3855.45 ^{a}	0.0153 **
- Forest related income	1296.27	2525.42	2057.13	0.3961
- PES income (1000 VND)	8529.41 ^{a}	9018.18 ^{a}	4715.15 ^{b}	0.0011 ***
Total cultivated area (ha)	0.70 ^{b}	1.47 ^{a}	1.27 ^{a}	0.0059 ***
Total area of homeland (ha)	0.26	0.35	0.44	0.0878 *
Area coffee production (ha)	0.48 ^{b}	1.06 ^{a}	0.94 ^{a}	0.0072 ***
Gross coffee production (kg)	1891.18 ^{b}	2757.58 ^{a,b}	4042.94 ^{a}	0.0123 **
Input investments per ha (1000 VND)	17,880.20	12,783.43	16,557.50	0.6793
Coffee production (kg) per ha	3836.37	2859.65	3561.37	0.7006
Average distance to land plots (km)	1.21	2.41	1.83	0.3202
Area cleared for agriculture (2000–2014)	0.23	0.42	0.40	0.0956 *
Area of uncertified land cleared (2000–2014)	0.11 ^{b}	0.34 ^{a}	0.40 ^{a}	0.0033 ***
PES participation (%)	88.2 ^{a}	100 ^{a}	63.6 ^{b}	<0.0001 ***

p-values estimate significance for differences between income groups: *** is significant at $p < 0.01$; ** is significant at $p < 0.05$; * is significant at $p < 0.1$; $^{a\text{–}c}$ Bonferroni test; groups with different letters are significantly different from each other ($p < 0.05$); USD 1 = VND 21,000.

4.2. PES Income and Land Use

There are large variations in the share of income by activity for the different income groups. The 'better off' have on average much higher off-farm income (83.0% of their total income and only 7.0% from agriculture) than the other two groups—the 'poor' and the 'medium' earn about 2.5% and 16.8%, respectively, off-farm. Off-farm income is derived from paid work, business activities, and remittances. In relation to 'paid work', the types of jobs vary—the 'poor' and 'medium' tend to be hired frequently for agricultural work on the land plots of other households' (such as for commercial flower and vegetable production, and coffee production), while the 'better off' work in the commune administration, as teachers, or in the tourism sector. Business-oriented activities are carried out almost exclusively by the 'better off' families. All business households in the sample were ethnic *Kinh*.

PES income are relatively high in the area. The findings from the survey data regarding linkages between PES and performance objectives are still mixed. In the surroundings of the Bidoup-Nui Ba National Park, anecdotal evidence shows that the number of 'violations' has decreased since the start-up of PES [21,23]. The positive correlation between income and land clearing found in the OLS may also demonstrate a link between the level of PES payments and households' expansion of agriculture into forest land. When we compared PES payments across income groups, we found that the 'poor' and 'medium' households receive the highest PES payments. This indicates that in terms of the poverty reduction objective of PES, PES has contributed to reducing income inequalities in the area. In other words, PES is relatively speaking more important for the 'poor' households than for the 'medium' or the 'better off' ones (respectively 65.2%, 38.7%, and 4.4% of total net income). Nevertheless, the 'medium' and the 'better off' households cultivate more land for coffee on average, and in this sense, PES has excluded an important segment of households that are more capable and more likely to expand coffee production.

A simple binary comparison of households in terms environmental knowledge and awareness also showed that there were insignificant differences between the PES (22.8) vs. non-PES beneficiaries (22.3) households ($p = 0.5237$). This partly also questions the positive effects of the knowledge and awareness related activities carried out as a part of the PES scheme.

4.3. Livelihoods, Land Use, and Coffee Production

The land-clearing data indicate that the households in Da Nhim cleared more land on average than in Da Chais in the 2000–2014 period (0.24 ha more on average). Much of the forest land clearing in Da Nhim is most likely attributable to the commune's distribution of forest land for coffee plantations. Here the authorities have deliberately chosen to distribute forest land certificates for coffee plantations to relieve the pressure for improved livelihoods, to provide more land, and to encourage the expansion of coffee production towards areas perceived as less vulnerable in terms of the provision of ecological services. If we subtract the certified legal land clearings organized by the authorities, this gives us an indicative measure of the level of what is officially perceived as illegal land clearings in the area (hereafter referred to as 'uncertified land clearing'). The average area per household across the two communes is 0.28 ha (Table 1), or 0.33 and 0.23 ha in Da Nhim and Da Chais, respectively. This may indicate that access to land is an even more pertinent political issue in Da Nhim than in Da Chais. Considering land use in relation to the various income groups, the 'medium' and the 'better off' households access and cultivate more land on average than the 'poor' (1.47 and 1.27 hectares vs. 0.70 hectares, respectively). On average, 72.8% of the cultivated land is used for coffee production. The 'medium' households cultivate the most land, both in total and for coffee production. The 'poor' households have less land than the other income groups. A modest significant difference in terms of land clearing in the 2000–2014 period was also detected between the groups (Table 3); that is, the 'poor' clear slightly less land than the other groups. Interestingly, however, if we compare income groups in terms of the level of uncertified land clearings in the 2000–2014 period, this is significantly lower for the 'poor' and highest for the 'better off' (though not significantly different from the 'medium'). No significant differences were found in input investments or in productivity between the groups. Further, in terms of output per input investment (that is kg of coffee per 1000 VND), there were no big differences between the groups (i.e., 0.21 for the 'poor', 0.22 for the 'medium', and 0.22 kg/1000 VND for the 'better off').

The data collected during the in-depth interviews indicate that illegal clearing of forest land and land transactions are more complex processes than revealed through the survey data. Land clearing and land transactions often take place in a step-wise process, where the poorer households are paid by the richer and more business-oriented households to clear new land. Thereafter, the land is sold to the other coffee producing households. This was also observed in other studies, such as Vu Tien Dien and Grais [40] and the Netherlands Development Organization (SNV) [62]. Such transactions most likely represent a substantial (illegal) off-farm income source for some of the 'business households' in the sample, a finding which was not well captured through the survey data. The numbers may have been reported as 'business activities', but they were not revealed as income from land transactions.

Business activities are carried out almost exclusively by the 'better off' households, and all of them are ethnic *Kinh*. They frequently own and run small or medium-sized businesses in the communes, typically shops where they sell food or agricultural articles, sometimes combined with cafés or restaurants. These households also provide other farmers with key agricultural inputs, and act as intermediaries for the sale of coffee. In these transactions, inputs are frequently sold on credit and the debt is then repaid in installments and interest in the form of a supply of coffee beans. In such cases, the debtors typically receive a lower price for their coffee compared to selling it directly in the markets (approximately 20% lower than the market price) [6]. This often results in the poorest households being caught in a vicious circle of debt and payments with inflated interest rates.

From a vulnerability point of view, concerns have been raised about the debt situation related to coffee production in the area [6]. The survey findings in this study demonstrate that the household debt levels are indeed high. Households have on average about 34 million VND in debt—some 70.5% of their total annual income on average (Table 3). Even though the differences in debt levels between households are statistically insignificant, the debt-to-income ratios for the 'poor' are much worse than for the 'medium' and 'better off' households (165% versus 77% and 58%, respectively). The levels of interest rates on debt repayments are similar between the different groups.

The better-off group hence has a stake in sustaining the current mode of production in the area, which may represent a major underlying factor for land-use change and forest encroachment in the area. Coffee production should hence be understood within a broader political, economic, and historic context, in which small-scale coffee producing farmers are part of a complex coffee economy, influenced locally by the business-oriented segment of the population that trades coffee, key inputs, and land. This system has been traditionally supported by the state, not only through input support and extension services, but also directly through the state-owned VINACAFE. This is the largest coffee trading and manufacturing company in Vietnam, accounting for 20% to 25% of the country's coffee exports.

4.4. Implications and Prospects for REDD+

While we agree in principle with the prominent role of the coffee sector as a driver of land-use change in the province, we challenge some of the predominant perceptions linked to the underlying poverty and livelihood-related explanation models found in key policy documents and discourses.

First, the findings question the role of smallholder coffee production in relation to deforestation in the study area. Considering the rate of land conversion evidenced through the survey gives reason to question the relative role of smallholder production in the overall picture of deforestation in the province. The land-use change data for the household plots in the 2000–2014 period indicate an increase in agricultural land of 0.35 ha (constituting about 31% of the total amount of agricultural land) (Table 1). This amounts to about 0.025 ha per household per year over this period. A simple calculation of the average total annual forest loss due to livelihood activities in the two communes yields an answer of 22.9 ha. With total forest cover of the two communes of 53,546 ha, the average loss due to household livelihood activities should be around 0.043% per year, which is well below the provincial average (about 0.5 ha per annum according to official figures from the Forest Inventory and Planning Institute [42]). These results are of course indicative figures, but considering that they are based on the current population of the communes, and since the population trend of the province has been increasing in recent years, the figures are probably not underestimating deforestation. It should also be noted that there are most likely differences in land-use practices and needs between locations in different contexts. The focus in REDD+ should hence also be on contexts where larger-scale plantation mode of coffee production is taking place.

Second, PES experiences should critically be reviewed and modified in REDD+. The way REDD+ has evolved is, to a large extent, symptomatic of how domestic REDD+ policies have evolved from global approaches and ideas, and in this case from an idea about results-based PES to more pragmatic national and local adaptations of broader policies and measures [2]. REDD+ in Vietnam in general, and Lam Dong in particular, is by-and-large viewed as a coordination mechanism for ongoing and future activities that may contribute to the reduction of GHG from forests. Hence, we foresee that the current PES setup will play a key role as a distribution mechanism for financial benefits, particularly in relation to forest conservation, thus making the experiences with PES since 2008 in Lam Dong particularly relevant for future REDD+ prospects.

When comparing PES income with coffee production per unit of land, the opportunity costs of coffee production very significantly outnumber those of PES. The net average income per hectare per year from PES is about 298,000 VND, while alternative agricultural land use would constitute about 5.7 million VND. Thus, it seems reasonable to assume that any causal links between the (low) level of PES payments and environmental performance are highly uncertain. In order for a performance-based PES or REDD+ mechanisms to succeed sustainably, the opportunity cost levels of current land uses must be addressed. If PES has had any impact, it is most likely attributable to increased control and patrolling, and information campaigns that reach all households (for example, newspapers, television, radio, etc.), rather than participation and performance-based payments (effects that were also partly questioned by comparing the environmental awareness scores between the beneficiary and the non-beneficiary households—see Section 4.3 above).

Third, the 'poor' deforest the least. Our findings demonstrate that the 'poor' households are the ones carrying out the least uncertified (illegal) land clearings, and are hence not the group of households causing most deforestation in the study area. The planned REDD+ activities seem to a larger extent to recognize some of the underlying issues of deforestation in the area than has been the case with PES. For example, the establishment of the Village Development Funds may have the potential to stimulate livelihood diversification and relieve poor farmers of debt dependency in the form of high interest rates charged by business households in the communes. Nevertheless, as with PES, the focus on the 'poor' ethnic minorities and their 'destructive' activities is also a key component of REDD+, and the lack of comprehensive analyses of the coffee sector at large is striking in plans and policies at all levels.

An underlying assumption of current REDD+ activities is also that training and capacity building for better and improved practices will increase households' agricultural investment returns, and in turn reduce the pressure on forests. Our livelihood analyses question the assumption that the production of 'poor' (and ethnic minority) households is economically and agronomically less efficient than that of wealthier households. Our findings do not reveal significant differences in levels of input investments, nor production per ha, and the data on the links between income levels and the expansion of production into forest land point to the higher income segments of the population. The effects of increasing the productivity of global commodity crops, such as coffee, are also uncertain in terms of decreasing pressure on more marginal lands, considering the high price elasticity of production levels and volumes [63].

Fourth, issues of tenure and access to productive land need to be addressed. Our findings point towards land being a major limiting factor for increased livelihood income. Issues related to sustainable access to productive land should be addressed in order to stop or constrain the illegal clearing of forest land. Improved land-use planning and securing legal and sustainable access to land for the poor and ethnic minority households should be a key component in all strategies aiming at relieving pressure on strategically important forest resources. This has been done in Da Nhim, to a certain extent, and could be a viable approach for directing coffee production strategically towards the less vulnerable areas in terms of environmental and carbon values. The interests of the business-oriented segment of wealthy households, who have a major stake in coffee-related businesses and (illegal) land transactions, seems to be a cementing factor for the current mode of production. These underlying factors need to be addressed before issues related to the expansion of coffee production into forest land can be solved in a sustainable way.

5. Concluding Remarks

The study has explored some of the predominant underlying policy 'orthodoxies' on the drivers of land-use change and forest encroachment related to small-scale coffee production amongst 'poor' and 'ethnic' households in a PES and REDD+ zone in the Lam Dong Province of Vietnam. Such perceptions are not endemic to Vietnam, but are frequently found in environmental policy schemes and discourses globally [20]. The study has explored drivers of land-use change by making use of a livelihood framework. We found that land is the main limiting factor for agricultural production and livelihoods, and most households in the study area need to supplement agricultural income with off-farm sources. Both the 'medium' and the 'better off' households have access to a more diversified set of off-farm income sources compared to the 'poor', who depend more upon agriculture for their livelihoods, both overall and in terms of coffee production. Policy discourses on the drivers of land-use change in the Central Highlands of Vietnam seem to neglect various underlying factors that drive the coffee production. The findings of this study also indicate that the coffee economy and land transactions related to the expansion of coffee production are controlled largely by the business-oriented households in the study area.

In terms of links between land-use change and poverty levels, the data showed that in absolute terms, the 'medium' and the 'better-off' households cultivate most land, both in total and for coffee production. The data also indicate that these households have cleared more forest land for agriculture over the years than the 'poor' households. Especially if we take the uncertified ('illegal') clearing into account, the 'poor' deforest the least. Thus, the focus on the linkages between poverty amongst the ethnic poor and coffee-related forest encroachment seems to be overrated in PES and REDD+ policies and discourses. The main argument of this paper is therefore that, in order to enable a more comprehensive understanding of land-use change and its management, the focus should be expanded beyond the poverty-environment nexus.

The expansion and development of the coffee sector in Vietnam must be seen in the historic and political context of marginalization of, and control over, land, resources, and people [21,30]. Historically, the general deforestation in the Lam Dong province has often been related to the large migrations of people from the north who came to populate and develop the province [27]. They represent a group of households that over the years have been encouraged by the Government of Vietnam to migrate, settle, and 'develop' the region. This study has demonstrated that business-oriented households, together with the state-owned coffee corporation (VINACAFE), are in control of the coffee sector in the region. Consequently, the current focus of PES and REDD+ on payments to 'poor' and 'ethnic' households, combined with education, information campaigns, and increased levels of forest patrolling, do not seem to be sufficient in trying to solve the underlying issues at stake.

Coffee production requires a long-term investment, and reverting to subsistence production is not a viable option for most households, considering their integration into a market and 'cash' economy. Solutions for attaining sustainable livelihoods need to address the critical lack of productive land in the area. Improved land rights and land-use planning seem to be warranted. The current development and infrastructure plans, for example, of larger-scale plantations of rubber in the Central Highlands also require more attention [64]. A focus on creating realistic and viable off-farm livelihood alternatives, other than being involved in the very dominant coffee sector, is most likely an important ingredient in any policy scheme that aims to reduce pressures on forests.

Acknowledgments: We extend special thanks to Manh Cuong Pham from the UN REDD Programme Office in Vietnam and Nam Pham Thanh from the SNV Office in Dalat, who provided invaluable support and advice during the fieldwork in Lam Dong, and comments regarding the data analyses. We thank Lam Ngoc Tuan at the University of Dalat for his comments and support during the fieldwork. We are also indebted to Professor Arild Vatn who provided valuable comments on the paper. Special thanks also go to all who assisted us during the fieldwork. We are also grateful to the two anonymous reviewers for very useful and constructive comments on this paper. This research was made possible with financial support from the Nansen Fund (managed by UNIFOR at the University of Oslo) and the Norwegian University of Life Sciences.

Contributions: Leif Tore Trædal carried out the data collection and analyses, and wrote the article. Pål Olav Vedeld provided inputs regarding the theoretical dimensions, the structure of the paper, and statistical interpretation of the data.

Conflicts of Interest: The authors declare no conflict of interest.

References

1. The Economics of Ecosystems and Biodiversity (TEEB). *The Economics of Ecosystems and Biodiversity for National and International Policy Makers—Summary: Responding to the Value of Nature*; TEEB: Geneva, Switzerland, 2009.
2. Angelsen, A. REDD+ as result-based aid: General lessons and bilateral agreements of Norway. *Rev. Dev. Econ.* **2016**, 1–28. [CrossRef]
3. Tan, N.Q. Environmental services in Vietnam: An analysis of the pilot project in Lam Dong Province. In *Forest Conservation Project, Occasional Paper*; Scheyvens, H., Ed.; Institute for Global Environmental Strategy: Kanagawa, Japan, 2011.
4. UN-REDD. *UN-REDD Viet Nam Phase II Programme: Operationalising REDD+ in Viet Nam*; UN-REDD Programme: Geneva, Switzerland, 2012.
5. Meyfroidt, P.; Vu, T.P.; Hoang, V.A. Trajectories of deforestation, coffee expansion and displacement of shifting cultivation in the Central Highlands of Vietnam. *Glob. Environ. Chang.* **2013**, *23*, 1187–1198. [CrossRef]
6. Hoang, H.C.; Giang, P.T.; Hai, B.V.; Boi, D.D.; Ha, L.V.H.; Binh, N.Q.; Phuong, D.H.; Tai, N.T.K.; Lan Phuong, N.T.; Thanh, N.D.; et al. *Participatory Rural Appraisal for the Project for Strengthening Community-Based Management Capacity of Bidoup-Nui Ba National Park*; Nong Lam University of Ho Chi Minh City, Department of Social Forestry and Agroforestry: Ho Chi Minh, Vietnam, 2011.
7. Dooley, K.; Griffiths, T.; Martone, F.; Osinga, S. *Smoke and Mirrors. A Critical Assessment of the Forest Carbon Partnership Facility*; FERN and Forest Peoples Programme: Belgium, UK, 2011.
8. DARD. *Action Plan on "Reduction of Greenhouse Gas Emissions through Efforts to Reduce Deforestation and Forest Degradation, Sustainable Management of Forest Resources, and Conservation and Enhancement of Forest Carbon Stocks" in Lam Dong Province, Period 2014–2020*; Department of Agriculture and Rural Development, Lam Dong People Committee: Da Lat, Vietnam, 2014.
9. Scoones, I. *Sustainable Rural Livelihoods: A Framework for Analysis*; IDS Working Paper; IDS: Brighton, UK, 1998.
10. Ellis, F. *Rural Livelihoods and Diversity in Developing Countries*; Oxford University Press: Oxford, UK, 2000.
11. Scoones, I. *Sustainable Livelihoods and Rural Development*; Agrarian Change and Peasant Studies Series; Fernwood Publishing: Black Point, NS, Canada, 2015; Volume 4.
12. Babigumira, R.; Angelsen, A.; Buis, M.; Bauch, S.; Sunderland, T.; Wunder, S. Forest clearing in rural livelihoods: Household-level global-comparative evidence. *World Dev.* **2014**, *64*, S67–S79. [CrossRef]
13. Hosonuma, N.; Herold, M.; De Sy, V.; De Fries, R.S.; Brockhaus, M.; Verchot, L.; Angelsen, A.; Romijn, E. An assessment of deforestation and forest degradation drivers in developing countries. *Environ. Res. Lett.* **2012**, *7*, 1–13. [CrossRef]
14. Kissinger, G.; Herold, M.; Sy, V.D. *Drivers of Deforestation and Forest Degradation: A Synthesis Report for REDD+ Policymakers*; Lexeme Consulting: Vancouver, BC, Canada, 2012.
15. Geist, H.J.; Lambin, E.F. Proximate causes and underlying driving forces of tropical deforestation. *BioScience* **2002**, *52*, 143–150. [CrossRef]
16. Hersperger, A.M.; Gennaio, M.; Verburg, P.H.; Bürgi, M. Linking land change with driving forces and actors: Four conceptual models. *Ecol. Soc.* **2010**, *15*, 1.
17. Carodenuto, S.; Merger, E.; Essomba, E.; Panev, M.; Pistorius, T.; Amougou, J. A Methodological Framework for Assessing Agents, Proximate Drivers and Underlying Causes of Deforestation: Field Test Results from Southern Cameroon. *Forests* **2015**, *6*, 203–224. [CrossRef]
18. Rudel, T.K.; Defries, R.; Asner, G.P.; Laurance, W.F. Changing drivers of deforestation and new opportunities for conservation. *Conserv. Biol.* **2009**, *23*, 1396–1405. [CrossRef] [PubMed]
19. McNally, R.; Phuong, V.T.; Chien, N.T.; Phuong, P.X.; Dung, N.V. *Analyses: Policies and Measures. Support for the Revision of Viet Nam's National REDD+ Action Programme (NRAP) (Draft Report)*; Forest Carbon: Jakarta, Indonesia; Vientiane, Laos, 2016.

20. Forsyth, T. *Critical Political Ecology: The Politics of Environmental Science*; Routledge: London, UK, 2003; p. 323.
21. McElwee, P. *Forests Are Gold. Trees, People, and Environmental Rule in Vietnam*; University of Washington Press: Seattle, WA, USA; London, UK, 2016; p. 283.
22. Forsyth, T.; Sikor, T. Forests, development and the globalisation of justice. *Geogr. J.* **2013**, *179*, 114–121. [CrossRef]
23. Tran, K.T. *Socio-Economic Survey for Assessing the VN Government Policy on Payment for Environmental Services for Lam Dong Province*; Asia Regional Biodiversity Conservation Program: Ho Chi Minh City, Vietnam, 2010.
24. Ogonowski, M.; Enright, A. *Cost Implications for Pro-Poor REDD+ in Lam Dong Province, Vietnam: Opportunity Costs and Benefit Distribution Systems*; IIED: London, UK, 2013.
25. Vietnam General Statistics Office. Population and Employment. 2014. Available online: https://www.gso.gov.vn/default_en.aspx?tabid=774 (accessed on 2 March 2016).
26. Lam Dong Portal. 2016. Available online: http://www.lamdong.gov.vn/en-us/home/Pages/default.aspx (accessed on 28 November 2016).
27. De Koninck, R. *Deforestation in Vietnam*; International Development Research Centre (IDRC): Ottawa, ON, Canada, 1999.
28. Wikipedia 2014: Location of Lam Dong within Vietnam. Available online: http://en.wikipedia.org/wiki/L%C3%A2m_%C4%90%E1%BB%93ng_Province#/media/File:Lam_Dong_in_Vietnam.svg (accessed in September 2016).
29. Déry, S. Agricultural colonisation in Lam Dong Province, Vietnam. *Asia Pac. Viewp.* **2000**, *41*, 35–49. [CrossRef]
30. Salemink, O. *The Ethnography of Vietnam's Central Highlanders: A Historical Contextualization 1850–1990*; RoutledgeCurzon: London, UK, 2003; p. 383.
31. Fortunel, F. *Le Café Au Viêtnam. De la Colonisation à L'essor d'un Grand Producteur Mondial*; L'Harmattan: Paris, France, 2000.
32. Ha, D.T.; Shively, G. Coffee Boom, Coffee Bust and Smallholder Response in Vietnam's Central Highlands. *Rev. Dev. Econ.* **2008**, *12*, 312–326. [CrossRef]
33. Doutriaux, S.; Geisler, C.; Shively, G. Competing for Coffee Space: Development-Induced Displacement in the Central Highlands of Vietnam. *Rural Sociol.* **2008**, *73*, 528–554. [CrossRef]
34. Nhan, D.T. Orientations of Vietnam coffee industry. In Proceedings of the International Coffee Conference, London, UK, 17–19 May 2001.
35. Marsh, A. Diversification by smallholder farmers: Viet Nam Robusta Coffee. In *Agricultural Management, Marketing and Finance Working Document*; FAO: Rome, Italy, 2007.
36. Stevie Ray Vaughan (SRV). *Decision No. 251-TTg on the 29th of April 1995 of the Prime Minister on the Establishment of the Vietnam National Coffee Corporation*; Socialist Republic of Vietnam: Hanoi, Vietnam, 1995.
37. Minot, N. *Competitiveness of Food Processing in Vietnam: A Study of the Rice, Coffee, Seafood and Fruit and Vegetables Subsectors*; International Food Policy Research Institute: Washington, DC, USA, 1998.
38. Petty, M. *Crippling Debts and Bankruptcies Brew Vietnam Coffee Crisis*; Reuters: London, UK, 2013.
39. Agergaard, J.; Fold, N.; Gough, K.V. Global–local interactions: socioeconomic and spatial dynamics in Vietnam's coffee frontier. *Geogr. J.* **2009**, *175*, 133–145. [CrossRef]
40. Vu, T.D.; Bay, P.N.; Stephen, P.; Van Chau, T.; Grais, A.; Petrova, S. *Land Use, Forest Cover Change and Historical GHG Emission. From 1990 to 2010. Lam Dong Province, Viet Nam*; Lowering Emissions in Asia's Forests (LEAF), Ed.; SNV, USAID: Hanoi, Vietnam, 2013.
41. Long, V.; Vu, T.P. *Report on Review of Forestry Policies in Vietnam*; Research Centre for Forest Ecology and Environment: Hanoi, Vietnam, 2011.
42. Federation of Image Professionals International (FIPI). Forest Inventory and Planning Institute. 2013. Available online: http://www.kemintran.vn (accessed in January 2015).
43. SRV. *Circular No. 34/2009/TT-BNNPTNT of June 10, 2009, on Criteria for Forest Identification and Classification*; Minsitry of Agriculture and Rural Development (MARD): Hanoi, Vietnam, 2009; p. 6.
44. SRV. *Decision No. 380 on the Pilot Policy for Payment for Forest Environmental Services*; Minister, T.P., Ed.; The Prime Minister Office: Hanoi, Vietnam, 2008.
45. Wunder, S.; The, B.D.; Ibarra, E. *Payment Is Good, Control Is Better: Why Payments for Forest Environmental Services in Vietnam Have So Far Remained Incipient*; CIFOR: Bogor, Indonesia, 2005.
46. VNS. Vietnam leads way in forest protection. *Viet Nam News*, 25 November 2014.

47. To, P.X.; Santiago, C. *Vietnam Leads Southeast Asia in Payments for Ecosystem Services*; Katoomba Group Ecosystem Marketplace: Washington, D.C. USA, 2010.
48. SRV. *Decree 99/2010/ND-CP, Payments for Forest Environment Services (PES) Policy*; Socialist Republic of Vietnam: Hanoi, Vietnam, 2010.
49. Chiramba, T.; Mogoi, S.; Martinez, I. Payment for Forest Ecosystem Services (PFES): pilot implementation in Lam Dong Province, Vietnam. In Proceedings of the UN-Water International Conference, Zaragoza, Spain, 15–17 January 2011.
50. SRV. *Viet Nam Forestry Development Strategy 2006–2020*; Socialist Republic of Vietnam: Hanoi, Vietnam, 2007; p. 53.
51. McElwee, P.D. Payments for environmental services as neoliberal market-based forest conservation in Vietnam: Panacea or problem? *Geoforum* **2012**, *43*, 412–426. [CrossRef]
52. To, P.X.; Dressler, W.H.; Mahanty, S.; Pham, T.T.; Zingerli, C. The prospects for Payment for Ecosystem Services (PES) in Vietnam: A look at three payment schemes. *Hum. Ecol.* **2012**, *40*, 237–249. [CrossRef] [PubMed]
53. Trædal, L.T.; Vedeld, P.O.; Pétursson, J.G. Analyzing the transformations of forest PES in Vietnam: Implications for REDD+. *For. Policy Econ.* **2016**, *62*, 109–117. [CrossRef]
54. UN-REDD. *Design of a REDD Compliant Benefit Distribution System for Viet Nam*; UN REDD Programme: Geneva, Switzerland, 2010.
55. SRV. *Vietnam Emission Reductions Program Idea Note (ER-PIN) for the Forest Carbon Partnership Carbon Fund*; Ministry of Rural Development and Agriculture: Hanoi, Vietnam, 2014.
56. ARBCP. *Avoided Deforestation Pilot Project in Da Nhim Watershed, Lam Dong Province, Vietnam. Project Desing Document (Working Draft)*; Asia Regional Biodiversity Conservation Programme; USAID and Winrock International: Little Rock, AR, USA, 2011.
57. Angelsen, A.; Kaimowitz, D. Rethinking the causes of deforestation: Lessons from economic models. *World Bank Res. Obs.* **1999**, *14*, 73–98. [CrossRef] [PubMed]
58. Hansen, M.C.; Potapov, P.V.; Moore, R.; Hancher, M.; Turubanova, S.A.; Tyukavina, A.; Thau, D.; Stehman, S.V.; Goetz, S.J.; Loveland, T.R.; et al. High-resolution global maps of 21st-century forest cover change. *Science* **2013**, *342*, 850–853. [CrossRef] [PubMed]
59. Thang, T.C.; Burton, M.P.; Brennan, D.C. Optimal replanting and cutting rule for coffee farmers in Vietnam. In Proceedings of the Australian Agricultural and Resource Economics Society (AARES) Annual Conference, Cairns, Australia, 11–13 February 2009; p. 34.
60. SRV. *Decision No. 09/2011/QD-TTg on Setting Norms on Poor Households and Households in Danger of Falling into Poverty for the 2011–2015 Period*; The Socialist Republic of Vietnam: Hanoi, Vietnam, 2011.
61. GSO. *General Statistics Office of Vietnam*; Government of Vietnam: Hanoi, Vietnam, 2014.
62. SNV. *Cat Tien—How Forbidden Is the "Forbidden Forest"?* SNV: Cat Tien District, Vietnam, 2010.
63. Lambin, E.F.; Meyfroidt, P. Global land use change, economic globalization, and the looming land scarcity. *Proc. Natl. Acad. Sci. USA* **2011**, *108*, 3465–3472. [CrossRef] [PubMed]
64. To, X.P.; Tran, H.N. *Rubber Expansion and Forest Protection in Vietnam*; Tropenbos International Vietnam: Hue, Vietnam, 2014.

Differentiation of Alternate Harvesting Practices Using Annual Time Series of Landsat Data

Lukas R. Jarron [1], Txomin Hermosilla [1,*], Nicholas C. Coops [1], Michael A. Wulder [2], Joanne C. White [2], Geordie W. Hobart [2] and Donald G. Leckie [2]

1 Department of Forest Resources Management, University of British Columbia, 2424 Main Mall, Vancouver, BC V6T 1Z4, Canada; lukas.jarron@alumni.ubc.ca (L.R.J.); nicholas.coops@ubc.ca (N.C.C.)

2 Canadian Forest Service (Pacific Forestry Centre), Natural Resources Canada, 506 West Burnside Road, Victoria, BC V8Z 1M5, Canada; mike.wulder@canada.ca (M.A.W.); joanne.white@canada.ca (J.C.W.); geordie.hobart@canada.ca (G.W.H.); don.leckie@canada.ca (D.G.L.)

* Correspondence: txomin.hermosilla@ubc.ca

Abstract: Sustainable forest management practices allow for a range of harvest prescriptions, including clearcut, clearcut with residual, and partial or selective cutting, which are largely distinguished by the amount of canopy cover removed. The different prescriptions are aimed to emulate natural disturbance, encourage regeneration (seed trees), or offer other ecosystem services, such as the maintenance of local biodiversity or habitat features. Using remotely sensed data, stand-replacing disturbance associated with clearcutting is commonly accurately detected. Novel time series-based change detection products offer an opportunity to determine the capacity to detect and label a wider range of harvest practices. In this research, we demonstrate the capacity of time series imagery, spectral metrics, and related attributed change products, to distinguish between different harvesting practices over a study area in central British Columbia, Canada. Producer's accuracy of harvest attribution was 79%, with 93% of harvest blocks >5 ha accurately identified. In relation to the amount of canopy cover removed, clearcut harvesting was the most accurately classified (84%), followed by clearcut with residual (79%), and partial cut (64%). Applying detailed spectral metrics derived from Landsat data revealed clearcut and partial cuts to be spectrally distinct. The annual nature of the Landsat time series also offers spatial harvest information within typical, often decadal, forest inventory update cycles. The statistically significant ($p < 0.05$) relationship between harvest practices and Landsat spectral information indicates a capacity to add increased attribution richness to remote sensing depictions of forest harvest.

Keywords: harvesting; remote sensing; disturbance analysis; Landsat; time series

---

1. Introduction

Approximately 65% of Canada is occupied by forest-dominated ecosystems [1], of which ~350 million hectares are represented by trees and other wooded land [2], with the remainder dominated by lakes and wetlands. Knowledge on the extent, condition, and management of forest resources is critical to meet both national [3,4] and international [5] expectations of sustainable forest management and to support reporting requirements. In addition, the magnitude of harvest events, and the type of harvest (clearcut or selective cut), has implications to carbon budget modeling. As such, information on the extent and type of forest harvesting is captured and maintained through spatial forest inventories to assist with management and reporting activities. Due to the nature of natural resource stewardship in Canada, provincial and territorial resource management agencies are responsible for managing forest resources. An element of forest resource management is the allocation of timber harvest permits, including the type of harvesting that may be undertaken [6,7]. Given the processes in place, many jurisdictions do not maintain spatial harvest records in a comprehensive and standardized fashion, with statistics often reported aspatially and/or on a jurisdictional basis (see [7]). Traditionally, forest inventories in Canada have been based upon the collection, delineation, and attribution of air photos on a periodic basis, often of a decadal interval. Ground data are used to calibrate and ensure the consistency of the information in the forest inventory. Jurisdictional forest inventories are strategic in nature, aimed at ensuring an understanding of the nature and extent of forest resources across a given jurisdiction [8]. At the operational level, new technologies for forest inventory are undergoing investigation [9] with lidar data showing the greatest promise and operational uptake [10]. Increasingly large areas are being flown to acquire lidar, yet provincial and territorial standards have yet to be altered to entrench lidar as the sole or favored data source. In contrast and at a different categorical and spatial scale, space-based remote sensing technology offers an opportunity to capture wall-to-wall information on forest harvesting. The opening of the Landsat archive has enabled the generation of time series informed, annual depictions of forest change by type [11]. Canada conducts a National Forest Inventory (NFI) to produce nationally consistent reporting on forest resources. Undertaken in partnership with provincial and territorial resource management agencies, the NFI is based upon a systematic grid of 2 × 2 km photo plots that are, in turn, supported by a subset of ground plots. The photo plots are analogous in information content to polygonal forest inventory data. Based upon the systematic grid and photo plot representation, a 1% sample of Canada's forest resources allows for statistically robust reporting. Given the 1% sample and a 10-year update cycle, there are numerous opportunities for the integration of wall-to-wall satellite derived information on forest dynamics [12]. The frequent and consistent acquisition of high quality, spatially precise, calibrated, analysis-ready remotely-sensed imagery, from sensors such as Landsat and Sentinel 2, facilitates improved management and stewardship, which subsequently supports integrated monitoring and reporting information needs [13].

Free and open access to the United States Geological Survey (USGS) Landsat archive [14] has led to an increase in the use of Landsat data for both large area and temporal analyses (e.g., [15–18]). Landsat imagery has imaging characteristics that allow for detection of human interactions with the landscape [19] based upon spatial, spectral, temporal, and radiometric qualities that enables implementation of robust and transparent analyses [19,20]. Further, as noted by Belward and Skøien [21], the temporal depth of the Landsat archive is unique among Earth observation satellites at the moderate or medium spatial resolution (10–100 m), typically required to capture anthropogenic activities. Utilizing the spatially complete and temporally deep database of Landsat data available for Canada [12] time series analyses have been conducted utilizing the 30 m spatial resolution data available from Landsat Thematic Mapper ™ and Enhanced Thematic Mapper Plus (ETM+) for the period 1984 to 2012 [18]. These data both define baseline knowledge of historic patterns of forest disturbances, as well as portray current forest disturbance and recovery trends that are needed to make informed management decisions. The 1984 baseline offered by Landsat is earlier than most existing spatial data records, hence otherwise unavailable historic information on forest harvesting is now available as well as a present day accounting that can augment jurisdictional records.

Furthermore, an opportunity exists to better understand the forest change attributable to harvest as reported by Hermosilla et al. [18], and to determine the capacity to increase the detail of the current attribution present in the change hierarchy used by Hermosilla et al. [11]. This potential is increasingly important as partial cuts become more commonplace and the ability to report on, or track the outcomes of, different harvest practices is desired. In 2014, only 12% of timber harvest on public land in British Columbia (by area) was clearcut, the remainder being some variation of partial cut technique or clearcut with reserves [22]. This increasing trend is expected to continue as large clearcut blocks are subject to public scrutiny [23]. The ability to distinguish these harvest practices using remotely sensed data presents a potential avenue to governments focused on implementation of sustainable forest management practices.

The science of detecting harvest events from remotely sensed data has evolved over time. Regression modeling using Landsat data by Healey et al. [24] highlights that the shortwave infrared bands and associated metrics are the best indicators for detecting forest removals. Before the opening of the Landsat archive, Kennedy et al. [25] focused on using Landsat metrics to detect the year of change for different harvest methods. While the year of change was accurate (overall accuracy = 90%), the classification of the harvest types was considerably lower (overall accuracy = 74%). Use of the Landsat time series is useful for differentiating forest disturbances, as demonstrated by Schroeder et al. [26], who classified clearcuts and fire events with an overall accuracy of 93%.

In this research, we compared ground truth information of specific harvest practices to disturbances detected using a time series of Landsat imagery [18]. In so doing, we determined the detection capacity for harvests implemented with differing levels of intensity (that is, canopy cover removal as related to a given harvest practice). Using spectral and temporal metrics, we determined the capacity of Landsat time series to relate how harvest practices manifest on a spectral level. We implemented statistical tests on the most descriptive spectral metrics to gain an improved understanding of the significance of the differences exhibited across the differing harvest practices present. We also investigated the accuracy of the time series disturbance classification against the type of harvest practice and assessed the influence of harvest polygon size on classification accuracy.

2. Study Area

The study area is a 50 × 35 km zone centered on the city of Prince George in north-central British Columbia, Canada (Figure 1). The region is situated in the interior plateau of the Western Cordillera, between the Coast and Columbian mountain ranges. Elevation in the area ranges between 550 and 800 meters with a gently rolling terrain. Classified within the Sub-boreal Spruce biogeoclimatic zone, the dominant climax tree species in the area are *Picea glauca × engelmannii*, *Picea glauca*, and *Abies lasiocarpa*. *Pinus contorta* is also common in mature stands and is the main pioneer species for the region along with *Populus tremuloides* [27].

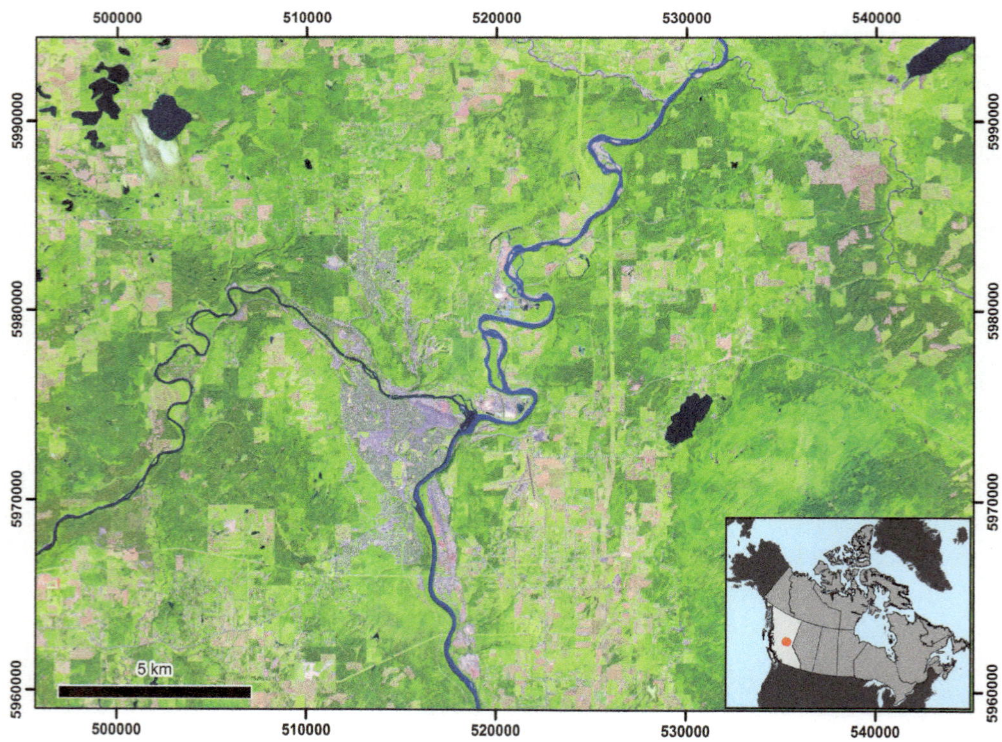

Figure 1. Location of the study area near Prince George, British Columbia, Canada overlaid on the C2C (Composite2Change) proxy composite image for 2003.

Prince George and the surrounding Cariboo developmental region are strongly invested in logging and forestry, accounting directly for 16.6% of all jobs in the region and 2.5% of GDP provincially [28]. Various harvest practices have been applied in the area. Harvest history in the region is reflective of trends occurring across the rest of North America, with clearcuts being the most common and partial or residual prescriptions being more preferential in recent years [29]. Forest management in the region has also been influenced by the recent mountain pine beetle outbreak, resulting in high mortality of the economically important *Pinus contorta* subsp. *contorta* [30].

3. Data

3.1. Landsat Time-Series Data

To identify harvest polygons, we utilized the Landsat-derived disturbance information dataset produced by Hermosilla et al. [18] following the Composite2Change (C2C) approach (Figure 2). This approach is based on the production of best-available pixel (BAP) image composites from Landsat imagery by selecting the most suitable observations for each pixel location within a specific date range (August 1, $\pm$30 days) based on the scoring functions defined by White et al. [13]. These functions rank every pixel based on the presence and distance to clouds and their shadows, the atmospheric quality, and the acquisition sensor. These image composites are further refined by removing noisy observations and infilling of data gaps using spectral trend analysis of pixel time series [31], which results in seamless annual surface-reflectance composites for all of Canada from 1984 to 2012, as well as the detection and characterization of forest change events greater than 0.5 ha (minimum map unit). Following the object-based image analysis approach presented in Hermosilla et al. [11], detected changes are attributed to a change type (i.e., fire, harvesting, road, or non-stand-replacing), based on their spectral, temporal, and geometrical characteristics using a Random Forests classifier. The overall accuracy of the change attribution process for all classes is 92%, with harvesting having a producer's and user's accuracy of 88%.

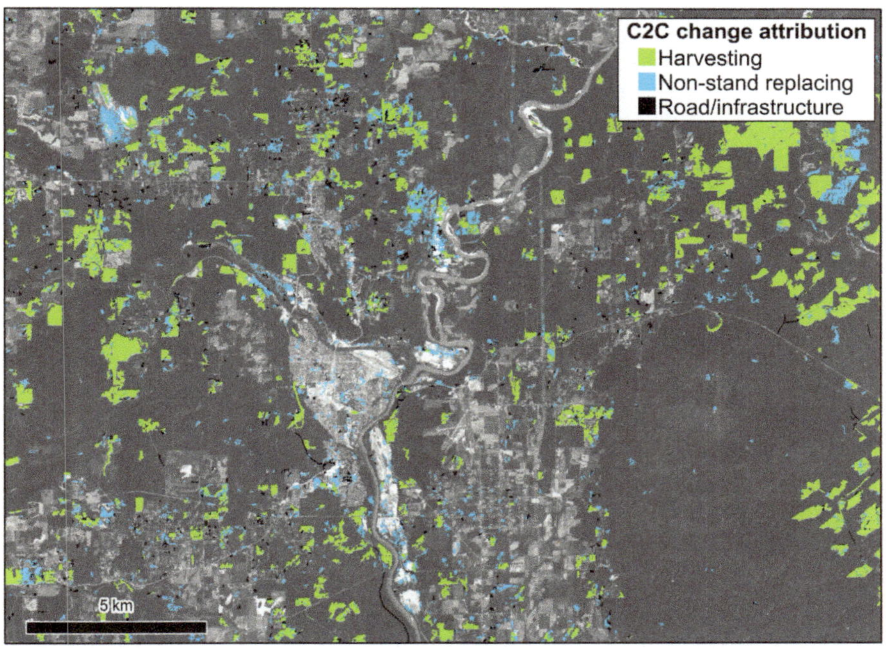

Figure 2. C2C (Composite2Change) forest change data between 1984 and 2003.

3.2. Validation Data

Information on the location, extent and type of harvest activity occurring within the region was compiled by the Pacific Forestry Center (PFC) of the Canadian Forest Service (CFS), and consists of a sample of harvest blocks selected from the Prince George forest region. Data were compiled from a combination of ground site visits (on 8–10 September 2003) and aerial photo interpretation. More details on the reference data are provided in Leckie et al. [32]. Ground visits were given priority at sites that were deemed to be the most likely to be misclassified from aerial photography. For this study, we utilized all reference harvest polygons that intersected with the C2C change objects, with more than 50% overlap, and were greater than the C2C minimum map unit (0.5 ha). The size of the analyzed harvest polygons derived from C2C varied from 0.5 to 96 ha, with an average size of 8.8 ha (σ = 13). A total of 160 reference harvest polygons were used for validation, distributed by harvest type as follows: clearcut (80 polygons), clearcut with residual (47), and partial cut (33). Clearcut implies a canopy closure less than 1% post-harvest. Clearcut with residual indicates a canopy cover between 1% and 10% post-harvest. Finally, partial cut implies a selective harvesting technique resulting in a stand with >10% canopy cover remaining (Figure 3). The average polygon size for clearcuts was 7.8 ha (σ = 10), while clearcuts with residual polygons were 9.5 ha on average (σ = 15), and partial cut polygons were 10.3 ha (σ = 15.6). The spatial distribution of the harvest polygons is shown in Figure 4.

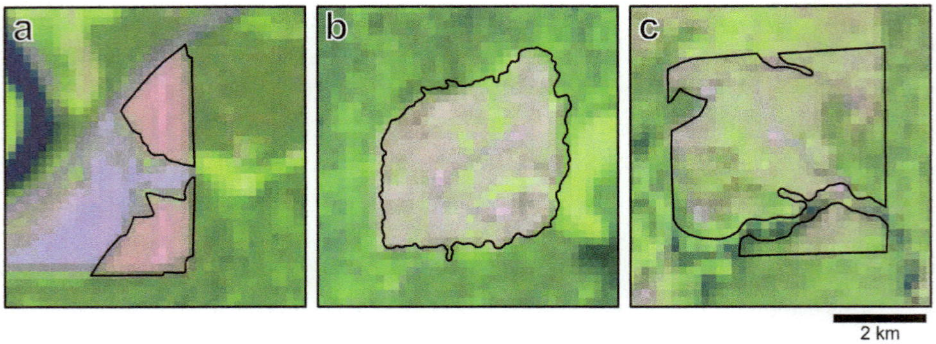

Figure 3. Examples of harvest types in Landsat false color (5-4-3) composition: clearcut (**a**); clearcut with residual (**b**); and partial cut (**c**).

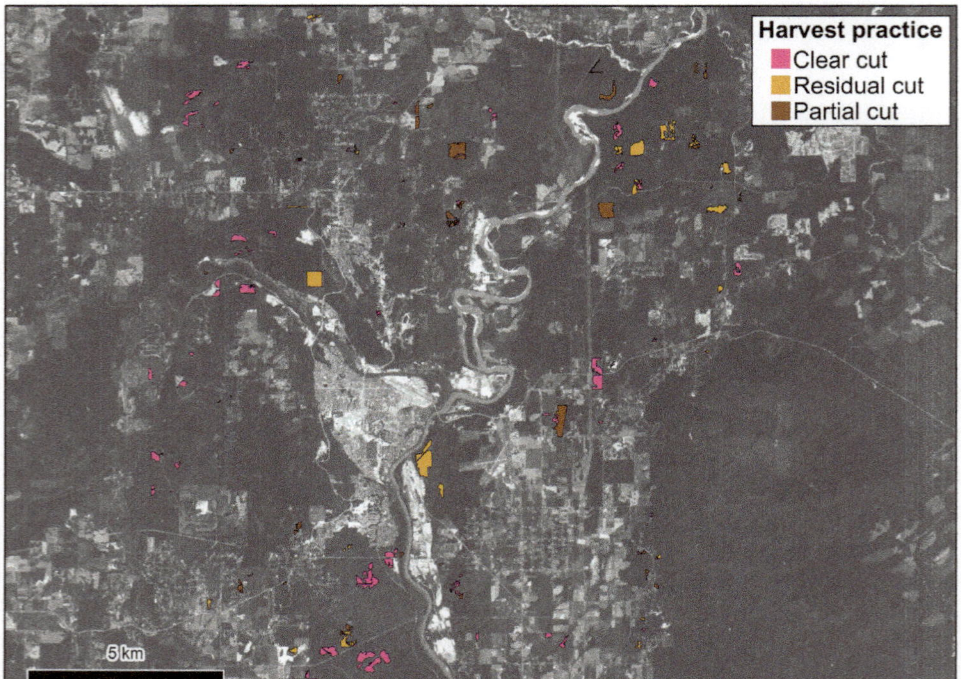

Figure 4. Reference forest harvest polygons compiled from a combination of ground site visits (on 8–10 September 2003) and aerial photo interpretation (Leckie et al. [32]).

4. Methods

To compare the changes detected from the Landsat time series and the forest harvest data, our approach was as follows. We assessed the capacity of the Landsat time-series-derived metrics to describe and discriminate between the three different harvest practices identified in the reference data. This was undertaken by matching the reference harvest polygons with the change pixels detected from the C2C Landsat time series. Then, we evaluated the C2C attribution of harvest, to assess (i) how accurately the C2C data identified harvest; and then (ii) how that accuracy varied by harvest type.

4.1. Spectral Characterization of Harvesting Practices

We analyzed the capacity of the Landsat time-series spectral metrics to discriminate between the three different harvesting practices. To do so, six spectral metrics, obtained from the time series, were calculated and extracted for each of the analyzed harvest polygons [31]. The six spectral metrics were change persistence (duration), average change magnitude variation, average spectral value post-change, average spectral value pre-change, range of pixel series values, and standard deviation of pixel series values (see full description in Hermosilla et al., [31]). Persistence represents the length in years following the harvest event that spectral values remain at post-change levels.

Magnitude variation refers to the spectral difference of Normalized Burn Ratio (NBR) [33] before and after the harvest event. The NBR is a spectral index used to map burn severity, and is calculated using Landsat TM/ETM+ bands 4 and 7 as follows:

$$NBR = \frac{B4 - B7}{B4 + B7} \tag{1}$$

Pre-change represents the average spectral NBR value of the years leading up to the year of greatest change and post-change represents the average NBR value directly following the harvest event. The standard deviation of pixel series values represents the spectral variation of NBR within each harvest practice. The time series range measured using tasseled cap greenness (TCG) [34] represents the variability of pixel values for each harvest practice. An ANOVA together with Tukey's post hoc test were performed on these spectral metrics in order to determine significant differences ($\alpha = 0.05$, $p < 0.05$) in their values between the harvesting practices. Additionally, we examined the complete trends of the spectral values centering the trends on the change year [35]. In this way, trends can be compared regardless of the year of disturbance, with trends referenced relative to the number of years before ($y - 1$, $y - 2$, etcetera) and after ($y + 1$, $y + 2$, etcetera).

4.2. Change Attribution

We evaluated the accuracy of the C2C change type attribution using the reference harvest polygons, and summarized these results in a modified confusion matrix [36]. As the reference data only characterizes harvest, we were not able to evaluate the accuracy of the other C2C change types. In cases of multiple C2C disturbance type classifications within the harvest polygon, the most frequently occurring class in terms of overlapping area was chosen as representative. Any change attribution value other than harvesting was considered as misclassified. Producer's accuracies were estimated by harvest practice. Lastly, the influence of the size of the harvesting polygons on the classification accuracy was analyzed.

5. Results

5.1. Discriminating Harvest Type

Figure 5 shows the distribution of the Landsat-derived spectral metrics for the three harvest practices (i.e., clearcut, clearcut with residual, and partial cut). Table 1 shows the significant differences between the spectral values for the different harvest types as determined with the ANOVA. The average pre-change NBR showed no statistical difference, and confirmed uniformity of the forest stands prior to harvest. Thus, significant differences in spectral metrics post-harvest were assumed to be the result of the harvest method applied and not influenced by different pre-harvest forest conditions.

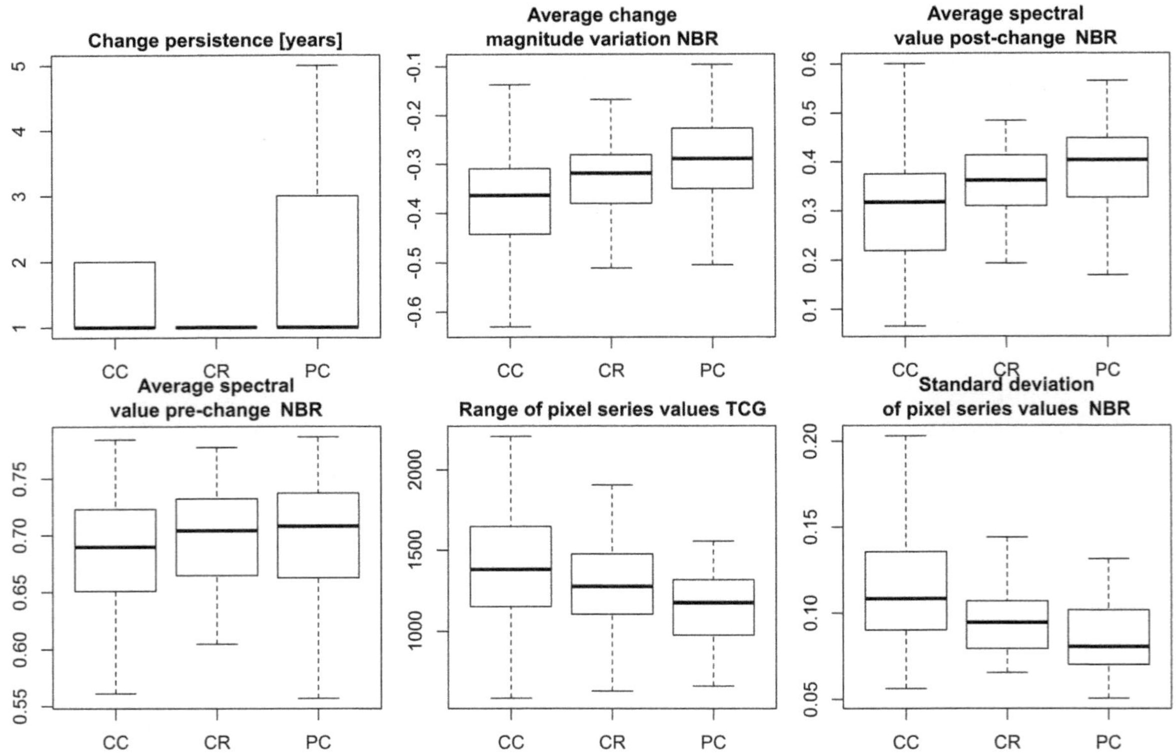

Figure 5. Boxplots representing median, interquartile range, and extreme values for selected Landsat-derived spectral metrics across harvest practices: clearcut (CC), clearcut with residual (CR), and partial cut (PC). Note that outliers are not displayed.

Table 1. Statistically significant different groups determined with Tukey's post hoc test ($p < 0.05$).

Metric	Clearcut	Clearcut with Residual	Partial Cut
Change persistence	a	a	b
Average change magnitude variation NBR	b	ab	a
Average spectral value post-change NBR	b	a	a
Average spectral value pre-change NBR	a	a	a
Range of pixel series values TCG	b	ab	a
Standard deviation of pixel series values NBR	a	a	b

Homogenous groups per metric are identified using the same letter. NBR, Normalized Burn Ratio; TCG, tasseled cap greenness.

The results of the ANOVA indicated that partial cuts were significantly different from clearcuts, and also different from clearcuts with residual. The average spectral magnitude variation of the change mirrors the expected intensity gradient in harvest practices. As shown in Figure 6, clearcuts show the largest spectral change magnitude variation, followed by clearcuts with residual, with partial cuts having the lowest spectral change values. Statistically, clearcut and partial cut categories are significantly different from each other, and the clearcut with residual statistically similar to either clearcut or partial cut, depending on the metric. The average spectral value post-change NBR complements the magnitude variation metric by indicating that more intense harvesting methods lead to significantly lower spectral values directly after the harvest. These intense changes with low spectral values are more likely to have a steeper recovery trend over the following years compared to less intense changes as shown by the persistence.

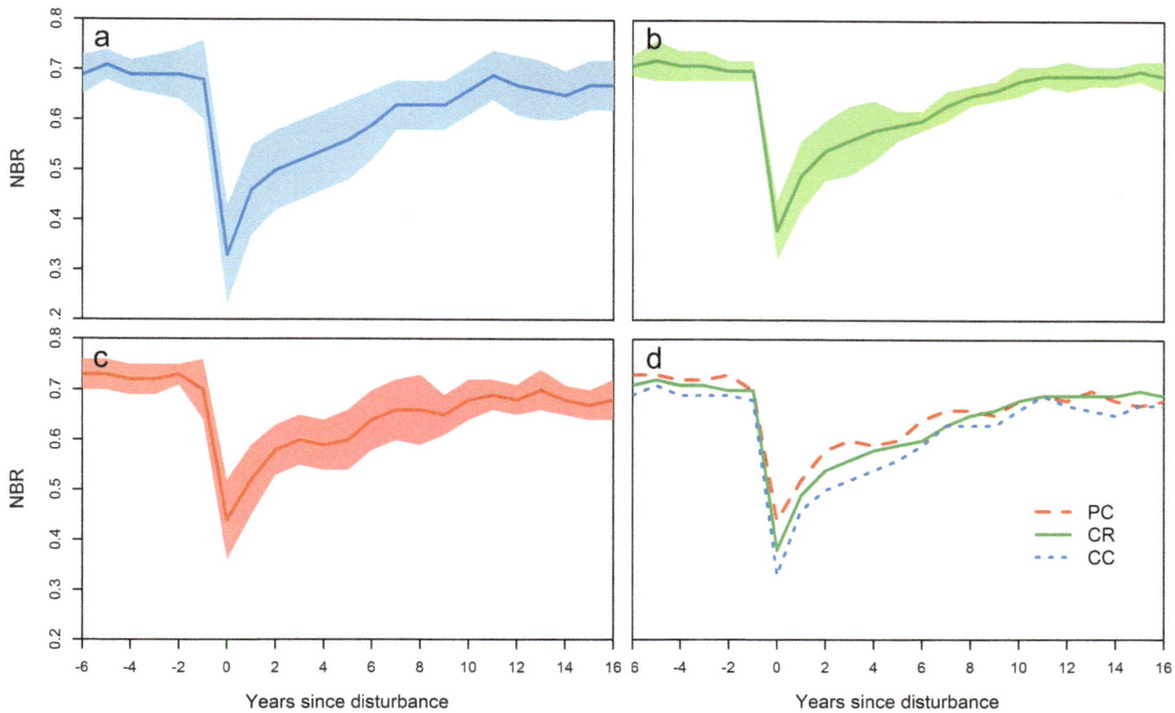

Figure 6. Average and standard deviation of Normalized Burn Ratio (NBR) spectral values per harvest practice: clearcut (**a**), residual cut (**b**) and partial cut (**c**), and comparison of the averaged spectral values: clearcut (CC), clearcut with residual (CR), and partial cut (PC) (**d**).

The change persistence (duration) shows significant differences between partial cut and the other harvest practices. Persistence for partial cuts ranged from 1 to 5 years, resulting in the recovery trend that is less steep and thus comparatively slower. All three harvest practices have a median persistence of 1 year, yet the partial cuts had a noticeably longer range of durations that can extend up to 5 years. The 5-year range reflects successive cuts removing mature trees to attain an even-aged new stand, which is the defining feature of partial cuts. The relative spectral heterogeneity of recovery time for partial cuts, combined with its criteria of being over 10% canopy cover (which may be difficult to accurately estimate in the field) explains the comparatively weaker classification percentage.

Another explanatory metric which detects difference between harvesting practices is the TCG. The results clearly show large magnitude reductions in TCG values as a consequence of clearcuts. This is in contrast to the smaller magnitude reductions caused by partial cuts. As a result, partial cuts are less likely to be classified as harvesting than clearcuts as the magnitude of the change is comparatively small and likely to be confused with other types of disturbance or missed completely. Once again, the smaller magnitude change can partially be attributed to the variable intensity (anything above 10% cover remaining) of the partial cut.

The mean of NBR values for each of the harvesting practices is shown in Figure 6a–c. Figure 6d shows the average NBR trajectory for each harvest type.

5.2. Change Attribution

Overall accuracy for all reference harvest polygons correctly classified as harvesting was 78.8% (Table 2). Table 3 displays the attribution accuracy of the harvest polygons based on size. The results showed that 93% of harvest polygons over 5 ha were accurately classified. As the size of the polygons decreases, the classification errors increase, leading to an accuracy of 66% for polygons less than 5 ha.

Of the different harvesting practices identified in the reference data, clearcut harvesting is the most accurately classified (83.8%), followed by clearcut with residual with 79.2% of correctly classified. Partial cut forestry practices showed the lowest attribution accuracy (63.6%). Clearcut with residual and partial cut practices had a higher percentage of misclassification as non-stand replacing changes. No harvest polygons were labeled by the C2C algorithm as fire.

Table 2. Accuracy of C2C (Composite2Change) harvest attribution.

	Reference Data			
	Clearcut	**Residual Cut**	**Partial Cut**	**Total**
Harvest	67	38	21	126
Non-stand replacing	7	5	9	21
Road/infrastructure	6	4	3	13
Fire	0	0	0	0
Total	80	47	33	160
Producer's accuracy (%)	83.8	79.2	63.6	78.8
Omission error (%)	16.2	20.8	36.4	

Table 3. Producer's accuracy, by object size.

Size	Polygons Misclassified	Accuracy (%)
>5 ha	5 of 70	93.0
2–5 ha	14 of 41	66.0
<2 ha	18 of 53	66.0s

6. Discussion

In this research, we assessed the capability of Landsat time series to identify and characterize different harvest practices reported in established harvest records, including clearcut, clearcut with residual, and partial cut. A range of different metrics were analyzed to examine which spectral properties differentiate the harvesting practices. The accuracy of the change attribution varied by harvest practice. Clearcuts were the most accurately classified harvesting practice. Clearcutting removes the majority of vegetation from site and exposes soil. Indeed, we found that the average spectral post-change value was the most statistically important metric for discriminating clearcuts, in keeping with the results of Lambert et al. [37] who identified clearcuts assessing the post-disturbance scenario. A large and fast spectral change in a long-term trajectory, depicted by a steep post change spectral evolution, is commonly representative of a major stand replacing disturbance [38]. Some clearcut plots were erroneously identified as roads, which also are characterized by a steep drop in the spectral response through time. In areas with harvest activity, however, roads are sometimes temporary and can have various surface materials (e.g., dirt, gravel) rendering them difficult to attribute due to lack of uniformity and a spectral similarity to clearcuts [31]. Specific spatial rules to distinguish between roads and other features based on geometric properties, such as linearity, are possible [39]. Moreover, roads tend to persist and will have different post-disturbance spectral characteristics that would allow them to be distinguished from harvesting, given additional observation years.

Clearcut with residual harvest practices were also correctly identified by the C2C algorithm in the majority of cases. The forest stands that are harvested in this manner typically undergo a near complete removal of trees, with between 1% and 10% of canopy cover retained. As a result, this harvesting practice is expected to have a similar spectral response to clearcuts. The remaining trees impact the averaged spectral values after the change event, thus reducing the magnitude, standard deviation, and other metrics. This is best exemplified through the range in greenness values of the time series, which show clearcut with residual practices to be not significantly different from partial cuts. As clearcut residual polygons retain large mature trees to aid in regeneration, the average greenness of the polygon has less deviation from the pre-change spectral response [40]. Depending on the configuration of the residual cover, it may confound the automated detection and attribution of the change event using the Landsat time series.

Partial cuts were the most challenging practice to identify, with lower accuracy relative to clearcuts and residual cuts. This detection and differentiation difficulty is explained by the spectral response of partial cut polygons, which show comparatively low change magnitude and standard deviation values through the time series. The majority of the confusion is from partial cuts being labeled as non-stand replacing changes, which are typically gradual changes in vegetation condition that do not lead to a change in land cover class (i.e., disease, insects, water stress, and decline; [41]). Partial cuts are defined as having >10 percent canopy cover post-harvest. Since the aim of partial cut practices is to mimic low severity natural disturbances [42,43] or maintain biodiversity [44] these changes have a high probability of being associated with non-stand replacing disturbances, such as drought stress or insects. The lower range of temporal TCG values for partial cuts made them distinguishable from clearcuts with this metric. This contrasts with Healey et al. [24], which found that TCG values were weakly correlated with forest removal using basal area as the forest-removal measurement, whereas we used canopy cover.

The analysis of the influence of polygon size on the change attribution accuracy indicates that smaller harvest polygons were more likely to be misclassified. The smallest polygons were 0.5 ha which is equivalent to 6 Landsat-pixels. The Landsat 30 m pixel would likely be influenced by factors outside the harvest plots, as exact delineation of polygon edges is difficult at this spatial resolution. These surrounding spectral values become more influential as polygon sizes decrease, increasing error from a proportional perspective. As expected, the large polygons were classified more reliably, due to less opportunity for edge effects and, more importantly, a greater number of pixels (and related spectral metric values) to sample per harvest event.

7. Conclusions

Despite some minor misclassification, the results of this study showed that partial cuts were significantly different from clearcuts in most spectral metrics characterizing the disturbance or the post change properties. This represents the potential ability to further distinguish between harvesting practices in the disturbance classification hierarchy presented in Hermosilla et al. [31]. Due to the opening of the Landsat archive, the analysis of spectral time series of Landsat data covering large jurisdictions is an increasingly operational approach for verifying harvest activity at a reduced cost compared to field visits or photo digitization approaches for traditional strategic forest inventories. Further differentiation of harvest practices will provide a rich baseline for forest management, analysis, and decision-making moving forward. Our results along with those of others [26] demonstrate the capability of utilizing Landsat time series to further classify forest change events into more specific categories. The methodological opportunities here applied are a direct consequence of the richness of the Landsat archive over Canada, in contrast to many nations and regions globally where the application of these methods would be more limited [12,45]. Spectral metrics derived from Landsat time series provide a quantitative and spatially explicit opportunity to generate refined attribution detail related to forest harvest practices. Traditional photo-based forest inventory techniques, while modernizing with increasingly digital methods, remain limited in their ability to provide a regularly updated and extensive representation of entire management areas. With the combination of the Landsat program's archival depth, calibrated sensors, analysis ready data, increased acquisition frequency, as well as growing synergies with complementary satellites (e.g., Sentinel 2), new opportunities exist for augmenting and developing forest inventories from regional to national scales.

Acknowledgments: This research was undertaken as part of the "National Terrestrial Ecosystem Monitoring System (NTEMS): Timely and detailed national cross-sector monitoring for Canada" project jointly funded by the Canadian Space Agency (CSA) Government Related Initiatives Program (GRIP) and the Canadian Forest Service (CFS) of Natural Resources Canada. We also acknowledge funding provided to Coops by NSERC (RGPIN 311926-13).

Author Contributions: The authors contributed equally to the overall research design, approach and analysis. The lead author wrote early drafts of the manuscript with contributions from the other authors through the editorial process.

Conflicts of Interest: The authors declare no conflict of interest.

References

1. Wulder, M.A.; White, J.C.; Cranny, M.; Hall, R.J.; Luther, J.E.; Beaudoin, A.; Goodenough, D.G.; Dechka, J.A. Monitoring Canada's forests. Part 1: Completion of the EOSD land cover project. *Can. J. Remote Sens.* **2008**, *34*, 549–562. [CrossRef]
2. Natural Resources Canada. *The State of Canada's Forests*; Annual Report 2016; Natural Resources Canada: Ottawa, ON, Canada, 2016.
3. Wulder, M.A.; Kurz, W.A.; Gillis, M.; Wulder, M. National level forest monitoring and modeling in Canada. *Prog. Plan.* **2004**, *61*, 365–381. [CrossRef]
4. Brandt, J.P.; Flannigan, M.D.; Maynard, D.G.; Thompson, I.D.; Volney, W.J.A. An introduction to Canada's boreal zone: Ecosystem processes, health, sustainability, and environmental issues. *Environ. Rev.* **2013**, *21*, 207–226. [CrossRef]
5. Kangas, A.; Maltamo, M. *Forest Inventory: Methodology and Applications*; Springer: Berlin/Heidelberg, Germany, 2006.
6. Wulder, M.A.; Campbell, C.; White, J.C.; Flannigan, M.; Campbell, I.D. National circumstances in the international circumboreal community. *For. Chron.* **2007**, *83*, 539–556. [CrossRef]
7. Masek, J.G.; Cohen, W.B.; Leckie, D.; Wulder, M.A.; Vargas, R.; de Jong, B.; Healey, S.; Law, B.; Birdsey, R.; Houghton, R.A.; et al. Recent rates of forest harvest and conversion in North America. *J. Geophys. Res.* **2011**. [CrossRef]
8. Gillis, M.D.; Leckie, D.G. Forest inventory update in Canada. *For. Chron.* **1996**, *72*, 138–156. [CrossRef]
9. White, J.C.; Coops, N.C.; Wulder, M.A.; Vastaranta, M.; Hilker, T.; Tompalski, P. Remote sensing technologies for enhancing forest inventories: A review. *Can. J. Remote Sens.* **2016**, *42*, 619–641. [CrossRef]
10. Wulder, M.A.; Coops, N.C.; Hudak, A.T.; Morsdorf, F.; Nelson, R.; Newnham, G.; Vastaranta, M. Status and prospects for LiDAR remote sensing of forested ecosystems. *Can. J. Remote Sens.* **2013**, *39*, S1–S5. [CrossRef]
11. Hermosilla, T.; Wulder, M.A.; White, J.C.; Coops, N.C.; Hobart, G.W. Regional detection, characterization, and attribution of annual forest change from 1984 to 2012 using Landsat-derived time-series metrics. *Remote Sens. Environ.* **2015**, *170*, 121–132. [CrossRef]
12. White, J.C.; Wulder, M.A. The Landsat observation record of Canada: 1972–2012. *Can. J. Remote Sens.* **2013**, *39*, 455–467. [CrossRef]
13. White, J.C.; Wulder, M.A.; Hobart, G.W.; Luther, J.E.; Hermosilla, T.; Griffiths, P.; Coops, N.C.; Hall, R.J.; Hostert, P.; Dyk, A.; et al. Pixel-based image compositing for large-area dense time series applications and science. *Can. J. Remote Sens.* **2014**, *40*, 192–212. [CrossRef]
14. Woodcock, C.E.; Allen, R.; Anderson, M.; Belward, A.; Bindschadler, R.; Cohen, W.; Gao, F.; Goward, S.N.; Helder, D.; Helmer, E.; et al. Free access to Landsat imagery. *Science* **2008**, *320*, 1011. [CrossRef] [PubMed]
15. Sexton, J.O.; Song, X.-P.; Feng, M.; Noojipady, P.; Anand, A.; Huang, C.; Kim, D.-H.; Collins, K.M.; Channan, S.; DiMiceli, C.; et al. Global, 30-m resolution continuous fields of tree cover: Landsat-based rescaling of MODIS vegetation continuous fields with lidar-based estimates of error. *Int. J. Digit. Earth* **2013**, *6*, 427–448. [CrossRef]
16. Hansen, M.C.; Potapov, P.V.; Moore, R.; Hancher, M.; Turubanova, S.A.; Tyukavina, A.; Thau, D.; Stehman, S.V.; Goetz, S.J.; Loveland, T.R.; et al. High-resolution global maps of 21st-century forest cover change. *Science* **2013**, *342*, 850–853. [CrossRef] [PubMed]
17. Ju, J.; Masek, J.G. The vegetation greenness trend in Canada and US Alaska from 1984–2012 Landsat data. *Remote Sens. Environ.* **2016**, *176*, 1–16. [CrossRef]

18. Hermosilla, T.; Wulder, M.A.; White, J.C.; Coops, N.C.; Hobart, G.W.; Campbell, L.B. Mass data processing of time series Landsat imagery: Pixels to data products for forest monitoring. *Int. J. Digit. Earth* **2016**, *9*, 1035–1054. [CrossRef]
19. Townshend, J.R.G.; Justice, C.O. Selecting the spatial resolution of satellite sensors required for global monitoring of land transformations. *Int. J. Remote Sens.* **1988**, *9*, 187–236. [CrossRef]
20. Wulder, M.A.; Masek, J.G.; Cohen, W.B.; Loveland, T.R.; Woodcock, C.E. Opening the archive: How free data has enabled the science and monitoring promise of Landsat. *Remote Sens. Environ.* **2012**, *122*, 2–10. [CrossRef]
21. Belward, A.S.; Skøien, J.O. Who launched what, when and why; trends in global land-cover observation capacity from civilian earth observation satellites. *ISPRS J. Photogramm. Remote Sens.* **2015**, *103*, 115–128. [CrossRef]
22. British Columbia Ministry of Forests and Range Silviculture Program Statistics. Available online: https://www.for.gov.bc.ca/hfp/silviculture/statistics/statistics.htm (accessed on 15 December 2016).
23. British Columbia Ministry of Forests, Mines and Lands. *The State of British Columbia's Forests*, 3rd ed.; Forest Practices and Investment Branch: Victoria, BC, Canada, 2010.
24. Healey, S.P.; Yang, Z.; Cohen, W.B.; Pierce, D.J. Application of two regression-based methods to estimate the effects of partial harvest on forest structure using Landsat data. *Remote Sens. Environ.* **2006**, *101*, 115–126. [CrossRef]
25. Kennedy, R.E.; Cohen, W.B.; Schroeder, T.A. Trajectory-based change detection for automated characterization of forest disturbance dynamics. *Remote Sens. Environ.* **2007**, *110*, 370–386. [CrossRef]
26. Schroeder, T.A.; Wulder, M.A.; Healey, S.P.; Moisen, G.G. Mapping wildfire and clearcut harvest disturbances in boreal forests with Landsat time series data. *Remote Sens. Environ.* **2011**, *115*, 1421–1433. [CrossRef]
27. British Columbia Ministry of Forests. *The Ecology of the Sub-Boreal Spruce Zone*; British Columbia Ministry of Forests: Victoria, BC, Canada, 1998.
28. Meyers Norris Penny. *British Columbia Forest Industry Economic Impact Study*; Meyers Norris Penny: Calgary, AB, Canada, 2015.
29. Forest Practices Board. *High Retention Harvesting and Timber Sustainability on the British Columbia Coast*; Forest Practices Board: Victoria, BC, Canada, 2009.
30. Dhar, A.; Parrott, L.; Heckbert, S. Consequences of mountain pine beetle outbreak on forest ecosystem services in western Canada. *Can. J. For. Res.* **2016**, *46*, 987–999. [CrossRef]
31. Hermosilla, T.; Wulder, M.A.; White, J.C.; Coops, N.C.; Hobart, G.W. An integrated Landsat time series protocol for change detection and generation of annual gap-free surface reflectance composites. *Remote Sens. Environ.* **2015**, *158*, 220–234. [CrossRef]
32. Leckie, D.; Tinis, S.; Henley, M.; Eichel, F.; Walsworth, N.; Hardman, D.; Burt, W.; Dechka, J.; Paradine, D.; Landry, R.; et al. *Deforestation and Forest Harvest Mapping with Integrated Automated and Manual Methods: Pilot Study Prince George, British Columbia, Canada*; Information Report Bc-X-438; Canadian Forest Service Pacific Forestry Centre: Victoria, BC, Canada, 2016.
33. Key, C.H.; Benson, N.C. *Landscape Assessment (LA) Sampling and Analysis Methods*; USDA Forest Service Fort: Collins, CO, USA, 2006.
34. Crist, E.P. A TM Tasseled Cap equivalent transformation for reflectance factor data. *Remote Sens. Environ.* **1985**, *17*, 301–306. [CrossRef]
35. Ahmed, O.S.; Franklin, S.E.; Wulder, M.A. Integration of lidar and Landsat data to estimate forest canopy cover in Coastal British Columbia. *Photogramm. Eng. Remote Sens.* **2014**, *80*, 953–961. [CrossRef]
36. Olofsson, P.; Foody, G.M.; Herold, M.; Stehman, S.V.; Woodcock, C.E.; Wulder, M.A. Good practices for estimating area and assessing accuracy of land change. *Remote Sens. Environ.* **2014**, *148*, 42–57. [CrossRef]
37. Lambert, J.; Denux, J.-P.; Verbesselt, J.; Balent, G.; Cheret, V. Detecting clear-cuts and decreases in forest vitality using MODIS NDVI time series. *Remote Sens.* **2015**, *7*, 3588–3612. [CrossRef]
38. Pickell, P.D.; Hermosilla, T.; Frazier, R.J.; Coops, N.C.; Wulder, M.A. Forest recovery trends derived from Landsat time series for North American boreal forests. *Int. J. Remote Sens.* **2016**, *37*, 138–149. [CrossRef]
39. Walter, V. Object-based classification of remote sensing data for change detection. *ISPRS J. Photogramm. Remote Sens.* **2004**, *58*, 225–238. [CrossRef]
40. Thorpe, H.C.; Thomas, S.C. Partial harvesting in the Canadian boreal: Success will depend on stand dynamic responses. *For. Chron.* **2007**, *83*, 319–325. [CrossRef]

41. Ahmed, O.S.; Wulder, M.A.; White, J.C.; Hermosilla, T.; Coops, N.C.; Franklin, S.E. Classification of annual non-stand replacing boreal forest change in Canada using Landsat time series: A case study in northern Ontario. *Remote Sens. Lett.* **2017**, *8*, 29–37. [CrossRef]
42. Lieffers, V.J.; Beck, J.A. A semi-natural approach to mixedwood management in the prairie provinces. *For. Chron.* **1994**, *70*, 260–264. [CrossRef]
43. Bergeron, Y.; Harvey, B. Basing silviculture on natural ecosystem dynamics: An approach applied to the southern boreal mixedwood forest of Quebec. *For. Ecol. Manag.* **1997**, *92*, 235–242. [CrossRef]
44. Burton, P.J.; Kneeshaw, D.D.; Coates, K.D. Managing forest harvesting to maintain old growth in boreal and sub-boreal forests. *For. Chron.* **1999**, *75*, 623–631. [CrossRef]
45. Wulder, M.A.; White, J.C.; Loveland, T.R.; Woodcock, C.E.; Belward, A.S.; Cohen, W.B.; Fosnight, E.A.; Shaw, J.; Masek, J.G.; Roy, D.P. The global Landsat archive: Status, consolidation, and direction. *Remote Sens. Environ.* **2016**, *185*, 271–283. [CrossRef]

11

Phosphorus in Preferential Flow Pathways of Forest Soils in Germany

Dorit Julich *, Stefan Julich and Karl-Heinz Feger

Institute of Soil Science and Site Ecology, TU Dresden, Pienner Str. 19, 01737 Tharandt, Germany; Stefan.Julich@tu-dresden.de (S.J.); karl-heinz.feger@tu-dresden.de (K.-H.F.)
* Correspondence: dorit.julich@tu-dresden.de

Academic Editors: Scott X. Chang and Xiangyang Sun

Abstract: The transport of nutrients in forest soils predominantly occurs along preferential flow pathways (PFP). This study investigated the composition of phosphorus (P) forms in PFPs and soil matrix in several temperate beech forests with contrasting soil P contents in Germany. The PFPs were visualized using dye tracer experiments. Stained and unstained soil was sampled from three profile cuts per plot and analyzed for P fractions. The results show that labile P concentrations were highest in the O-layer and had the same range of values at all sites (240–320 $mg \cdot kg^{-1}$), although total P (TP) differed considerably (530–2330 $mg \cdot kg^{-1}$). The ratio of labile P to TP was significantly lower in the P-rich soil compared to the medium and P-poor soils. By contrast, the ratio of moderately labile P to TP was highest at the P-rich site. The shifts in P fractions with soil depth were generally gradual in the P-rich soil, but more abrupt at the others. The contents of labile and moderately labile P clearly differed in PFPs compared to soil matrix, but not statistically significant. The studied soils are characterized by high stone contents with low potential for P sorption. However, indications were found that labile organically bound P accumulates in PFPs such as biopores.

Keywords: phosphorus fractions; forest soil; preferential flow; PFP

---

1. Introduction

Phosphorus (P) in forest soils is distributed highly variably, both between and within soil types, with respect to contents, speciation, availability, and sources of P [1]. Soils contain various P forms, generally differentiated in organically bound (P_o) and inorganic (Pi) P. Organic P occurs as orthophosphate monoesters (e.g., inositols, mononucleotides, sugar phosphates), orthophosphate diesters (phospholipids, teichoic acids, nucleic acids, and their degradation products), and phosphonates. Inorganic P is bound in minerals (e.g., apatite) and is present in dissolved form as ortho-, pyro-, and/or polyphosphates [2,3]. The total content and stock of soil P varies from very low (<100 $mg \cdot kg^{-1}$ and <10 $g \cdot m^{-2}$) to P-rich (>3000 $mg \cdot kg^{-1}$ and >1000 $g \cdot m^{-2}$), and the mineral P supply and P speciation is strongly dependent on the parent material and pedogenesis [4,5]. Whereas P supply from the mineral phase is the main source of available P in P-rich soils, P availability for plants in P-poor soils is considerably influenced by P speciation [4]. These differences in P supply drive P fluxes in the systems and determine the dominating strategy of ecosystem nutrition [6]. Besides those internal factors, external environmental and anthropogenic impacts (as forest management, N deposition, etc.) substantially influence the distribution of P and nutrient cycling in temperate forest soils [7–10].

The predominant transport pathway for nutrients in forest soils is along preferential flow pathways (PFP), bypassing large parts of the soil matrix [11–13]. An established method to investigate PFPs and their role in water and nutrient transport are tracer experiments [14–17]. Previous tracer studies indicated that PFPs in forest soils can be stable for decades, resulting in differences in biological and chemical parameters between PFPs and the soil matrix, particularly when preferential flow along root channels dominates the flow system [12,18]. Accumulation in stable flow paths has been found in previous studies for radionuclides, SOC, N [15,18], and also for C, Ca, Mg, and Fe [12]. Backnäs et al. [19] investigated P fractions in PFPs as compared to the soil matrix in a Podzol under forest in Southern Lapland, Finland. The results indicated that the leaching of P through PFPs occurred primarily along stone surfaces, where PFPs are larger than PFPs related to coarse grains or roots. Thus, P sorption may be limited due to potentially faster water flow, weaker lateral mass exchange and less sorption interaction than in smaller PFPs [20]. By contrast, P accumulation might occur in small preferential pores with decelerated water flow and longer contact time between the solution and inner pore surfaces [19]. As a result, such PFPs with high potential to accumulate P increase the likelihood for elevated P concentrations in export fluxes [21]. However, there are very few data available regarding P accumulation and transport in forest soils. Therefore, this study investigated the composition of P forms in PFPs and the soil matrix in four temperate forest ecosystems in Germany. Furthermore, the influence of different initial conditions regarding the content of total and available P at the sites on the distribution of P forms in the soil profiles and in PFPs was examined. The correlation between P fractions and chemical parameters was tested to identify the most important factors influencing the occurrence of P fractions in the investigated soils.

2. Materials and Methods

2.1. Site Description

Field experiments and sampling were conducted at four beech dominated forests (*F. sylvatica* L.) across Germany with similar stand age (110–120 years). The four study sites Lüss (LUE), Vessertal (VES), Mitterfels (MIT), and Bad Brückenau (BBR) are core sites of the German DFG Priority Program SPP 1685 which aims at testing the overall hypothesis that the P-depletion of soils drives forest ecosystems from P acquiring systems (efficient mobilization of P from the mineral phase) to P recycling systems (highly efficient cycling of P) [6]. Further, all sites are long-term monitoring sites in Germany and part of the EU-ICP Forests Level II Program (International Co-operative Programme on the Assessment and Monitoring of Air Pollution Effects on Forests). The sites were selected with regard to their considerable different availability of mineral P following the sequence BBR > MIT > VES > LUE. The BBR site was defined as P-rich site, LUE as P-poor site, and MIT and VES as medium P-supplied sites. Three sites (VES, MIT, and BBR) are located in the low mountain ranges, whereas LUE is in the lowlands. The sites differ in parent material, soil type, and climate (Table 1). Soil properties are summarized in Table 2. For better comparisons, the soil horizons were aggregated for each site considering their physical and chemical properties to: O—organic layer, A—topsoil horizon, B—subsoil horizon, C—lower subsoil horizon containing weathered parent material.

Table 1. Overview over the study sites.

	LUE	VES	MIT	BBR
Location	52°50′32″ N 10°16′06″ E	50°36′26″ N 10°46′20″ E	48°58′35″ N 12°52′49″ E	50°21′38″ N 9°55′71″ E
Elevation (m a.s.l.)	115	810	1023	809
MAT (°C)	8.0	5.5	4.5	5.8
MAP (mm)	779	1200	1229	1031
Parent material	Pleistocene sands	Trachyandesite	Gneiss	Basalt
Soil type (WRB *)	Dystric skeletic Cambisol	Hyperdystric skeletic folic Cambisol	Hyperdystric skeletic chromic Cambisol	Hyperdystric folic Cambisol

LUE: Lüss (Lower Saxony); VES, Vessertal (Thuringian Forest); MIT, Mitterfels (Bavarian Forest); BBR, Bad Brückenau (Bavarian Rhön Mountains); MAT, Mean annual temperature; MAP, Mean annual precipitation; * World Reference Base for Soil Resources 2014, World Soil Resources Reports 106, FAO, Rome, Italy.

Table 2. Soil properties at four study sites in beech forests in Germany.

Study Site/ Horizon	Depth (cm)	Texture	pH ($CaCl_2$)	C (%)	N (%)	Al_{ox}	Al_{di}	Fe_{ox}	Fe_{di}	P_{ox}	TP
						($g \cdot kg^{-1}$)					
Lüss (LUE)											
O	9 to 0		2.8	36.63	1.59	0.73	0.79	0.87	1.30	0.15	0.49
A	0 to −12	Loamy sand	3.0	1.69	0.07	0.18	0.37	0.29	2.17	<0.04	0.06
B	−12 to −52	Loamy sand	3.7	0.90	0.04	1.04	1.00	1.40	3.21	<0.04	0.08
C	−52 to −100	Sand	4.3	0.12	0.01	0.50	0.41	0.15	1.01	<0.04	0.04
Vessertal (VES)											
O	4 to 0		3.4	32.17	1.72	4.11	1.20	1.28	1.75	0.30	0.86
A	0 to −3	Sandy loam	3.0	7.49	0.45	2.75	1.68	3.14	7.46	0.53	0.70
B	−3 to −65	Sandy loam	4.1	3.03	0.20	9.44	7.40	5.13	9.23	0.83	0.98
C	−65 to −100	Loamy sand	4.1	0.37	0.03	4.64	2.83	0.59	5.55	0.16	0.32
Mitterfels (MIT)											
O	4 to 0		3.5	44.12	2.19	0.95	0.87	1.28	1.45	0.34	0.88
A	0 to −4	Silt loam	3.0	35.50	2.05	3.32	3.83	3.58	4.97	0.47	1.14
B	−4 to −80	Silt loam	3.9	5.45	0.34	6.26	5.62	8.05	12.17	0.53	0.91
C	−80 to −100	Loamy sand	4.3	1.09	0.09	5.73	4.63	3.61	10.89	0.35	0.66
Bad Brückenau (BBR)											
O	3 to 0		4.6	27.33	1.68	5.53	5.51	15.27	21.85	1.29	2.24
A	0 to −7	Silty clay	4.2	17.57	1.28	7.77	8.30	21.15	31.52	1.51	2.45
B	−7 to −30	Silt clay loam	4.3	7.45	0.54	10.58	10.83	24.46	35.41	1.84	2.62
C	−30 to −65	Loam	4.5	4.60	0.34	10.70	9.57	18.87	25.92	1.63	2.39

Horizon: O—organic layer, A—topsoil horizon, B—subsoil horizon, C—lower subsoil horizon containing weathered parent material. Values for C and N are total contents (for C this equals organic C); Al_{ox}, Fe_{ox}, and P_{ox}—ammonium oxalate-extractable contents; Al_{di} and Fe_{di}—dithionite-extractable contents; TP—total P contents.

2.2. Field Experiments

At each study site a test plot of 1 m^2 was established at a minimum distance of 2 m from any trees. On the first day, the plots were irrigated with 30 L of tracer-water solution containing 3 $g \cdot L^{-1}$ of Brilliant Blue FCF (within 1 hour) to activate rapid flow paths (PFPs) in the soils. Brilliant Blue dye was used due to its good visibility, retardation behavior in soil, and its non-toxicity to the environment [16,17]. After the dye tracer irrigation, the plots were sprinkled with 2 L of water to wash off the dye residues from the organic layer [19], and then covered overnight with a tarpaulin to exclude disturbances (e.g., rainfall). The irrigated plots were excavated on the following day by successively preparing five horizontal profile cuts per plot (at 20, 35, 50, 65, and 80 cm from the plot border) down to 1 m depth and 1 m width (see Figure 1). Stained (flow region) and unstained (non-flow region) soil was sampled respectively from the three inner cuts at 35, 50, and 65 cm from all morphological horizons. The amounts of soil material per sample varied because the thickness of soil horizons and stained/unstained parts per horizon differed considerably. The samples were sieved (<2 mm), dried at 40 °C, and then stored in plastic bottles until analysis. For the analysis of C, N, and total P the samples were finely grounded.

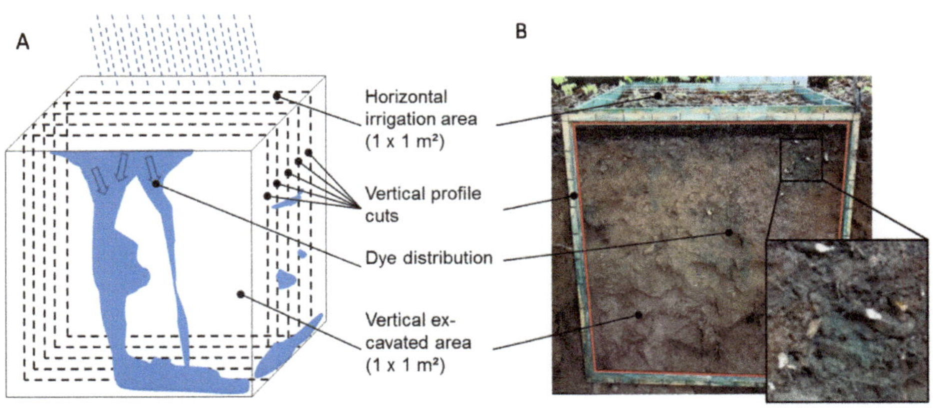

Figure 1. Scheme of tracer experiments (**A**): Dye application at the soil surface (1 m^2), covering overnight, and excavation of five vertical (1 m^2) profile cuts (example on the right (**B**)). The black framed enlargement shows staining of flow pathways along roots and stone surfaces.

2.3. Chemical Analyses

The P fractions in stained and unstained soil samples were analyzed using the Hedley fractionation method [22,23]. This method uses increasingly strong extractants to determine P forms with differing availability for plants. All samples were extracted with a replicate. Additionally, certified reference soil material from the Wageningen Evaluating Programs for Analytical Laboratories, International Soil-analytical Exchange (WEPAL-ISE 865 and 884) and blanks were treated equally to the samples for quality control of the extraction procedure.

In the fractionation procedure, a subsample of 0.5 g soil was first shaken in distilled water containing an anion exchange resin (Dowex 22, 16–50 mesh, Sigma-Aldrich, Taufkirchen, Germany) in netting bags (PP-405/230-41, Bückmann, Mönchengladbach, Germany) [24]. The following extractants applied were: 0.5 M sodium bicarbonate ($NaHCO_3$-P), 0.1 M sodium hydroxide (NaOH-$_i$), 1 M hydrochloric acid (HCl_{dil}-P), hot concentrated 3.5 M hydrochloric acid (HCl_{conc}-P), and finally nitric acid digestion of the residual P fraction (residual P). After each extraction step, the amount of inorganic P in the extract solution was measured using the method of Murphy and Riley [25] and O'Halloran and Cade-Menun [26]. For the determination of total P (TP) in $NaHCO_3$, NaOH, and HCl_{conc}, the extracts were oxidized in an autoclave using ammonium persulfate. Organically bound P (P_o) was calculated as the difference between TP and P_i [23]. The P concentrations in all extracts were measured colorimetrically with a spectrophotometer (UV-mini 1240, Shimadzu Deutschland GmbH, Duisburg, Germany). For a detailed description of the procedure applied here see Tiessen and Moir [23].

According to the concept of Hedley fractionation, the different P fractions represent (1) surface-adsorbed readily available P, (2) strongly-bonded or adsorbed moderate available P, and (3) very strongly-bonded or inaccessible mineral P with very low availability [22,23,27]. Therefore, the analyzed P fractions were classified for further data analyses as labile P (resin-P_i + $NaHCO_3$-$P_{i,o}$), moderately labile P (NaOH-$P_{i,o}$ + HCl_{dil}-P_i), and stable P (HCl_{conc}-$P_{i,o}$ + residual-P) (cf. [28]). The initial TP content of all samples was determined using the same method as the residual P fraction, by digestion with concentrated nitric acid and measurement with ICP-OES (CIROS, SPECTRO Analytical Instruments GmbH, Kleve, Germany). Total C and N contents were measured after dry combustion of grounded samples with a CN analyzer (Vario EL III Elementar GmbH, Hanau, Germany). The soils contained no inorganic C, thus soil organic C (SOC) equals total C. The concentrations of C in PFPs were not corrected for any C added as part of the Brilliant Blue dye. The C content of the dye is 56%, and the amount of total added C in the experiments was 50 g C m^{-2}. Bundt et al. [15] conducted comparable experiments with a dye-derived addition of 75 g C m^{-2} and found a resulting increase of C with a maximum of 2.5 % SOC in in PFPs in 0–9 cm depth, and a maximum of 3.5 % SOC in PFPs in 50–100 cm depth. In the current study, less dye was applied to the plots, and thus the change in C content in PFPs is expected to be very low. Amorphous metal sesquioxides (Al_{ox}, Fe_{ox}, Mn_{ox}) and P_{ox} were determined by extraction with acid 0.2 M ammonium oxalate (100 mL added to a subsample of 2 g). Dithionite-extractable (Fe_{di}, Al_{di}, Mn_{di}) was analyzed with 50 mL Na-citrate-dithionite in a subsample of 0.5 g (modified after Holmgren, 1967). The measurement of element contents in the extracts was conducted with ICP-OES.

2.4. Statistical Analysis

The statistical significance of differences between P contents of different chemical fractions in flow pathways and matrix were tested using the R software package (version R i386 3.1.0, R Foundation for Statistical Computing 2014, Vienna, Austria). Given that most of the data sets exhibited non-normal distributions (tested by Shapiro-Wilk test and q-q-plot), the non-parametric Wilcoxon rank-sum test was used to statistically compare mean values. The statistical tests were conducted pair-wise including all data of the four sites, and at horizon level due to the hierarchical structure of the data (profiles, cuts, horizons). For correlations between chemical parameters and P fractions, the Spearman's rho correlation coefficients were calculated and tested for significance.

3. Results and Discussion

The soil P fractions classified as labile, moderately labile, and stable P for PFP and soil matrix samples are shown in Figure 2. The values represent mean values for the three vertical cuts per profile. For the forest floor (O-layer), it was not possible to distinguish between stained and unstained material, due to the homogeneous distribution of the dye tracer from the irrigation. At the sites VES and MIT, the tracer did not reach the C horizon. Thus, the presented results contain no soil matrix data for the O-layer and no PFP data for the lower subsoil.

The TP values differed at the four study sites in their contents (45–2700 $mg \cdot kg^{-1}$) and distribution within the soil profiles (Figure 2, TP = total bar length). Whereas TP at the P-poor site (LUE) was clearly highest in the O-layer and strongly reduced in the mineral soil, the P-rich soil (BBR) showed the lowest values in the O-layer and had a slight increase in the subsoil. However, the amount of TP in all horizons at the P-rich site was much higher compared to the other sites, and was more or less equally distributed through the profile. The summarized P fractions were compared to the initial TP contents in order to determine the recovery rate of the analytical method. Considering all samples, the TP recovery was very high, with a correlation coefficient of 0.98.

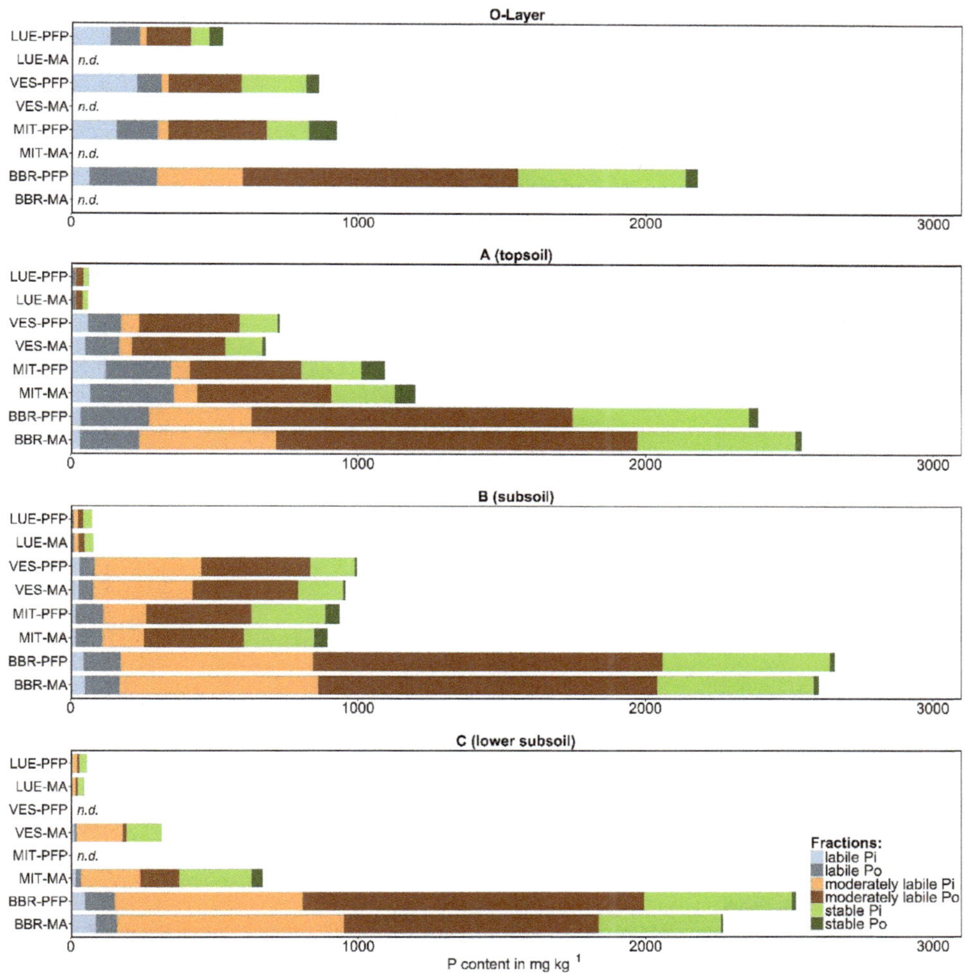

Figure 2. Phosphorus fractions summarized as labile (resin-P_i + $NaHCO_3$-$P_{i,o}$), moderately labile (NaOH-$P_{i,o}$ + HCl_{dil}-P_i), and stable P (HCl_{conc}-$P_{i,o}$ + residual-P) for PFP (preferential flow pathways) and MA (soil matrix) samples; n.d., no data available; study sites in the sequence P-poor to P-rich are: LUE, Lüss, Lower Saxony; VES, Vessertal, Thuringian Forest; MIT, Mitterfels, Bavarian Forest; BBR, Bad Brückenau, Bavarian Rhön Mountains.

3.1. Labile P Fractions

Despite the large difference in TP between the sites, labile P (here defined as resin-P_i + $NaHCO_3$-$P_{i,o}$) was in the same range of values, at least for the medium P-supplied (VES, MIT), and the P-rich (BBR) sites. In the O-layer, the P-poor site (LUE) also had comparable values for labile P as the other sites. The share of labile P to TP in the O-layer was highest at the P-poor site (45 %), slightly lower at the medium sites (36 % and 33 %), and lowest in the P-rich site (14 %) (Figure 3a.1). There was a steep decrease in the ratio labile P to TP with soil depth at the P-poor to medium sites; and a more gradual decrease with depth at the P-rich site.

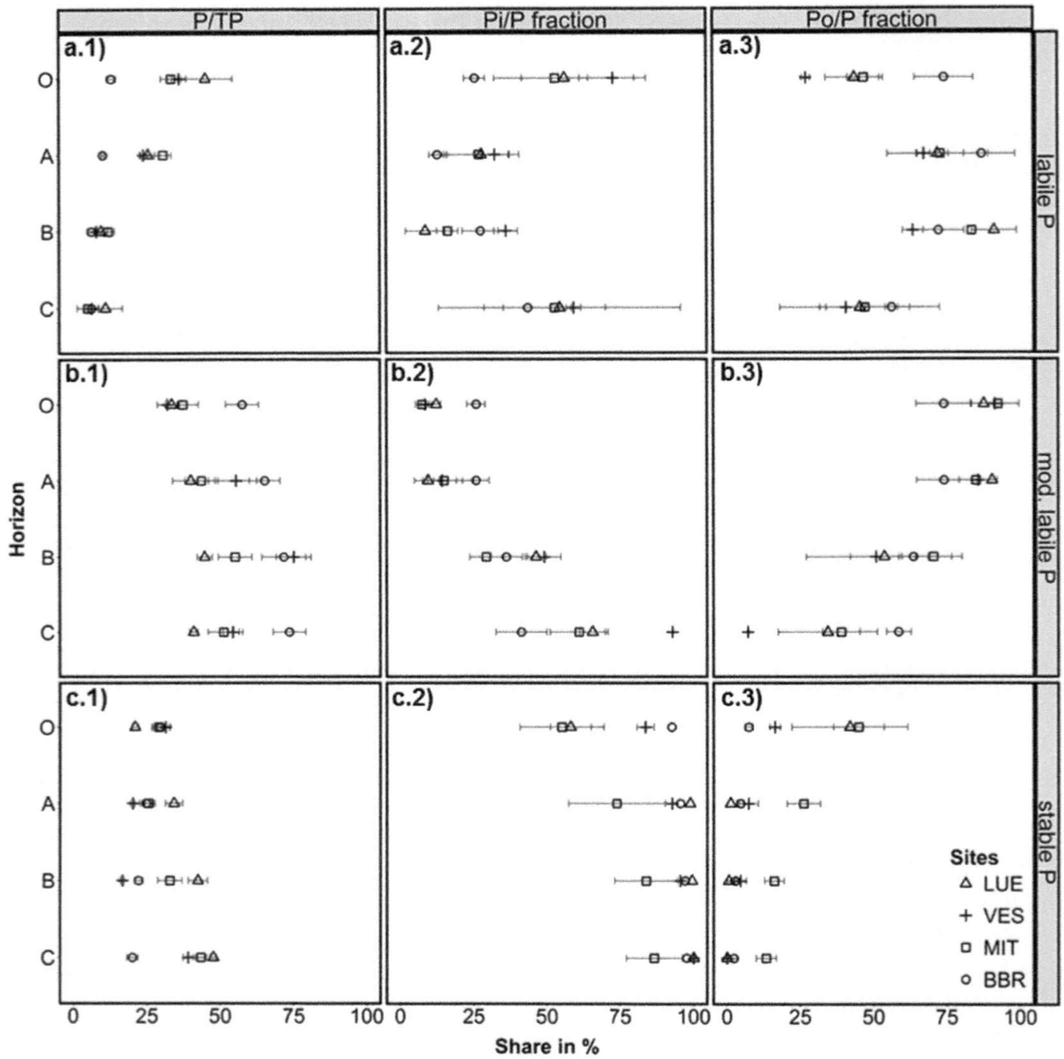

Figure 3. Share of **a.1**) the labile (resin-P_i + $NaHCO_3$-$P_{i,o}$), **b.1**) moderately labile (NaOH-$P_{i,o}$ + HCl_{dil}-P_i), and **c.1**) stable (HCl_{conc}-$P_{i,o}$ + residual-P) fractions on total P (TP) in organic layer (O) and soil (A, B, C horizons). Share of inorganic P (**a.2**: labile P_i, **b.2**: moderately labile P_i, **c.2**: stable P_i) and organically bound P (**a.3**: labile P_o, **b.3**: moderately labile P_o, **c.3**: stable P_o) on the particular pools. Error bars: standard deviation of three samples (profile cuts). Study sites in the sequence P-poor to P-rich are: LUE, Lüss, Lower Saxony; VES, Vessertal, Thuringian Forest; MIT, Mitterfels, Bavarian Forest; and BBR, Bad Brückenau, Bavarian Rhön Mountains.

Organically bound P dominated the labile P pool in the A and B horizon for all sites, and also in the O-layer for BBR. At VES, 27% of labile P in the O-layer consisted of P_o, while MIT and LUE had approximately equal amounts of P_i and P_o (Figure 3a.2,a.3). Thus, most of the easily plant-available P_i

was located in the O-layers at the medium and P-poor sites, whereas labile P was clearly dominated by P_o at the P-richest site. This finding was unexpected, as very low P supply from the mineral phase is assumed to enhance tight P recycling in forest ecosystems [6] (and therefore the fast uptake of P_i) which would result in a higher share of P_o at the P-poor site. In a recent study conducted at the same P-rich and P-poor study sites, Bünemann et al. [29] found that biological/biochemical processes were the dominant control of P_i availability in the low-P-sorbing sandy soil of the P-poor site. By contrast organic P mineralization was negligible when compared to the high P_i availability at the P-rich study site, which has a high-P-sorbing soil and very high TP content. Further, the high contents of amorphous and crystalline sesquioxides at the P-rich site (Table 2) may act as a sink for labile P_i, when it is adsorbed onto those surfaces and precipitated (depending on pH) (cf. [30,31]).

The differences in labile P between PFP and soil matrix were marginal considering the sum of P_i and P_o of this pool (Figure 2), but the ratio of P_i to P_o differed. Thus, the medium P-supplied site showed clearly higher contents of labile P_i and less labile P_o in PFPs compared to soil matrix. By contrast, labile P_o was significantly higher in PFPs of the P-rich site. However, the variation of results for labile P_i and P_o was quite high between the three profile cuts within the soil plot (see error bars in Figure 3), which indicates that conclusions from the results have to be drawn carefully.

3.2. Moderately Labile P Fractions

The sum of moderately labile P forms (NaOH-$P_{i,o}$ + HCl_{dil}-P_i) was the dominant fraction in mineral soil at the P-rich sites (Figures 2 and 3). At the P-poor site, this pool was equal or slightly lower than the stable P pool. At all sites, moderately labile P was predominantly organically bound in the O-layer and topsoil. In the subsoil, the ratio of P_i to P_o shifted towards P_i (Figure 3b.2,b.3). The shifts with depth were relatively gradual in the P-rich soil, but were more abrupt at the medium and P-poor sites.

The NaOH-extracted P represents the moderately available P forms related to Al, Fe, and humic substances. In acidic forest soils on silicate parent material, NaOH-P was observed to be involved in P_i adsorption (mainly to Al-oxyhydroxides) as well as P_o adsorption (to Fe-oxyhydroxides or Al-saturated soil organic matter) [4]. The share of HCl_{dil}-extractable P (apatite-bound) was very small for most samples, which indicates highly weathered soils [32]. In the P-rich soil, HCl_{dil}-P was found to be in the range of 150–300 $mg{\cdot}kg^{-1}$, which confirms that the supply with mineral P is still high at this site. These results are in line with findings of Prietzel et al. [4] who identified apatite-P in this P-rich soil using P *K*-edge X-ray absorption near-edge structure (XANES) spectroscopy.

The sum of moderately labile P showed a variable distribution between PFP and soil matrix (Figure 2). In the topsoils, the P-poor to medium sites (LUE and VES) had slightly higher contents of this pool in PFPs, whereas clearly higher contents in soil matrix were observed at the medium to P-rich sites (MIT and BBR). Here, both higher P_i as well as higher P_o values contributed to the increased pool contents. For the subsoils, slightly higher contents of moderately labile P in PFPs as compared to soil matrix were detected at all sites which was mainly caused by higher P_o contents, especially at the P-rich site. For the P-poor site, the differences are hardly interpretable because the chemical analyses of the very low P concentrations led to high data uncertainty.

3.3. Stable P Fractions

Stable P in the soils (HCl_{conc}-$P_{i,o}$ + residual-P) followed the sequence of decreasing initial TP in the order of: BBR > MIT > VES > LUE, and was characterized by low contents of P_o compared to P_i in all soils (Figure 2). Elevated P_o portions in the O-layer were observed at the P-poor site and at the medium site MIT (Figure 3c.3), which indicates the presence of P associated with very stable organic compounds. However, the ratio of stable P to TP in the O-layer was comparable at all sites with around 30 % of TP at the P-rich to medium sites, and slightly lower at the P-poor site with 21 %. The depth distribution of this relation showed a slight decrease with depth in the P-rich soil, but an increase in the medium P-supplied soils and particularly increased in the P-poor soil. Here, almost 50 % of the

P in the subsoil is stable bound and not available for plants, which is in line with the assumption of decreasing availability of mineral-bound P from the P-rich site to the P-poor site.

The contents of stable P in PFPs and soil matrix were on the same level. Only for the P-rich soil, slightly higher values in PFPs were observed.

The distribution of P fractions revealed for the P-rich site comparable labile P concentrations as for the other sites, but a clearly lower ratio of labile P to TP, and much more P_o than P_i in this pool. Considering the assumption of negligible P_o mineralization [29], it can be assumed that the P-rich soil has enough labile P available from mineral sources and has no need for increased P_o mineralization. P_o therefore remains stable, but P_i is consumed and reduced thus relative to the amount of labile P_o. This may also explain the higher ratio of moderately labile P to TP at the P-rich site compared to the other sites. In this pool, P is moderately bound to inorganic and organic compound, and can be acquired if necessary.

3.4. Factors Affecting the Distribution of P Fractions in the Soil Profiles

The statistical test of correlations between parameters included in the first step all data from the four study sites and covered a wide range of values of chemical soil properties. The results confirmed a close relationship between labile P and the contents of soil organic C and total N, which is in accordance with previous studies [4,30,33,34] (for the correlation matrix and scatter plots see Table S1 of the supplementary material). This correlation is stronger for labile P_o (N: $R^2 = 0.98$, C: $R^2 = 0.97$) than for labile P_i (N: $R^2 = 0.94$, C: $R^2 = 0.93$). Also in agreement with previous studies [30,34], the easily available resin-P was strongly correlated with $NaHCO_3$-P_i ($R^2 = 0.79$) and $NaHCO_3$-P_o ($R^2 = 0.75$) indicating the contribution of inorganic as well as organic labile P fractions to the resin-P_i pool.

Moderately labile P was strongly related to sesquioxides (Al, Fe, Mn). In this fraction, P_i showed close correlations to Al_{ox}, Al_{di}, Mn_{ox}, Mn_{di}, and had slightly lower correlations with Fe_{ox} and Fe_{di}. Moderately labile P_i was also positively correlated with pH, but negatively correlated to the C/N ratio. By contrast, the strongest correlation of P_o in this fraction was found with Fe-oxyhydroxides, and to a lesser extent with Al- and Mn-oxyhydroxides. This strong contribution of Fe compounds to P_o adsorption, but the adsorption of mainly P_i to Al-oxyhydroxides, is in line with the study of Prietzel et al. [4] for acidic forest soils.

For stable P, a close correlation of P_i with Fe, Al, and Mn compounds was identified, whereas P_o was strongly related to C and N. This shows that P_o is occluded in stable organic compounds or particulate organic matter.

To indicate the dominant processes influencing the distribution of P_i and P_o in soils, the ratio C to P_o was calculated (Table 3). Low C/P_o values (<200) indicate the net mineralization of P, whereas high ratios (>300) are indicative for immobilization as result of microbial P_i utilization during decomposition of organic matter [35,36]. When comparing the test sites with considerable differences in soil P contents, the C to P_o ratio was highest at the P-poor site and lowest at the P-rich site. The strong reduction of C/P_o with depth at all sites can be attributed to the sharp decrease of organic C compared to P. In the O-layers, C/P_o was > 300 at the P-poor and medium sites, which indicates P immobilization (incorporation into biomass). C/P_o-values between 200 and 300, which occurred in the O-layer of the P-rich site, indicate a stable rate of cycling between P_i and P_o [35]. These findings support the hypothesis of intense P recycling at P-poor sites [6] leading to P depletion and high C to P_o ratios. The C/P_o ratio also differed considerably between PFP and soil matrix samples at all site with mostly clearly higher values in PFPs (Table 3). For the P-rich site, the difference between PFP and matrix was more pronounced in the topsoil and diminished with soil depth. In the P-poor soil, the contrasting behavior was observed: the difference increased with depth indicating stronger P_o depletion in the subsoil PFPs.

Table 3. Ratios of organic C to sum of organic P (Po) at four beech forest study sites in Germany for PFP (preferential flow pathways), Ma (soil matrix), and RC (root channel-PFP) samples.

C/P_o	LUE		VES		MIT		BBR	
Horizon	PFP	Ma	PFP	Ma	PFP	Ma	PFP	Ma
O	1143		840		910		223	
A	512	494	155	174	593	371	139	97
B	356	274	89	66	101	93	58	51
C	204	88		170		59	42	43

Horizon: O—organic layer; A—topsoil horizon; B—subsoil horizon; C—lower subsoil horizon containing weathered parent material. Study sites in the sequence P-poor to P-rich are: LUE—Lüss, Lower Saxony; VES—Vessertal, Thuringian Forest; MIT—Mitterfels, Bavarian Forest; and BBR—Bad Brückenau, Bavarian Rhön Mountains.

For all chemical parameters, the differences between PFP and the soil matrix were tested statistically (for test results see Table S2 of the supplementary material). The most statistical differences were found for the P-rich site. Significantly higher values in PFPs occurred here for C, labile P_o, and C to TP ratio with $p < 0.01$, and for N, stable P_o and stable P_i with $p < 0.05$. For the other sites, almost no statistically significant differences of P fractions and other parameters were identified. At the P-poor site, the ratio C to TP and the C content were statistically higher in the PFPs ($p < 0.05$). At the medium sites, the content of metals differed substantially, with high contents of Fe_{ox}, Mn_{ox}, Al_{di} and Mn_{di} in PFPs, but lower values of Fe_{di} compared to matrix samples. However, these statistical tests included all data per site and have to be considered carefully due to the hierarchical structure of the data. For the same tests at the horizon level per site, the data sets were much smaller (three samples for PFP and matrix respectively) and revealed no significant differences between PFP and matrix for all parameters (Table S3). In previous studies, Hagedorn and Bundt [18] suggested that PFPs in forest soils can be stable for decades, resulting in biological and chemical gradients between PFPs and soil matrix [15]. Bogner et al. [12] confirmed this hypothesis for PFPs along root channels and found an enrichment of C, N, Ca, Mg, and Fe in PFPs compared to the soil matrix. Such an accumulation in PFPs was also documented for different P fractions in flow pathways along stones by Backnäs et al. [19]. Our study sites are characterized by very high stone contents, resulting in very heterogeneous conditions for water and solute flow. PFPs were therefore located mostly along stone surfaces with low potential for P sorption [19], and preferential flow occurred less in root channels or other biopores where P accumulation would be expected to be more prevalent. Single biopores could be identified and sampled separately, which contained decayed root material. Indications were found in these biopores of an accumulation of labile organically bound P (P_o), but could not be statistically confirmed. With respect to the distribution of subsurface flow pathways at hillslopes (midrange mountain sites), it has been shown that a predominant surface-near lateral flow does not support chemical gradients between PFPs and soil matrix. Thus, non-significant results in the statistical test are a result of specific site conditions, spatial soil heterogeneity, and analytical uncertainties with regard to very low P concentrations at the P-poor site.

Nevertheless, several factors affecting the distribution of P within the tested soils became apparent. The results showed that the labile $P_{i,o}$ pool and stable P_o were strongly related to C and N, whereas moderately labile $P_{i,o}$ as well as stable P_i correlated with metal-oxides and -oxyhydroxides (Al, Fe, Mn). At the medium to P-rich sites, where higher contents of sesquioxide in topsoil matrix compared to PFPs were observed, moderately labile P_i and P_o also was clearly higher in soil matrix. The higher P_o contents of this moderately labile pool in subsoil PFPs could be related to Fe_{ox} accumulated in the flow pathways.

4. Conclusions

As expected, considerable differences were found for the four study sites in both the P contents and the distribution of P fractions within the soil profiles. Despite these large differences, the labile P contents in the O-layers were in the same range of values at all sites. At the site with very low supply of P from the mineral phase, the predominant portion of labile P was available in the O-layer, but was greatly reduced in the mineral soil. This indicates that for this site intense P recycling happens in the O-layer, which is also supported by a high C to P_o ratio in this layer. By contrast, labile P was available throughout the whole profile at the P-rich site. Here, the supply of plant-available P seems to be sufficient enough that there is no need for tight P recycling, and may therefore explain the very low share of labile P to TP, when compared to the P-poor site. The contents of labile and moderately labile P clearly differed in PFPs compared to soil matrix, and indicate that labile organically bound P (P_o) accumulates in PFPs with a high content of organic matter. However, these results could not be proved statistically at the tested sites. Therefore, more sites particularly for soils with a higher proportion of preferential flow through biopores should be studied to verify the indications.

Supplementary Materials: Table S1:
Correlation matrix with Spearman's rho coefficients to test the correlation between chemical soil parameters (dataset: $n = 123$, includes data of four forest soils in Germany), Table S2: Results (p-values) of the non-parametric Wilcoxon rank-sum test for differences between PFPs and soil matrix samples (pair-wise test for all data per site), Table S3. Results (p-values) of the non-parametric Wilcoxon rank-sum test for differences between PFPs and soil matrix samples (pair-wise test at horizon level per sites).

Acknowledgments: We thank Gisela Ciesielski and Manuela Unger for technical support in the laboratory. Furthermore, our thanks go to Maximilian Kirsten, Jianhong Liang, Peer Appelfelder, and Nils Heide for their assistance in the field and laboratory works. Finally, we thank Dan Hawtree for helpful comments and linguistic help. This work was funded within the framework of Priority Program SPP 1685 "Ecosystem Nutrition: Forest Strategies for Limited Phosphorus Resources" by the German Research Foundation (DFG), grant No. JU 2940/1-1. We further acknowledge support by the Open Access Publication Funds of the TU Dresden.

Author Contributions: D.J. designed and performed the field experiments, and did the data analyses; all authors wrote the manuscript.

Conflicts of Interest: The authors declare no conflict of interest.

References

1. Bol, R.; Julich, D.; Brödlin, D.; Siemens, J.; Kaiser, K.; Dippold, M.A.; Spielvogel, S.; Zilla, T.; Mewes, D.; von Blanckenburg, F.; et al. Dissolved and colloidal phosphorus fluxes in forest ecosystems—An almost blind spot in ecosystem research. *J. Plant Nutr. Soil Sci.* **2016**, *179*, 425–438. [CrossRef]
2. Newman, R.H.; Tate, K.R. Soil phosphorus characterisation by ^{31}p nuclear magnetic resonance. *Commun. Soil Sci. Plant Anal.* **1980**, *11*, 835–842. [CrossRef]
3. Cade-Menun, B.; Liu, C.W. Solution phosphorus-31 nuclear magnetic resonance spectroscopy of soils from 2005 to 2013: A review of sample preparation and experimental parameters. *Soil Sci. Soc. Am. J.* **2014**, *78*, 19. [CrossRef]
4. Prietzel, J.; Klysubun, W.; Werner, F. Speciation of phosphorus in temperate zone forest soils as assessed by combined wet-chemical fractionation and XANES spectroscopy. *J. Plant Nutr. Soil Sci.* **2016**, *179*, 168–185. [CrossRef]
5. Walker, T.W.; Syers, J.K. The fate of phosphorus during pedogenesis. *Geoderma* **1976**, *15*, 1–19. [CrossRef]
6. Lang, F.; Bauhus, J.; Frossard, E.; George, E.; Kaiser, K.; Kaupenjohann, M.; Krüger, J.; Matzner, E.; Polle, A.; Prietzel, J.; et al. Phosphorus in forest ecosystems: New insights from an ecosystem nutrition perspective. *J. Plant Nutr. Soil Sci.* **2016**, *179*, 129–135. [CrossRef]
7. Prietzel, J.; Kaiser, K. De-eutrophication of a nitrogen-saturated Scots pine forest by prescribed litter-raking. *J. Plant Nutr. Soil Sci.* **2005**, *168*, 461–471. [CrossRef]
8. Neary, D.G.; Ice, G.G.; Jackson, C.R. Linkages between forest soils and water quality and quantity. *For. Ecol. Manag.* **2009**, *258*, 2269–2281. [CrossRef]
9. Van Miegrot, H.; Johnson, D.W. Feedbacks and synergism among biogeochemistry, basic ecology, and forest soil science. *Forest Ecol. Manag.* **2009**, *258*, 2214–2223. [CrossRef]
10. Santos, R.M.B.; Sanches Fernandes, L.F.; Pereira, M.G.; Cortes, R.M.V.; Pacheco, F.A.L. A framework model for investigating the export of phosphorus to surface waters in forested watersheds: Implications to management. *Sci. Total Environ.* **2015**, *536*, 295–305. [CrossRef] [PubMed]

11. Uchida, T.; Tromp-van Meerveld, I.; McDonnell, J.J. The role of lateral pipe flow in hillslope runoff response: An intercomparison of non-linear hillslope response. *J. Hydrol.* **2005**, *311*, 117–133. [CrossRef]
12. Bogner, C.; Borken, W.; Huwe, B. Impact of preferential flow on soil chemistry of a podzol. *Geoderma* **2012**, *175*, 37–46. [CrossRef]
13. Van der Heijden, G.; Legout, A.; Pollier, B.; Bréchet, C.; Ranger, J.; Dambrine, E. Tracing and modeling preferential flow in a forest soil—Potential impact on nutrient leaching. *Geoderma* **2013**, *195–196*, 12–22. [CrossRef]
14. Van Verseveld, W.J.; McDonnell, J.J.; Lajtha, K. The role of hillslope hydrology in controlling nutrient loss. *J. Hydrol.* **2009**, *367*, 177–187. [CrossRef]
15. Bundt, M.; Jäggi, M.; Blaser, P.; Siegwolf, R.; Hagedorn, F. Carbon and nitrogen dynamics in preferential flow paths and matrix of a forest soil. *Soil Sci. Soc. Am. J.* **2001**, *65*, 1529–1538. [CrossRef]
16. Flury, M.; Flühler, H. Brilliant Blue FCF as a dye tracer for solute transport studies—A toxicological overview. *J. Environ. Qual.* **1994**, *23*, 1108–1112. [CrossRef]
17. Weiler, M.; Flühler, H. Inferring flow types from dye patterns in macroporous soils. *Geoderma* **2004**, *120*, 137–153. [CrossRef]
18. Hagedorn, F.; Bundt, M. The age of preferential flow paths. *Geoderma* **2002**, *108*, 119–132. [CrossRef]
19. Backnäs, S.; Laine-Kaulio, H.; Kløve, B. Phosphorus forms and related soil chemistry in preferential flowpaths and the soil matrix of a forested podzolic till soil profile. *Geoderma* **2012**, *189*, 50–64. [CrossRef]
20. Jarvis, N.J. A review of non-equilibrium water flow and solute transport in soil macropores: principles, controlling factors and consequences for water quality. *Eur. J. Soil Sci.* **2007**, *58*, 523–546. [CrossRef]
21. Jensen, M.B. Subsurface transport of phosphorus in relation to its mobilization and immobilization in structured soil (steady state flow experiments). *Acta Agric. Scand. Sect. B-Soil Plant Sci.* **1998**, *48*, 11–17.
22. Hedley, M.J.; Stewart, J.W.B.; Chauhan, B.S. Changes in inorganic and organic soil phosphorus fractions induced by cultivation practices and by laboratory incubations. *Soil Sci. Soc. Am. J.* **1982**, *46*, 970. [CrossRef]
23. Tiessen, H.; Moir, J.O. Characterization of Available P by Sequential Extraction. In *Soil Sampling and Methods of Analysis*; Carter, M.R., Gregorich, E.G., Eds.; Canadian Society of Soil Science: Boca Raton, FL, USA, 2008; Chapter 25; pp. 293–306.
24. Sibbesen, E. A simple ion-exchange resin procedure for extracting plant-available elements from soil. *Plant Soil* **1977**, *46*, 665–669. [CrossRef]
25. Murphy, J.; Riley, J.P. A modified single solution method for the determination of phosphate in natural waters. *Anal. Chim. Acta* **1962**, *27*, 31–36. [CrossRef]
26. O'Halloran, I.P.; Cade-Menun, B.J. Total and Organic Phosphorus. In *Soil Sampling and Methods of Analysis*; Carter, M.R., Gregorich, E.G., Eds.; Canadian Society of Soil Science: Boca Raton, FL, USA, 2008; pp. 265–291.
27. Crews, T.E.; Brookes, P.C. Changes in soil phosphorus forms through time in perennial versus annual agroecosystems. *Agric. Ecosyst. Environ.* **2014**, *184*, 168–181. [CrossRef]
28. Negassa, W.; Leinweber, P. How does the Hedley sequential phosphorus fractionation reflect impacts of land use and management on soil phosphorus: A review. *J. Plant Nutr. Soil Sci.* **2009**, *172*, 305–325. [CrossRef]
29. Bünemann, E.K.; Augstburger, S.; Frossard, E. Dominance of either physicochemical or biological phosphorus cycling processes in temperate forest soils of contrasting phosphate availability. *Soil Biol. Biochem.* **2016**, *101*, 85–95. [CrossRef]
30. Tiessen, H.; Stewart, J.W.B.; Cole, C.V. Pathways of phosphorus transformations in soils of differing pedogenesis. *Soil Sci. Soc. Am. J.* **1984**, *48*, 853. [CrossRef]
31. Renneson, M.; Dufey, J.; Legrain, X.; Genot, V.; Bock, L.; Colinet, G. Relationships between the P status of surface and deep horizons of agricultural soils under various cropping systems and for different soil types: A case study in Belgium. *Soil Use Manag.* **2013**, *29*, 103–113. [CrossRef]
32. Hashimoto, Y.; Watanabe, Y. Combined applications of chemical fractionation, solution ^{31}P-NMR and P K-edge XANES to determine phosphorus speciation in soils formed on serpentine landscapes. *Geoderma* **2014**, *230*, 143–150. [CrossRef]
33. Frizano, J.; Johnson, A.H.; Vann, D.R.; Scatena, F.N. Soil phosphorus fractionation during forest development on landslide scars in the Luquillo Mountains, Puerto Rico. *Biotropica* **2002**, *34*, 17–26. [CrossRef]
34. Thomas, S.M.; Johnson, A.H.; Frizano, J.; Vann, D.R.; Zarin, D.J.; Joshi, A. Phosphorus fractions in montane forest soils of the Cordillera de Piuchué, Chile: Biogeochemical implications. *Plant Soil* **1999**, *211*, 139–148. [CrossRef]

35. McDowell, R.W.; Stewart, I. The phosphorus composition of contrasting soils in pastoral, native and forest management in Otago, New Zealand: Sequential extraction and ^{31}P NMR. *Geoderma* **2006**, *130*, 176–189. [CrossRef]
36. Stutter, M.I.; Shand, C.A.; George, T.S.; Blackwell, M.S.A.; Dixon, L.; Bol, R.; MacKay, R.L.; Richardson, A.E.; Condron, L.M.; Haygarth, P.M. Land use and soil factors affecting accumulation of phosphorus species in temperate soils. *Geoderma* **2015**, *257*, 29–39. [CrossRef]

12

Using REDD+ Policy to Facilitate Climate Adaptation at the Local Level: Synergies and Challenges in Vietnam

Pamela McElwee [1,*], Van Hai Thi Nguyen [2], Dung Viet Nguyen [2], Nghi Huu Tran [3], Hue Van Thi Le [4], Tuyen Phuong Nghiem [4] and Huong Dieu Thi Vu [4]

1 Department of Human Ecology, School of Environmental and Biological Sciences, Rutgers, The State University of New Jersey, New Brunswick, NJ 08901, USA
2 People and Nature Reconciliation (PanNature), Hanoi 10000, Vietnam; van@nature.org.vn (V.H.T.N.); dungnv@nature.org.vn (D.V.N.)
3 Tropenbos International Vietnam (TBI), Hue 530000, Vietnam; nghi@tropenbos.vn
4 Centre for Natural Resources and Environmental Studies (CRES), Vietnam National University, Hanoi 10000, Vietnam; thivanhue@gmail.com (H.V.T.L.); tuyennghiem_cres@yahoo.com (T.P.N.); huongvudieu@yahoo.com (H.D.T.V.)
* Correspondence: pamela.mcelwee@rutgers.edu

Abstract: Attention has recently been paid to how REDD+ mitigation policies are integrated into other sectoral policies, particularly those dealing with climate adaptation at the national level. But there is less understanding of how subnational policy and local projects are able to incorporate attention to adaptation; therefore, we use a case study in Vietnam to discuss how REDD+ projects and policies address both concerns of mitigation and adaptation together at subnational levels. Through stakeholder interviews, focus groups, and household surveys in three provinces of Vietnam with REDD+ activities, our research sought to understand if REDD+ policies and projects on the ground acknowledge that climate change is likely to impact forests and forest users; if this knowledge is built into REDD+ policy and activities; how households in forested areas subject to REDD+ policy are vulnerable to climate change; and how REDD+ activities can help or hinder needed adaptations. Our findings indicate that there continues to be a lack of coordination between mitigation and adaptation policies in Vietnam, particularly with regard to REDD+. Policies for forest-based climate mitigation at the national and subnational level, as well as site-based projects, have paid little attention to the adaptation needs of local communities, many of whom are already suffering from noticeable weather changes in their localities, and there is insufficient discussion of how REDD+ activities could facilitate increased resilience. While there were some implicit and coincidental adaptation benefits of some REDD+ activities, most studied projects and policies did not explicitly target their activities to focus on adaptation or resilience, and in at least one case, negative livelihood impacts that have increased household vulnerability to climate change were documented. Key barriers to integration were identified, such as sectoral specialization; a lack of attention in REDD+ projects to livelihoods; and inadequate support for ecosystem-based adaptation.

Keywords: REDD+; household livelihoods; climate adaptation; vulnerability; forest policy; land

---

1. Introduction

There has been increasing concern in recent years for the need to link climate mitigation and adaptation policies together, particularly with regard to forests [1,2]. Forest policies to respond to climate change often involve either mitigation actions, such as biological carbon sequestration, or adaptation actions, such as promoting resilience of ecosystems, but rarely are both considered together. Although the Intergovernmental Panel on Climate Change (IPCC) since 2007 has called for combined approaches, little has happened to facilitate mitigation and adaptation policies within the UN Framework Convention on Climate Change (UNFCCC), and individual country policies vary considerably in how much they integrate both approaches [3,4]. The IPCC defines adaptation as "adjustments in practices, processes, or structures" that can "moderate or offset the potential for damage or take advantage of opportunities created by a given change in climate" [4], and thus can encompass a wide range of potential policies for both forests and forest-using peoples.

Reducing Emissions from Deforestation and Degradation (REDD+) is the most well-known forest mitigation strategy to lower land-use generated greenhouse gas (GHG) emissions; the fundamental premise of REDD+ is that if households and governments are given payments and other types of rewards that equal or exceed what is earned from deforestation, then forests will be better protected, carbon emissions will be reduced, and these areas can serve as greater sinks for future GHG mitigation [5]. The rollout and implementation of REDD+ policies in various countries over the past decade has received much scholarly attention [6–10], although these has been less attention to how adaptation policies have been integrated into REDD+. Most existing studies of REDD+ and adaptation have been at the national level and have assessed how different ministries and sectors have coordinated through REDD+ projects to incorporate adaptation and mitigation concerns [11–17], while a smaller number of studies have looked at how voluntary forest carbon projects include adaptation measures [18–20]. However, we know less about how sub-national policymakers are treating adaptation in the development of forest carbon policies, and how households in areas with REDD+ projects, particularly those already vulnerable to climate impacts, are affected in terms of their adaptation options by REDD+ activities.

This article uses a case study in Vietnam to explore how REDD+ projects and policies link both concerns of mitigation and adaptation together at subnational levels in both policy and household impacts, and if not, what the barriers to doing so are. We build off a previous assessment for Vietnam that determined there was potential to address adaptation in REDD+ at national levels, as many stakeholders recognized the importance of integration of this sector [13]. Our project follows up at local levels to see if these potentials have been realized by exploring two main questions:

(1) To what degree do REDD+ policies and projects on the ground at subnational levels acknowledge that climate change is likely to impact forests and forest users, and how is this built into REDD+ policy and activities?
(2) How are households in forested areas subject to REDD+ policy also vulnerable to climate change, and how can REDD+ activities help or hinder needed adaptations?

Overall, we find that there continues to be a lack of coordination between mitigation and adaptation policies for forests in Vietnam, particularly with regard to REDD+. Policies at both the national and provincial level, and site-based projects, have paid little attention to the adaptation needs of local communities, and how REDD+ activities could facilitate increased resilience in livelihoods. While there were some implicit and coincidental adaptation benefits of REDD+ activities, most of the projects and local policies that we examine did not explicitly consider their activities to touch on adaptation or resilience, and in at least one case, negative livelihood impacts that have increased household vulnerability to climate change were documented. We conclude the article with insights into the barriers that continue to exist that keep REDD+ and adaptation from being considered more holistically.

2. Background: Intersections between REDD+ and Climate Adaptation in International and National Policy and Practice

REDD+ policies have been discussed as part of the UN Framework Convention on Climate Change (UNFCCC) since 2005, and has been on the agenda of all subsequent Conference of the Party

(COP) meetings, as technical subcommittees have worked out elements of how REDD+ might be implemented [21]. COP 19 in Warsaw in 2013 adopted a number of important technical decisions on REDD+, including on results-based finance, coordination of support, forest monitoring systems, safeguards, reference levels, measuring, reporting and verifying (MRV), and addressing the drivers of deforestation [22]. Pilot programs to prepare countries for "REDD+ readiness" have been underway in many nations, funded by bilateral and multilateral donors, and involving new institutions like the United Nations' UN-REDD+ program and the Forest Carbon Partnership Facility (FCPF) of the World Bank [23].

Very few of the decisions taken at various COPs have explicitly linked REDD+ and adaptation approaches. For example, the 2015 Paris Agreement that was negotiated at COP 21 entered into force in November 2016, following ratification by at least 55 parties accounting for 55% of total global emissions (Vietnam ratified the agreement in November 2016). REDD+ is explicitly mentioned in article 5 of the agreement, but it does not state how countries are to implement forest sinks, or how results-based payments will be made, and leaves such decisions up to individual countries. These actions will be clarified by states in their submissions of "nationally determined contributions" (NDCs) that the Paris Agreement regularly requires. Article 7 notes that adaptation actions are similarly to be decided at the country level, through National Adaptation Plans (NAPs) also to be regularly filed with the UNFCCC, and such actions should *"follow a country-driven, gender responsive, participatory and fully transparent approach, taking into consideration vulnerable groups, communities and ecosystems, and should be based on and guided by the best available science and, as appropriate, traditional knowledge, knowledge of indigenous peoples and local knowledge systems, with a view to integrating adaptation into relevant socioeconomic and environmental policies and actions, where appropriate"* [24].

Within the international literature, there is increasing reference to how forest mitigation and adaptation activities might intersect [1]. In some countries, the same actors are in charge of both REDD+ and adaptation plans [11]. (However, this is not the case in Vietnam where REDD+ activities and actors are centered in the Ministry of Agriculture and Rural Development (MARD), and adaptation activities are mainly driven by the Ministry of Natural Resources and Environment (MONRE)). Some privately financed forest carbon projects, such as those certified by the Climate, Community and Biodiversity Alliance (CCBA) standard, have required attention to social adaptation as part of their certifications [19]. Other REDD+ policies, at national and local levels, have had little to say about adaptation. Some authors have assumed that attention to safeguards under REDD+ (a requirement from the Warsaw COP in 2013) demonstrates a sufficient approach to livelihoods, which will result in adaptation benefits [11], while other authors are more skeptical that safeguards and local participation (which might lead to adaptation benefits) are actually happening in REDD+ projects on the ground [25,26]. Of the existing reports on combined mitigation and adaptation policy, very few assess the complicated relationships between livelihoods, climate impacts, and REDD+ through on-the-ground surveys or interviews, leaving most discussions at higher levels of policy. One of the few papers to tackle this problem did find that while many forest mitigation projects in a case study in Belize did not have specific adaptation actions embedded in them, nearly half did have adaptation-related outcomes, such as improved livelihoods, and 90% of adaptation-explicit projects also reported mitigation-related outcomes, such as enhanced carbon stocks [18].

This topic deserves further study, as a lack of integration of REDD+ and adaptation into subnational and local policy could have serious consequences. Several authors have noted that activities taken to increase mitigation of land-based GHGs under REDD+ might have unintended impacts on the livelihoods of forest-dependent people [27,28], and therefore also have impacts on the ability of these households to adapt to climate change. A hypothetical example might be a forest plantation created for carbon sequestration that reduced water availability for nearby households, who then might become more vulnerable to climate-change induced droughts [18]. Overall, communities' and individuals' ability to cope with many forecasted climate changes, like localized changes in rainfall timing and amounts, among other impacts, are likely to be strongly conditioned on their ability to access and mobilize resources like land, trees, water, fish and other means of livelihood [29,30]. If access rights to forest change under REDD+ projects, this could render communities and households more vulnerable to the effects of climate change at local levels if traditional assets like forests that are used for adaptation responses (e.g., as a source of quick cash or as food) become inaccessible [25,31]. Alternatively, REDD+ could potentially strengthen local access rights to forests through increased financing to ensure their protection from outside deforestation pressures, thus possibly increasing communities' resilience to climate change [32,33]. The aforementioned assessment of forest carbon mitigation projects in Belize show the positive benefits of attention to adaptation which resulted in diversified livelihoods, strengthened land tenure, and more robust local forest management [19].

Other reasons to combine mitigation and adaptation approaches include the need to maximize limited climate financing [11,18,34]; to harmonize sectoral policy and avoid institutional duplication and overlap in approaches [35]; and to potentially "climate-proof" mitigation and other development projects [36,37]. For example, the planting of biofuels (a mitigation policy) might be impacted by climate-induced changes in the future and would need to be planned for, otherwise projects' contributions to overall mitigation might decrease without adaptation measures [38].

However, numerous challenges face any attempt to integrate adaptation and mitigation together in policy and projects, at both national and local levels. For example, examination of existing REDD+ development at national scales has revealed major challenges in coordination across sectors already, with both duplications and gaps in how REDD+ works with other development policies [39–41]. Additionally, there are often mismatches between time scales for projects, with mitigation usually being more immediate and adaptation more longer term [19]. In cases where adaptation has explicitly been linked into a REDD+ mitigation project, there are often difficulties in financing and extended time spans for projects [42]. Further, future climate change impacts on both households and forests are variable and often depend on localized context, making generalizations about adaptation difficult to put into policy [43–46]. For example, one study that assessed climate change forecasts for the provisioning of ecosystem services from forests in Finland found a series of complicated impacts, some positive, some negative, with no clear direction for policy actions to increase adaptation [44].

On the positive side, one potentially promising new approach has been the concept of ecosystem-based adaptation (EBA), promoted for "the use of natural capital by people to adapt to climate change impacts, which can also have multiple co-benefits for mitigation, protection of livelihoods and poverty alleviation" [47]. EBA is often presented as a win-win for both mitigation and adaptation [48], and activities under this label include such activities as restoration or protection of coastal mangroves, which sequester carbon as well as helping coastal communities withstand the impacts of storms [49]. A recent review of EBA noted that forests can support human climate adaptation through: (1) provision of goods to vulnerable communities; (2) regulation of microclimates, especially for agriculture; (3) regulation of soil and water to buffer climate impacts; (4) coastal forest protection against storms; and (5) urban trees that regulate temperature and water [50,51]. However, there have been few assessments of the degree to which EBA approaches are integrated into either national or subnational REDD+ or other policies [52,53], and country experience shows that many challenges remain in operationalizing EBA [54].

Vietnam is a particularly appropriate country in which to look at both climate adaptation and forest-based mitigation, as it has been an early adopter of REDD+ activities, through the UN-REDD+ programme and the FCPF, as well as a number of voluntary projects. Slightly more than half of Vietnam's 2010 greenhouse gas emissions were from the agriculture, forestry, or land use sectors, indicating a high priority for investments in emissions reductions in this category [55]. Vietnam has also been identified as one of the top fifteen countries in the world vulnerable to natural hazards like drought and storms in terms of the number of people and scale of exposure [56], and forecasted temperature increases will exacerbate this condition to levels previously not experienced. The forecasted climate impacts to 2100 will likely be an increase in rainfall in wet seasons and decrease in dry of around 10% or more, increased intensity and frequency of storms and floods, and a likely sea level rise of at minimum one meter [57]. In order to minimize climate change impacts on Vietnam, adaptation projects to reduce vulnerability have been increasing in scale and importance in recent years, including in water management, health care provisioning, and land use planning, such as resettlement away from vulnerable zones [58–60]. Many of these actions have been combined with disaster-risk reduction strategies and aimed at increasing resilience of households to a multitude of climate related effects [61–64]. However, there have been relatively fewer adaptation actions directed at forest-dwelling and using communities, which tend to be located in mountainous areas of the country, while more adaptation attention and financing focuses on coastal and delta areas [57,65].

3. Materials and Methods

3.1. Fieldsites

This study was carried out in 3 provinces of Dien Bien, Kon Tum, and Kien Giang (see Figure 1a). These sites were selected as representative of the North, Center and South of the country and were sites in which preliminary REDD+ readiness projects were on-going, all sponsored by different donors, which gave us a range of project types to explore. In each study province, local communes were selected for in-depth study based on where existing projects for REDD+ or other forestry-focused projects have been operating (see Figure 1b–d).

Dien Bien is a mountainous area located in the Northwest along the border with Laos. The total natural area of this province is 956,290 ha, with 41.1% of the total area classified as forest. The total population is 547,785, and 47% of the province's households were considered under the government poverty line in 2016, the highest rate in the entire country. Livelihoods primarily consist of agricultural production, livestock husbandry, and forestry exploitation and development activities. Around 50,000 households, mostly consisting of ethnic minorities like Hmong, Thai, Dao, Kho Mu, and others, have participated in government payments for environmental services (PES) programs since 2011. A Japan International Cooperation Agency (JICA) REDD+ project was piloted in two districts from 2012 to 2013 and included activities such as raising awareness of REDD+, FPIC (Free, Prior and Informed Consent) agreements, agro-forestry extension, and development of the province's overall REDD+ policy.

Kon Tum is a mountainous area located in the Central Highlands, with a total forest area of 603,814 ha, 58.5% of the province. There is a large and vulnerable ethnic minority population (54% of total), dependent on both cash crop and subsistence agriculture, and 26% of the province's households were considered under the government poverty line in 2016. The major crops grown in this area are primarily cassava for subsistence, with only a little rice, corn and rubber for supplemental income. Households located in areas of the province with basalt soils have been able to transition into cash crop agriculture, particularly rubber but also coffee, tea, cashew and litsea in the past 10 years, but these activities have been faulted for deforestation and forest degradation. A REDD+ project has been piloted in Kon Plong district of Kon Tum in 11 villages of Hieu commune by Fauna and Flora International (FFI), Kon Tum Provincial People Committee's, Kon Tum's Agriculture & Rural Development Department (DARD), and PanNature (an NGO) in the period 2011–2014. Hieu commune has around 20,500 ha that is nearly 90% forest, and 660 households, mostly of the M'nam ethnicity, have participated in project activities, such as building capacity for local authorities and communities in order to directly implement REDD+ activities (including setting up a new local community-based institution which would have locally derived and implemented rules), with the hopes of providing financial benefits to forest-dependent local and indigenous people from selling carbon credits in the voluntary carbon market.

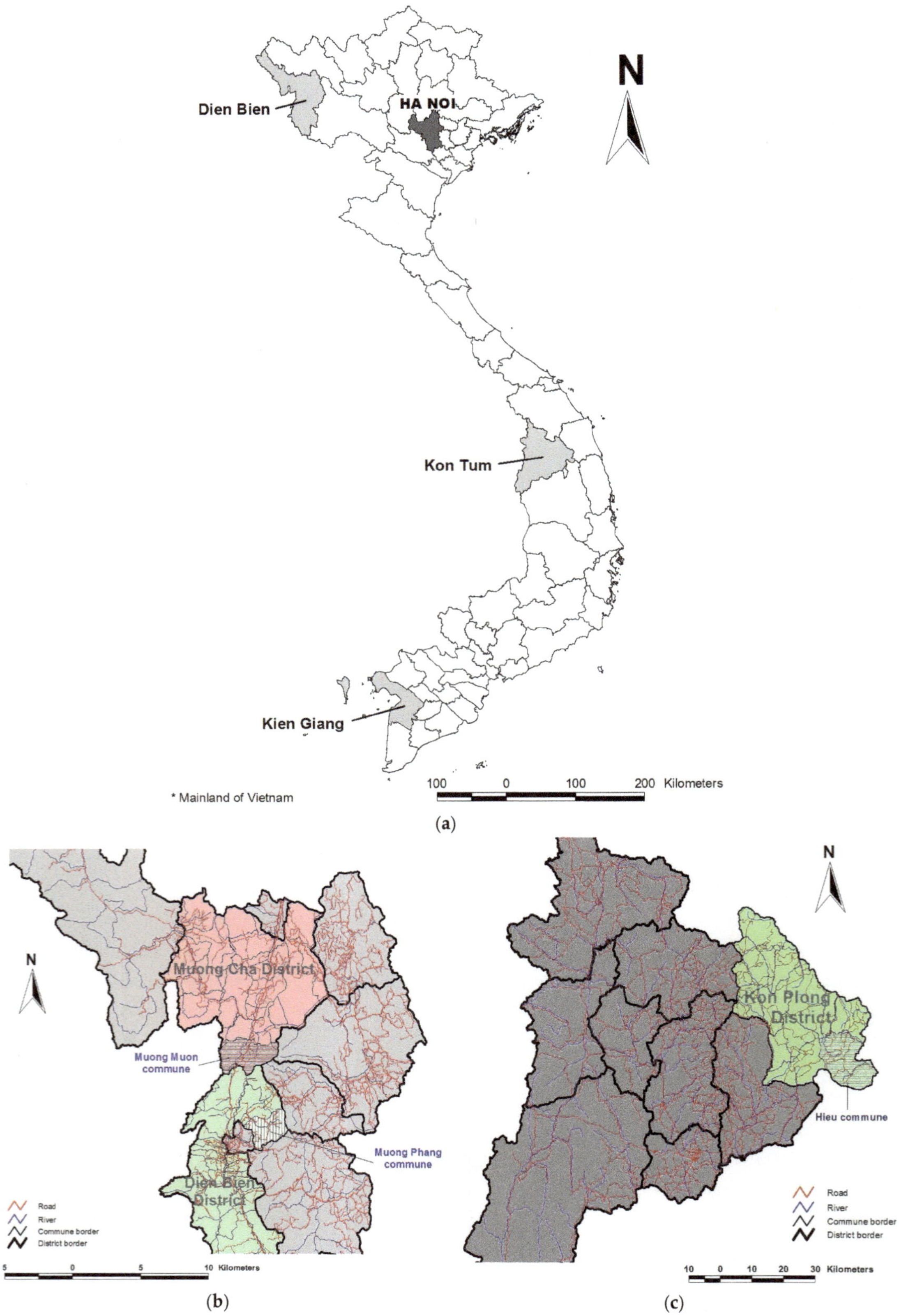

Figure 1. *Cont.*

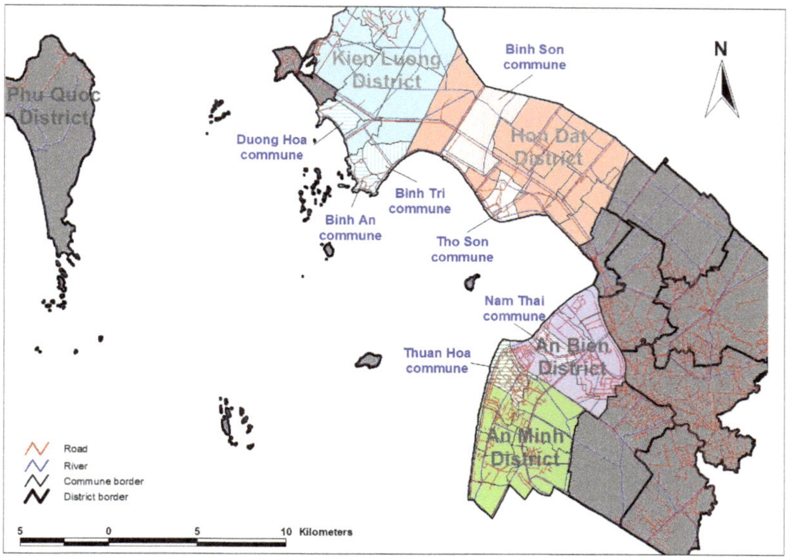

(**d**)

Figure 1. (**a**) Map of Vietnam showing provinces where research took place; (**b**) Dien Bien Province; (**c**) Kon Tum Province; (**d**) Kien Giang Province. Communes labelled in blue were sites of REDD+ projects and our research.

Kien Giang is located in the Mekong Delta area in the South. The total area of the province is 634,853 ha, in which forestry land is only 13.6% of the total area, mostly coastal mangroves. The total population is about 1.7 million people, of which ethnic minorities (primarily Khmer) are about 16% of the total, and the provincial poverty rate is 9.7%. Two-thirds of the population live in rural areas and have activities related to agriculture and aquaculture near mangrove forests. In 2008, with support from AUSAID and GTZ, Kien Giang began to implement a Conservation and Development of the Kien Giang Biosphere Reserve project. Within this project, activities have included surveys of mangrove species diversity; mangrove and coastline mapping via remote sensing and satellite image interpretation; studying biomass, carbon stocks, and biological diversity, including an assessment of forest regeneration needs and potential; and developing a REDD+ feasibility study. Local people have been involved in training courses and awareness raising activities held by the project and the local government agencies on the topic of PES and reducing emissions due to deforestation and agriculture production in the province.

3.2. Data Collection

Within each province, we carried out a mixed methods approach to collecting social and environmental data, making several field trips to each province throughout 2012, 2013 and 2014, spending from two to three weeks collecting data in each site. In each province a standard questionnaire was administered to a sample of households (selected at random from a village census by choosing every *k*th household on list) proportionately spread across villages to generate 100 households per province, for a total sample size of 300 households. Households are usually the main units making land-use and livelihood decisions, and this project has used the standard Vietnamese government definition of households. The survey asked questions about livelihoods, income, assets, participation in forest projects, climate vulnerability and adaptation measures. The data from the surveys was entered into SPSS for analysis.

In addition, focus groups were carried out with small numbers of local residents in each study area to help us to build histories of resource use, determine how residents learned from one another and set up institutions for managing forests and reducing climate risk and vulnerability, how these institutions functioned in different situations, and how such institutions interacted with official forest policies like

REDD+. In each research site, four focus groups (of approximately ten people per group) were run. A general focus group consisting of representatives of local civil society and government groups (such as the women's union, youth union, veteran's union, Communist Party, etc.) were asked to discuss the general issues of village such as main livelihoods, household economic status ranking, and seasonal calendars of the village, among other topics. A forest user focus group including families who were allocated forest land and who joined village forest patrolling groups discussed forest issues, changes of land and forest during the past ten to fifteen years, and participatory land-mapping exercises. For women's focus groups, representative women from different household types (poor, average, rich), and woman-headed households were selected to join for discussion. Finally, one focus group of officially designated "poor" households was run as well, and focused on risk-mapping, climate impacts on the poor, and other topics. In each case invitations were issued to attendees with the advisement of the village head and snowball sampling (e.g., asking invitees to bring along neighbors with related knowledge). We also targeted knowledgeable people in each community for lengthier, unstructured key informant interviews to collect life histories aimed at understanding social and climate vulnerability, as well as changes in resource-use patterns and access over time, among other topics.

Finally, we used stakeholder interviews with government officials and policymakers in each field site to gather information on the development of local forest policies. We asked how they were responsible for designing a forest policy incorporating social considerations; types of social data that policy makers use; and local input and participation to forest policy to their locality. We interviewed 15 policymakers at district and provincial levels in each fieldsite, and several national level stakeholders involved with REDD+ as well, for 60 policy interviews total.

4. Results

In the first part of this results section we discuss the national policy frameworks that have developed for REDD+ and for climate adaptation, and the degree to which they are integrated at subnational levels, while later we present data from a household survey undertaken in REDD+ project areas that aimed to understand forest-based livelihoods, climate vulnerabilities and adaptation, and the impacts of REDD+ projects on these.

4.1. National Policy Development for REDD+ and Adaptation

Vietnam developed a National Strategy on Climate Change in late 2011, which addresses both adaptation and mitigation, and REDD+ activities are a key element of the strategy [66]. It is estimated that through REDD+, 88.2 million tCO_2 emissions per year could be reduced [67]. Vietnam has recently submitted an intended "nationally determined contribution" (NDC) outlining plans to carry out the Paris Agreement, and the submission calls for an 8% reduction in greenhouse gas emissions by 2030 as compared to the business-as-usual scenario, which could be increased to a 25% reduction depending on international support. Specific activities in Vietnam's NDC related to forests and REDD+ include actions to:

- "Review and identify the areas and objects to apply sustainable forest management, afforestation and reforestation, biodiversity conservation, including special priority for regions with large forests that are important for forestry production and livelihoods of local communities of people;
- Develop and improve policies to promote sustainable forest management; mechanisms and policies to attract private sector investment for sustainable forest management, afforestation, reforestation, biodiversity conservation and livelihood development;
- Integrate and effectively use domestic and international resources for implementation of programmes and projects related to forest management and development, livelihoods and biodiversity conservation such as REDD+, the policy of payment for forest environmental services, etc.

- Strengthen and expand international cooperation for investment, technical assistance and capacity building, information and experience sharing on the sustainable forest management and development, biodiversity conservation and livelihood development" [68].

The development and governance structures for REDD+ in Vietnam have been reviewed by others, to which we refer readers for additional details [69–71]. We primarily focus on the adaptation linkages at national level in this paper. Most of the REDD+ activities in Vietnam have been carried out by various department of MARD, and have focused on traditional forestry activities like inventories and forest land use planning. However, one potential institutional linkage exists to facilitate adaptation into REDD+; a national steering network was set up in 2009 to coordinate REDD+ activities, and the Department of Meteorology and Climate Change of the Ministry of Natural Resources and Environment (MONRE) sits as a member the network, one of three non-MARD government departments (the other two are the Agro-economic Department of the Ministry of Planning and Investment and the Government Office, a prime-ministerial level coordinating office).

The steering network was charged with developing a National REDD+ Action Plan (NRAP), which was first completed in 2012 (and which is now under revision). The goals of the original 2012 NRAP were to (1) Build a national REDD+ Program and provincial REDD+ plans; (2) Enhance institutional capacity and coordination between ministries; (3) Raise awareness to stakeholders in forestry; (4) Improve technical capacity in reference levels, and monitoring; and (5) Develop benefit sharing systems and an information system on safeguards [72]. However, this initial NRAP made no mention of adaptation in any systematic way [13], although there are implied adaptation benefits from some of the REDD+ activities. For example, MARD is currently focused on a revision of forest criteria and classification for Vietnam to help clarify forest tenure agreements, which will likely have impacts on both REDD+ and adaptation projects in the future. The NRAP also affirms that by 2016–2020, there should be attention to "diversification and improvement of livelihoods of the forest owners and the people at large" [72], which would likely have positive adaptation benefits for households. Although previous researchers found limited policy support for adaptation in the forestry field at the national level in their analysis in 2014 [13], there has been some progress since then, particularly in research activities. For example, current climate adaptation focused activities in the forest sector include forest breeding of trees resistant to climate change; developing a national plan for adaptation to climate change in the forestry sector; and coordinating with the World Bank to conduct research on climate change adaptation in forestry [73].

Vietnam's NDC clearly states that adaptation will also be an important part of the country response. However, Vietnam has not yet filed a National Adaptation Plan (NAP) with the UNFCCC, though it intends to do so in the future. The NDC has a lack of specificity with regard to funding and priorities for adaptation beyond stressing its importance. Although the NDC states that costs of adaptation in Vietnam are estimated to exceed 3%–5% of GDP by 2030, how funding will be mobilized and for what specific activities is not yet clear, beyond a few priorities of: (1) responding pro-actively to disasters and improving climate monitoring; (2) ensuring social security, including EBA and community-based adaptation; and (3) responding to sea level rise and urban inundation [68].

While the ways adaptation may be carried out in the forest sector are not explicitly referenced in the NDC, the National Strategy on Climate Change refers more specifically to the ways that forestry can contribute to "preventing and coping with natural disasters, flash floods and landslides in mountain areas", and that policy will be needed to "improve quality of forests and afforestation, to turn bare lands and hills green, to effectively exploit different kinds of forest to secure and improve resistance against natural disasters, preventing desertification, land erosion and degradation; to enhance protection, management and development of mangrove forests and flooded ecosystems; to raise the forest coverage to 45% by 2020". There is also a stated goal to "preserve biodiversity, protect and develop ecosystems and species which can well resist climatic changes; to protect and preserve genes and species endangered by impacts of climate change". In addition, the National Strategy explicitly refers to "managing forest in a sustainable way, preserving and improving forests' absorption of carbon, and maintaining and diversifying local people's livelihood as well as helping them to adapt to climate change" [66].

The primary approach for livelihood improvements under REDD+ will likely be household payments, and it has been calculated that there will likely be a national-level payment rate from REDD+ of around 265,000 (12$US) VND/ha/year in the future. Together with the general financial support from the state budget for forest protection measures (100,000–400,000 VND (5–20$US) /ha/year), as well as the average payment for environmental services (PES) available for forest protection in some upland forests (on average 250,000 VND (11$US)/ha/year) [74], participation in forest protection and management under a combination of REDD+ and other programs can contribute to incomes of local people, especially for poor and vulnerable groups in remote areas. Existing PES payments to households in forested areas of Vietnam account for on average 6%–7% (and up to 30%) of household income in participating areas, and are often used to enable school fees and healthcare bills to be paid, to help ensure food security through purchases of rice or seedlings, and other forms of investment [75,76]. When payments can be made to community funds (done in some areas but not others), they can be used to build community infrastructure (e.g., roads, bridges, community houses, etc.); upgrading and buying common assets for the community (e.g., school supplies); paying for a village forest protection group; or even can setting up micro-loans for diversifying livelihood activities and other activities [77,78]. Through these ways, households and communities could be helped via REDD+ payments to adapt and recover following climatic shocks or disasters, although more research will need to be carried out once these payments begin to follow how these monies are actually being used (no areas of Vietnam have received national REDD+ payments as of 2016).

4.2. Subnational Policy on REDD+ and Adaptation and Interlinkages

The primary purpose of the NRAP is to set out key legal and institutional roles as well as priority interventions in REDD+ for the period 2011–2020 [72]. However, the document is mainly an enabling document rather than one providing detailed guidance to develop REDD+ interventions on the ground. Thus, the NRAP is to be supplemented by Provincial REDD+ Action Plans (PRAPs) and Site-based REDD+ Action Plans (SiRAPs). Depending on the particular context of sub-national levels, the PRAPs will help to develop mechanisms and set out suggested REDD+ prioritized interventions that are suitable for the local political, social and environmental conditions in order to support local actors to participate in REDD+ implementation more effectively and sustainability. Currently, fifteen provinces are developing PRAPs; ten are completed while five others are in the development process. (They include: Ca Mau, Lam Dong, Binh Thuan, Ha Tinh, Bac Kan and Lao Cai (supported by the UN-REDD+ program); Thanh Hoa, Ha Tinh, Nghe An, Quang Binh, Quang Tri and Thua Thien Hue (supported by the World Bank Forest Carbon Partnership Facility and the USAID-funded Vietnam Forests and Delta Program) and Dien Bien, Son La, and Hoa Binh provinces (supported under the aforementioned JICA project)). Of our three study provinces, only Dien Bien has an approved PRAP (finalized in May of 2014). In addition to the development of subnational PRAPs, pilot REDD+ activities for REDD+ readiness or voluntary carbon market accession have been implemented in many of Vietnam's provinces, funded by multilateral or bilateral donors or NGOs.

Many provinces have used these donor pilot projects to build off of for development of PRAPs. Because all provinces already were required to create Forest Protection and Development Master Plans to 2020 as required by MARD, REDD+ programs have piggybacked onto this process. This means that the PRAPs usually include both general forest protection and development activities, as well as specific activities for REDD+ pilot areas, divided into 5 different approaches: reducing deforestation, reducing degradation, sustainable forest management, conserving and enhancing forest carbon stocks. In addition, depending on the characteristic of the activities, PRAP activities also can be broken down into direct investment interventions (type I) and the supporting interventions (type II). Type I are defined as direct investment or activities for forest protection, management and development. Type II are supporting activities, which are mainly focused on providing incentives to forest owners, local communities and other relevant actors to carry out the type I activities. The supporting activities include, depending on the local demands, provision of community development funds, financial and technical support for sustainable local livelihoods, or capacity building on various aspects of forest or livelihoods activities. All these activities have been required to be considered in the development of PRAPs, as per government decision 5414/QD-BNN-TCLN dated on 25 December 2015. Although many of these activities likely have adaptation benefits, throughout the national guidance issued for PRAP development, the words "climate change", "mitigation" and "adaptation" are rarely used or not mentioned directly. Part of this inattention is likely due to the fact that PRAPs are primarily being developed by forestry departments at local levels, with less of the sector integration and multi-stakeholder engagement that has characterized the NRAP process at the national level.

We analysed the Dien Bien PRAP in some detail to understand the potential for synergies and linkages with climate adaptation policy. In addition, team members attended several workshops during the process of the development of the Dien Bien PRAP. Dien Bien was the first province in Vietnam to have completed and launched a PRAP with the support from JICA under the project "*Sustainable Forest Management in the Northwest Watershed Areas (SUSFORM-NOW)*". The PRAP was designed to cover the period 2013–2020 and was approved by the Dien Bien Provincial People's Committee in early 2014. Because the Dien Bien PRAP was developed and approved even before the national guidelines for PRAP development were issued, there are some now required activities that are not mentioned in this specific PRAP (see Table 1; the left hand column are activities now required to be addressed in most PRAPs). In particular, some interviewees noted that in fact Dien Bien's PRAP is more similar to a general Forest Protection and Development Plan rather than a specific REDD+ plan focused on forest carbon mitigation.

Key activities to be implemented in the future under the Dien Bien PRAP include forest patrolling and monitoring; forest land allocation; livelihood support; and forest plantations, but there are few details on how these will be accomplished yet. Similar to the NRAP, the Dien Bien PRAP does not mention adaptation directly, and is only implied indirectly through the livelihood development activities to support forest protection in implementing REDD+. For example, the Dien Bien PRAP has highlighted that forestland allocation to organizations, households, individuals and communities should be completed, as one of the supporting activities to reduce deforestation and forest degradation drivers at the local level. Such activities are considered to have the potential to indirectly improve the adaptive capacity of local people to tackle other climate impacts by providing more secure access rights to forests (see Table 1, far right column), although how this will play out in reality will need further research. Additionally, the PRAP does place emphasis on the need to incorporate climate-resilient forest species in future forest plantations, and has several other foci of potential adaptation benefits to forests themselves (Table 1, second column from right).

In addition to helping support the creation and implementation of PRAP for Dien Bien, JICA developed a guidebook for other provinces carrying out PRAP development as well [79]; however, the handbook makes no explicit mention of adaptation support or activities. The handbook outlines a 13-step process that is necessary to develop a PRAP, of which some of the key activities are forest and socio-economic surveys; calculations of reference emission levels; searches for potential funding of emissions reductions; surveys on forest monitoring; and policies on safeguards. The handbook does identify the global Adaptation Fund as a source of REDD+ financing, but does not identify what adaptation activities related to REDD+ should be incorporated into PRAPs. Additionally, the socio-economic surveys only indirectly touch on climate vulnerabilities, as primarily these surveys are to "identify forest status such as forest distribution & stock at present and past forest change, the driver of forest decrease and increase, socio-economic conditions such as demography and agriculture & forestry production, and to assess past programs & policies relating forest protection and development" [79]. Thus, there appears to be much more room for PRAPs and PRAP development guidance to focus explicitly on adaptation.

Table 1. Key sectors and activities in subnational Provincial REDD Action Plan (PRAP) in Dien Bien Province.

I. General Forest Sectors and Plans	**In Dien Bien PRAP?**	**Potential Adaptation Benefit to Forests?**	**Potential Adaptation Benefit to Households?**
1.1. Forest protection and natural forest regeneration and enrichment	x	Y	Y
1.2. Forest plantation	x	Unclear	Unclear
1.3. Forest rehabilitation and reforestation	x	Y	Y
1.4. Silvicultural-related construction projects			
II. Prioritized Activities carried out at the potential REDD+ areas			
2.1. Activities to reduce deforestation			
a. Reviewing and analyzing overall land-use planning and forest/forestland planning			
b. Reviewing and analyzing the overlap between forest/forestland planning and socio-economic development planning			
c. Strictly control the conversion of natural forests into other purposes	x	Y	Unclear
d. Supporting the forestland allocation, contracts and lease to households, individuals and local communities	x	Unclear	Y
e. Livelihood improvements			
f. Establishing small-scale local livelihood credit development funds			
g. Enhancing the forest protection and development law enforcement			
2.2. Activities to reduce forest degradation			
a. Reducing/ preventing illegal timber logging and utilization			
b. Establishing the administrative and technical monitoring system on timber legality assurance			
c. Promoting sustainable non-timber forest product (NTFP) models			
d. Market orientation to agro-forest products	x		Y
e. Developing village forest protection and development conventions	x		Y
f. Encouraging job creation in the REDD+ implementation sites	x		Y
2.3. Sustainable forest management			
a. Support to develop and implement the sustainable forest management plan and SFM certificates			
b. Improving the forest governance ability for forest owners	x		Y
2.4. Conserving carbon stocks			
a. Improving the quality of forests: forest enrichment, or diversification of crops to adapt impacts of climate change			
b. Biodiversity conservation in special-use forests and protection forest and payments for forest environmental services	x	Y	Y
2.5. Enhancing carbon stocks			
a. Technical support for forest plantation: climate change resilient seedlings, etc.	x	Y	Unclear
b. Forest enrichment in exhausted forest areas			
c. Afforestation by land-use conversion projects			

4.3. Project Based Interlinkages with Adaptation

The REDD+ projects in our three study sites differ from the three projects analysed previously by other reports on adaptation and REDD+ in Vietnam [13]. The REDD+ Community Carbon Pool Program (REDD+ CCP) is the first and only REDD+ project to be implemented in Kon Tum province and is focused at a site-based level in Hieu commune, Kon Plong District; a 3-year SiRAP had been developed and approved by the Kon Tum Provincial People's Committee. Under the REDD+ CCP project, 18,700 ha forests have been re-allocated with community land use rights, which is intended to lead to local access and control over forestland and forest resources. The development of local community forest management institutions that comprised equitable, easy to understand, locally devised and implemented rules was seen as a way to effectively reduce emissions and provide benefits to forest-dependent local people, and the project was hopeful to meet the requirements for CCBA certification, which explicitly includes projects that generate benefits for climate change adaptation.

In Dien Bien, the JICA SUSFORM-NOW project aimed to test REDD+ activities in two pilot communes; develop capacity for forest rangers and commune level field officers; and prepare the PRAP and related technical guidance documents as noted previously. Overall, the project aimed to build efficient models and develop capacity for provincial cadres to implement REDD+ themselves so that those models would be replicated to other areas. As such, the project was aimed to contribute to better forest management at the local level and respond to climate change in the forestry sector, although adaptation was not explicitly mentioned in project documents.

In Kien Giang, the GIZ/AusAid Conservation and Development program was originally designed as a climate change adaptation, mitigation and integrated coastal zone management project. In recent years it has also supported research into mangrove forests, awareness raising of forest protection, and training on REDD+ readiness in Kien Giang and Ca Mau provinces. One of the major activities funded by the project has supported local communities to develop "green fences" along the coastal area through mangrove plantations to protect against landslides and coastal erosion due to strong waves from the seas and future sea level rise. Local people were also supported by the project to establish mangrove nurseries. The project provided training courses for local people, especially women, on primary health care services and rural sanitation, for example building toilets and using clean water, among others, which have adaptation benefits. Kien Giang province also has had since 2005 an initiative of benefit sharing in mangrove forest protection that enables local people to combine forest protection and aquaculture development to secure local livelihoods. People are allocated two to three ha of protection mangrove forest, and they are allowed to use 30% of this area to raise shrimp, as long as they maintain 70% mangrove forest. By 2013, there were 1076 households participating who were allocated mangrove forest with "green books" (a long-term contract for forest protection) along 200 km of the coast of Kien Giang province. The program aims to give local people incentive to protect mangrove forest while developing shrimp production for their livelihood; however, according to GIZ/AusAid project officers' assessments, the expectation from local people is for a higher percentage of land devoted to shrimp production due to low profits and higher costs of maintaining mangrove forest as compared to non-participating shrimp farms.

4.4. Household Level Impacts from Climate Changes and REDD+ Projects

Our household survey provided a general summary of the characteristics of households in the three sites: around 31% of surveyed households were classified as poor by the government (based on guidelines of the Ministry of Labor, Invalids, and Social Affairs, as every household in Vietnam receives regular assessments of their economic status to determine some social benefits); 49% as average; and 11% better off (the rest did not know or did not answer). The average household generated 78 million (around 3580$US) per HH per year in subsistence and cash income (see Table 2), with great variation between the three sites, as Kien Giang households had over four times the income as those in Kon Tum. In each site, a different income source was the primary contributor to local livelihoods: business and trading in Dien Bien; livestock in Kon Tum; and aquaculture in Kien Giang.

The primary uses of forests by households were for fuelwood and forest product collection; as sites for aquaculture in the case of mangroves in Kien Giang (where households generated on average 4758$US per household from aquaculture); as sources of lands for fields in swidden cultivation systems in Dien Bien and Kon Tum; and as supplies of wood for house building, particularly for poorer or ethnic minority households. Forest-based subsistence and cash income was highest in Dien Bien, while it was surprisingly low in Kon Tum, despite having the most extensive forest cover of all three sites (and which we explain below was a result of restrictions on forest use instituted by the CCP project).

The living conditions of the populations in the research sites were generally low with poor infrastructure, making areas difficult to access in the rainy season, leading to limited access to information regarding markets and services. Local communities were all sensitive to changes in weather and had been very much impacted by natural disasters that appeared with higher frequency and more intensity in recent years. In general, local people at all fieldsites had concerns about climate changes and risk, although the specific type of impacts varied (see Table 3). In the Northwest mountainous area of Dien Bien, people were concerned about droughts, with 59% of surveyed households noting decreases in rainfall, and 53% noting longer dry seasons. In Kon Tum, people were more concerned with increases in storms and rainfall, leading to localized flooding. Households in the Mekong Delta area of Kien Giang were particularly concerned about salinity intrusions farther from the coast in the dry season. During group meetings, communities often linked these climate changes explicitly to forest policy and activities; for example, in Kon Tum, local people explained that the higher impacts of storms and flooding were due to deforestation in the last few decades in the area. In Dien Bien, local people noted that higher temperature and severe droughts had caused forest fires to increase as well (focus group data).

When asked to rank the most serious climate-related risk to their property and livelihoods, households choose a variety of answers (Figure 2a). Typhoons and storms were ranked as the most serious risk by nearly 30% of survey respondents, with landslides and drought chosen as the most serious risk by fewer than 15% and 10% of households respectively. Although households overall understood climate risks to be important and in most cases increasing, these risks were put in the context of other challenges that households had to face (Figure 1b). Health problems were considered the most serious and frequently encountered risk, with natural disasters and pests ranked second and third. Problems with access and quality and outcomes of education of their children were also considered risks, as people were worried they could not afford for their children go to school; these concerns are reflected as "children's schooling" in the Figure 2b. Other risks included labour shortages due to health and other problems (such as alcoholism or drug use). Poor infrastructure in the study areas, especially in health care, education, and public services, increased households' feeling of risky livelihoods. This findings confirmed that vulnerability to shocks, be it climate or health or unemployment, have long been one of the major challenges for the poor in Vietnam [57]. In particular, while these climate risks have the potential to impact all income groups, the poor tend to have less resilience, such as less access to insurance and less ability to rebuild or move away from affected areas [80,81].

Table 2. Average household livelihood sources and values in VND.

Province	Agriculture	Livestock	Aquaculture/Fishing	Forestry	Business	Total Average HH Subsistence and Cash Income
Dien Bien	14,493,982	24,124,279	4,464,852	12,008,551	30,811,538	79,772,542 (3626$US)
Kon Tum	5,352,315	20,136,242	0	3,754,609	10,415,938	30,078,040 (1367$US)
Kien Giang	55,395,454	19,965,853	104,679,297	5,951,666	35,293,787	126,419,650 (5746$US)
Average across all households	18,390,243	21,955,750	53,053,674	7,594,240	25,113,156	78,756,744 (3580$US)

* At the time of the survey, 1 USD = approximately 22,000 Vietnam Dong (VND).

Table 3. Household perceptions of weather and climate changes in recent years (% of households citing each reason).

Province	Increase in Rainfall	Decrease in Rainfall	Longer Rainy Season	Longer Dry Season	Increase in Number of Storms	Increase in Storm Strength	Salinity Intrusion	Higher Temperatures
Dien Bien	14%	59%	5%	53%	3%	24%	n/a	30%
Kon Tum	37%	3%	26%	17%	74%	43%	n/a	9%
Kien Giang	48%	38%	70%	30%	23%	33%	100%	61%

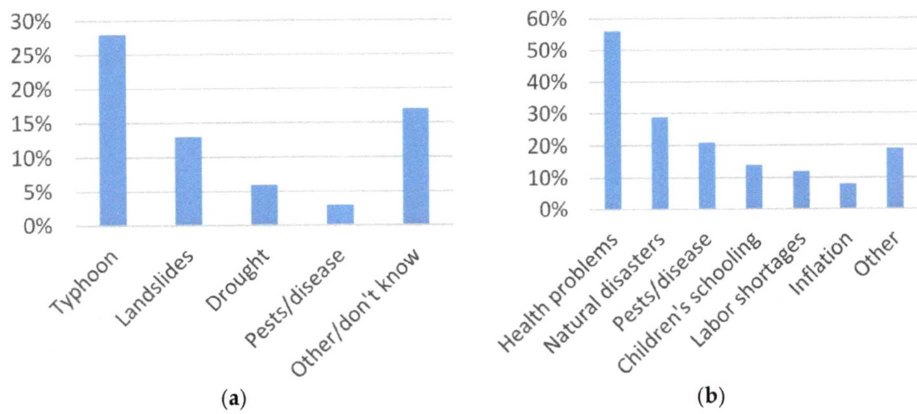

Figure 2. (**a**) The most serious climate/weather risk as identified by households; (**b**) The most serious of all risks as identified by households.

4.4.1. Impacts of REDD+ on Livelihoods and Vulnerability

In one of our studied projects, household level REDD+ activities had been confined to awareness raising of REDD+ and there were no significant livelihood or other activities directed at the household (Dien Bien). In Kon Tum however household activities had been taking place, and showed the potential negative impacts of REDD+ on both livelihoods and climate vulnerability. The commune where REDD+ activities were taking place is one of the 300 poorest communes in Vietnam [82]. Most of the local people are of the M'Nam ethnic group and have long directly depended on forestland and forest resources for livelihoods, such as rice or cassava cultivation on shifting cultivation plots in designated forestlands, firewood collection, and gathering and commercialization of diverse NTFPs. These forest activities are important because the local climate is highly variable, with rainy and cold weather affecting agricultural production. As a result, the growing season in Hieu commune is very short, around 6 months per year, from February until the end of September. Due to this weather, if households want to increase their rice or cassava production, they can only expand their cultivation areas (as they cannot diversify out of season), but this conflicts with REDD+ project priorities of conserving forest lands for emissions reductions [83].

In 2011 the initial stage of the REDD+ CCP project began, and the message that households could "protect forest for selling carbon" that the project delivered to local people through FPIC awareness-raising activities raised their hopes and expectation of gaining benefits and improving their livelihoods. The household survey and interviews conducted showed that the local people in Hieu commune perceived REDD+ as a new type of income source that would be used to replace existing practices. Therefore, the local people accepted the need to stop expanding their traditional swidden plots as well as restricting NTFP extraction of *"la kim cuong"* (*Anoectochilus setaceus*), a medicinal plant found in forests, as trade-offs in order to get income from forest carbon in the future. The project drew a lucrative picture about the benefits of REDD+ and forest carbon, but in reality, difficulties have already emerged from the project, which include scarcity of cultivable land as swiddening has stopped or been discouraged, and the loss of income from NTFP extraction. Data from the survey and interviews show that most people already changed their household livelihood strategies several years ago (household surveys showed lower forest income in Kon Tum than the other two sites, at only 170$US worth of timber, fuelwood and NTFPs extracted per household), but are still waiting for forest carbon benefits, which have not yet been paid. While the project finished at the end of 2014, it could not get carbon certificates to sell in the market as per the initial objective, and the project designer, FFI, has been trying to obtain further funding and access some carbon market to get emissions reductions certificates.

In the meantime, however, households with differences in cultivable land sources, capital and labor, have had to adjust their livelihood strategies in different ways. The poor and landless households are the most vulnerable group in this situation, and their way to adapt to the new context varies within this group. Some households have returned to make use of their old swidden plots or tried to find some small pieces of land near their paddy fields, or even have planted field crops like cassava in the garden surrounding their houses. Some other households, particularly those who do not have cultivable land, cannot open new swidden plots, and who also cannot wait two to three years for carbon credits, have decided to give up their rights to participate and get benefits from REDD+ in the future by leaving the villages or continuing to do restricted activities. For example, some husbands in these families have decided go to other communes or cities to find new jobs to compensate for the loss of income due to participating in REDD+, which has increased their families' overall vulnerability to risk and shocks. Further, "leakage" has been an on-going issue. Statistics by the local authority shows there has been no more forest clearance since the middle of 2013 in Hieu commune. However, there were some households who decided to leave and clear some forest of Bo E commune (a commune nearby but not in the REDD+ project) in order to establish new swidden plots illegally.

4.4.2. Household Adaptation Actions

At the household level in all research sites, some spontaneous adaptation actions are being taken to cope with climate risks that are already being felt. Households were most proactive in the agriculture sector, using adaptation strategies to adjust cropping patterns, harvesting time, selecting salt-resistant varieties, and so on, as to reduce damages to livelihoods. For example, 13% of households in Dien Bien and 12% of households in Kien Giang stated they had changed a crop variety, while only 3% of households in Kon Tum had done so. Changing crop calendars and harvesting crops early to avoid flood and disaster losses was another strategy, one that had been taken by 100% of households in Kien Giang, 90% of households in Kon Tum, and 60% of households in Dien Bien at least once. However, most households felt they were not doing enough to adapt to climate change, and needed more policy support. In particular, households rarely mentioned adaptation actions they were taking with regard to forestry, indicating that households felt less knowledge about this sector and needed guidance as to steps to take.

In group discussions, those residents who stated that they had not taken any adaptation actions explained that they knew that adaptation was necessary but they had a lack of resources. Poor households in particular stated they did not have enough human, physical or financial resources to protect their fields and homes from climate hazards, therefore they tended to lose relatively more when hit by floods and storms than wealthier households, and had a lower capacity to cope with and adapt to shocks due to lower access to savings. While acknowledging these challenges, local government officials interviewed stated that they lacked budgets to support households to carry out climate change strategies, plans and adaptation actions at community levels. These stakeholders at district and commune levels noted that national adaptation strategies and action plans were normally very ambitious but often not feasible due to lack of funding and low participation from local people, since there was very little funding to distribute.

5. Discussion and Conclusions

This study has confirmed that there continues to be a lack of coordination between mitigation and adaptation policies in Vietnam, particularly with regard to REDD+. While much lip service has been paid to combining approaches, in reality, policies at the national and provincial level, and site-based projects, have paid little attention to the adaptation needs of local communities, and how REDD+ activities could facilitate increased resilience in livelihoods. While there were some potential implicit adaptation benefits of REDD+ projected activities (such as promises of future activities to improve livelihoods, or to increase forest tenure security which might help local communities have better access to forest resources), local REDD+ planning through development of PRAPs has not systematically considered activities that focus on adaptation or resilience. This is despite the fact that many national policymakers, donors, and NGOs interviewed a few years ago expressed strong support for integrated attention to adaptation [13]. There was also little discussion in either PRAPs or site-specific REDD+

projects regarding EBA (with the exception of Kien Giang), again despite the fact that the idea has been enthusiastically discussed with regard to Vietnam [84,85].

These are missed opportunities, as our household data shows a great deal of need for assistance in reducing climate vulnerability (particularly in coping with drought and floods) and increasing adaptive capacity in many REDD+ project sites. There are potentials for integration with agricultural adaptation in particular. REDD+ activities in agroforestry or financial support for climate-smart agriculture could help households deal with the increasing climate risks they are facing, particularly those areas that are experiencing either drought or flooding as noted in household surveys. Tree shelterbelts that could be planted to enhance carbon stocks could help to reduce wind and drought pressures on nearby agricultural fields. Similarly, using trees to secure landslide and flood prone areas near fields can help reduce impacts on agricultural livelihoods from flooding [50]. Other examples of adaptation needs might include: a stronger focus on useful multipurpose tree species for reforestation and carbon stock enhancement that could provide for both carbon and increased livelihoods (e.g., food or products for sale); policies to increase value and marketing of forest goods that can be harvested sustainability (such as NTFPs); and policies to reduce woodfuel use but increase energy access (e.g., improved cookstoves or small-scale hydro-powered electrification projects), among others [19,86,87]. Poor households in particular showed needs for financial support to undertake adaptation actions in agriculture, which REDD+ payments could potentially be useful for, but households stated these needed to be coupled with training, education, and other forms of support so they would know what to invest in to increase resilience. Payments alone will not be sufficient.

Further, within both national and subnational REDD+ policy approaches, our analysis notes a lack of attention to the potential consequences of climate change on forest structure and composition, and the implications of this for REDD+ activities into the future. Several of the most serious climate impacts from the literature on forest vulnerability in Asia include forest fires; pest outbreaks; shifts in species distributions; higher tree mortality; changes in forest composition; or loss of wood volume [51,88–90]. Very few of these possible climate vulnerabilities of forests are discussed in the National Strategy on Climate Change (which only mentions the need to "increase capacity and efficiency of systems for evaluating, forecasting, preventing, monitoring, supervising and urgently responding to forest fires" [66]); in the NRAP; or the in the PRAP for Dien Bien. Future REDD+ projects to address forests' climate vulnerabilities might focus on reducing fire hazard and risk (e.g., community supported fire watches) or reforestation projects that prioritize drought and fire resistant native species (e.g., rather than introduced eucalyptus or acacia, which have been primary species in reforestation in the past but which are both drought and windfall-prone) [51].

Several pilot REDD+ projects examined also did not explicitly acknowledge how climate-induced changes might impact household livelihoods, and what role these vulnerabilities may play in REDD+ participation. In the one project we examined that did specifically have a focus on adaptation (Kien Giang), livelihood activities were developed that extended beyond typical REDD+ activities, such as addressing water scarcity and shoreline erosion, adaptation responses that were considered useful by households facing water and land erosion risks, and which may increase positive feelings and household participation in other parts of the project (e.g., tree planting). However, in another project site (Kon Tum), restrictions on livelihood activities had taken place under REDD+ and had caused negative consequences, particularly for the poorer households, due to restrictions on swidden agriculture and NTFP collection, the lack of suitable alternative livelihood plans, and delays in seeking carbon financing. Other studies in Vietnam have shown the importance of NTFPs as "insurance" for poor households, particularly for buffering unpredictable shocks like disasters or health problems [91]. Yet the Kon Tum project did not consider these roles of NTFPs, and how loss of access to these forests might create unforeseen negative impacts. Indeed, it appears that the project has made households more vulnerable to the risks of poverty and climate change impacts than before, particularly in the cases where REDD+ participation seems to have triggered male outmigration in some families.

Despite the negative outcomes of one of the examined projects, there appear to be opportunities to promote synergies with adaptation, where REDD+ could improve local actors' adaptive capacity. For example, in Kien Giang, the approach to forest management which allows some limited production activities in mangrove areas has proven moderately successful, and shows that co-management that allows for some livelihood activities, rather than complete abandonment of forest use as was the case in Kon Tum, can provide benefits for both people and forests. The fact that livelihoods are supported, and that mangroves provide useful protection against storms, landslides and river erosion, makes for a double adaptation benefit. Kien Giang could address concerns from households about low profitability by expanding the area of mangroves allowed under the 70/30 model from the current small pilot area to allow for each participating household to manage a larger area. Despite these successes, however, there have been shortfalls in incorporating mangroves into other REDD+ and PES policies. Although many reports within Vietnam have noted that mangroves are important for both mitigation and adaptation [92–94], there have been limited on-the-ground mangrove projects linked to REDD+, and there are currently no mangrove areas receiving national PES money due to an inability to determine who the buyers of mangrove ecosystem services are [95]. These are challenges that should be immediately prioritized in future REDD+ development.

All of these potential synergies rest on removing barriers to integration between REDD+ and adaptation. Key barriers to integration can be seen in both the PRAP development process and in individual site projects. These include:

(1) Sectoral specialization: the PRAP process, as seen in Dien Bien, was led by forestry officials and primarily focused on narrow interpretations of forests, and did not involve much input from agricultural or climate adaptation offices. The minimal inputs of the Ministry of Environment and Natural Resources, the key climate adaptation ministry, in REDD+ policy development below the national level is further evidence of this disengagement.

(2) A primary focus of REDD+ policy on technical measures rather than livelihoods: in the PRAP for Dien Bien and in national PRAP development guidance, there was far more attention devoted to reference levels of deforestation and carbon emissions equations than to outlining ways to involve local people in participatory forestry projects with livelihood benefits. The Kon Tum CCP project similarly spent time focused on meeting technical requirements for selling emissions reductions on the market, without interim livelihood activities.

(3) A focus primarily on emissions rather than co-benefits or multipurpose trees: in the NRAP and in Dien Bien's PRAP, more attention was paid to maintaining large intact forest areas or plantations rather than support for small scattered tree plantings, such as in agroforestry or shelterbelts that would benefit farmers. A focus on maximizing emissions levels in REDD+ through extensive forestry thus might create a disincentive for more adaptive measures in local household-based forestry.

As we have shown, the potentials for integration of adaptation activities into REDD+ in Vietnam are there. Climate vulnerable households in our study sites, like many place in Vietnam, are already feeling the effects of some climate and weather changes now [57], and see natural disasters as serious risks threatening their families' wellbeing. REDD+ activities that provide ways to strengthen the adaptive capacity of these households would therefore be extremely useful and welcome. Future PRAP development in the remaining forested provinces of Vietnam would do well to consider better integration of adaptation considerations in future planning, and site-specific projects clearly need to learn from previous lessons with regard to the necessity of considering both livelihoods and forest outcomes simultaneously to avoid increasing the climate vulnerability of participating households.

Acknowledgments: This research was made possible by a grant from the National Science Foundation Geography and Regional Science Division for the project "Downscaling REDD+ policies in developing countries: Assessing the impact of carbon payments on household decision-making and vulnerability to climate change in Vietnam" (grant #11028793) to McElwee, support to McElwee from Hatch funding of the US National Institute for Food and Agriculture (NIFA), and a US Agency for International Development (USAID) Partnerships for Enhanced Engagement in Research grant to the CRES, Tropenbos and PanNature for the project: "Research and capacity building on REDD+, livelihoods, and vulnerability in Vietnam: developing tools for social analysis of development planning". Dao Minh Truong contributed the figures. Additional people who contributed to the collection of data include Dao Minh Truong, Le Trong Toan, Ha Thi Thu Hue, Ha Thi Tu Anh, and Nguyen Xuan Lam. Thanks are due to Lo Quang Chieu, Vice-Director of Department of Agriculture and Rural Development in Dien Bien, the Hieu Commune People's Committee in Kon Tum, and Doan Van Thanh, Le Thi Kim Anh, Ho Tuan Quang, Nguyen Minh Tan and Nguyen Van Thieu in Kien Giang who facilitated fieldwork in these provinces.

Author Contributions: All authors conceived and designed the research project; V.H.T.N., D.V.N. N.H.T., H.V.T.L., T.P.N., and H.D.T.V. conducted the fieldwork; all authors contributed to writing the paper.

Conflicts of Interest: The authors declare no conflict of interest. The funding sponsors had no role in the design of the study; in the collection, analyses, or interpretation of data; in the writing of the manuscript, and in the decision to publish the results.

References

1. Ravindranath, N.H. Mitigation and adaptation synergy in forest sector. *Mitig. Adapt. Strateg. Glob. Chang.* **2007**, *12*, 843–853. [CrossRef]
2. Matocha, J.; Schroth, G.; Hills, T.; Hole, D. Integrating climate change adaptation and mitigation through agroforestry and ecosystem conservation. In *Agroforestry—The Future of Global Land Use*; Advances in Agroforestry; Springer: Dordrecht, The Netherlands, 2012; pp. 105–126.
3. Füssel, H.; Klein, R. Climate change vulnerability assessments: An evolution of conceptual thinking. *Clim. Chang.* **2006**, *75*, 301–329. [CrossRef]
4. Intergovernmental Panel on Climate Change. *Climate Change 2007: Synthesis Report*; Intergovernmental Panel on Climate Change and Cambridge University Press: Cambridge, UK, 2008.
5. Pistorius, T. From RED to REDD+: The evolution of a forest-based mitigation approach for developing countries. *Curr. Opin. Environ. Sustain.* **2012**, *4*, 638–645. [CrossRef]
6. Olander, L.P.; Galik, C.S.; Kissinger, G.A. Operationalizing REDD+: Scope of reduced emissions from deforestation and forest degradation. *Curr. Opin. Environ. Sustain.* **2012**, *4*, 661–669. [CrossRef]
7. Gupta, J.; van der Grijp, N.; Kuik, O. *Climate Change, Forests and REDD: Lessons for Institutional Design*; Routledge: London, UK, 2013.
8. Minang, P.A.; van Noordwijk, M.; Duguma, L.A.; Alemagi, D.; Do, T.H.; Bernard, F.; Agung, P.; Robiglio, V.; Catacutan, D.; Suyanto, S.; et al. REDD+ Readiness progress across countries: Time for reconsideration. *Clim. Policy* **2014**, *14*, 685–708. [CrossRef]
9. Fischer, R.; Hargita, Y.; Günter, S. Insights from the ground level? A content analysis review of multi-national REDD+ studies since 2010. *For. Policy Econ.* **2016**, *66*, 47–58. [CrossRef]
10. Mbatu, R.S. REDD+ research: Reviewing the literature, limitations and ways forward. *For. Policy Econ.* **2016**, *73*, 140–152. [CrossRef]
11. Somorin, O.A.; Visseren-Hamakers, I.J. Integration through interaction? Synergy between adaptation and mitigation (REDD+) in Cameroon. *Environ. Plan. C Gov. Policy* **2016**, *34*, 415–432. [CrossRef]
12. Somorin, O.A.; Brown, H.C.P.; Visseren-Hamakers, I.J.; Sonwa, D.J.; Arts, B.; Nkem, J. The Congo Basin forests in a changing climate: Policy discourses on adaptation and mitigation (REDD+). *Glob. Environ. Chang. Part A* **2012**, *22*, 288–298. [CrossRef]
13. Pham, T.T.; Moeliono, M.; Locatelli, B.; Brockhaus, M.; Gregorio, M.; Mardiah, S. Integration of adaptation and mitigation in climate change and forest policies in Indonesia and Vietnam. *Forests* **2014**, *5*, 2016–2036.
14. Pramova, E.; Di Gregorio, M.; Locatelli, B. *Integrating Adaptation and Mitigation in Climate Change and Land-Use Policies in Peru*; CIFOR: Bogor, Indonesia, 2015.
15. Locatelli, B.; Evans, V.; Wardell, A.; Andrade, A.; Vignola, R. Forests and climate change in Latin America: Linking adaptation and mitigation. *Forests* **2011**, *2*, 431–450. [CrossRef]
16. West, S. *REDD+ and Adaptation in Nepal*; REDDNet Case Study; REDDNet: Bangkok, Thailand, 2012.
17. McFarland, W. *Synergies between REDD+ and Adaptive Capacity to Climate Change at the Local Level—A Ghana Case Study*; REDDNet Case Study; REDDNet: Bangkok, Thailand, 2012.
18. Kongsager, R.; Locatelli, B.; Chazarin, F. Addressing climate change mitigation and adaptation together: A global assessment of agriculture and forestry projects. *Environ. Manag.* **2015**, *57*, 271–282. [CrossRef] [PubMed]

19. Kongsager, R.; Corbera, E. Linking mitigation and adaptation in carbon forestry projects: Evidence from Belize. *World Dev.* **2015**, *76*, 132–146. [CrossRef]
20. Takacs, D. Carbon into gold: Forest carbon offsets, climate change adaptation, and international law. *Hastings West-Northwest J. Int. Environ. Law* **2009**, *15*, 39.
21. Fry, I. Reducing emissions from deforestation and forest degradation: Opportunities and pitfalls in developing a new legal regime. *Rev. Eur. Community Int. Environ. Law* **2008**, *17*, 166–182. [CrossRef]
22. Voigt, C.; Ferreira, F. The Warsaw Framework for REDD+: Implications for national implementation and access to results-based finance. *Carbon Clim. Law Rev. (CCLR)* **2015**, *2*, 113–129.
23. Cerbu, G.A.; Swallow, B.M.; Thompson, D.Y. Locating REDD: A global survey and analysis of REDD readiness and demonstration activities. *Environ. Sci. Policy* **2011**, *14*, 168–180. [CrossRef]
24. UNFCCC. The Paris Agreement. Available online: http://unfccc.int/files/essential_background/convention/application/pdf/english_paris_agreement.pdf (accessed on 31 October 2015).
25. Ingalls, M.L.; Dwyer, M.B. Missing the forest for the trees? Navigating the trade-offs between mitigation and adaptation under REDD. *Clim. Chang.* **2016**, *136*, 353–366. [CrossRef]
26. Poudyal, M.; Ramamonjisoa, B.S.; Hockley, N.; Rakotonarivo, O.S.; Gibbons, J.M.; Mandimbiniaina, R.; Rasoamanana, A.; Jones, J.P.G. Can REDD+ social safeguards reach the "right" people? Lessons from Madagascar. *Glob. Environ. Chang.* **2016**, *37*, 31–42. [CrossRef]
27. Jagger, P.; Sills, E.O.; Lawlor, K.; Sunderlin, W.D. *A Guide to Learning about Livelihood Impacts of REDD+*; Center for International Forestry Research (CIFOR) Occasional Paper 56; CIFOR: Bogor, Indonesia, 2010.
28. Corbera, E.; Brown, K. Offsetting benefits? Analyzing access to forest carbon. *Environ. Plan. A* **2010**, *42*, 1739–1761. [CrossRef]
29. Leach, M.; Mearns, R.; Scoones, I. Environmental entitlements: Dynamics and institutions in community-based natural resource management. *World Dev.* **1999**, *27*, 225–247. [CrossRef]
30. Ribot, J.C.; Peluso, N.L. A theory of access. *Rural Sociol.* **2003**, *68*, 153–181. [CrossRef]
31. Osborne, T. Tradeoffs in carbon commodification: A political ecology of common property forest governance. *Geoforum* **2015**, *67*, 64–77. [CrossRef]
32. Atela, J.O.; Minang, P.A.; Quinn, C.H.; Duguma, L.A. Implementing REDD+ at the local level: Assessing the key enablers for credible mitigation and sustainable livelihood outcomes. *J. Environ. Manag.* **2015**, *157*, 238–249. [CrossRef] [PubMed]
33. Kashwan, P. Forest policy, institutions, and REDD+ in India, Tanzania, and Mexico. *Glob. Environ. Polit.* **2015**, *15*, 95–117. [CrossRef]
34. Ayers, J.M.; Huq, S. The value of linking mitigation and adaptation: A case study of Bangladesh. *Environ. Manag.* **2008**, *43*, 753–764. [CrossRef] [PubMed]
35. Fujisaki, T.; Hyakumura, K.; Scheyvens, H.; Cadman, T. Does REDD+ ensure sectoral coordination and stakeholder participation? A comparative analysis of REDD+ national governance structures in countries of Asia-Pacific region. *Forests* **2016**, *7*, 195. [CrossRef]
36. Fankhauser, S.; Schmidt-Traub, G. From adaptation to climate-resilient development: The costs of climate-proofing the Millennium Development Goals in Africa. *Clim. Dev.* **2011**, *3*, 94–113. [CrossRef]
37. Thornbush, M.; Golubchikov, O.; Bouzarovski, S. Sustainable cities targeted by combined mitigation–adaptation efforts for future-proofing. *Sustain. Cities Soc.* **2013**, *9*, 1–9. [CrossRef]
38. Stromberg, P.M.; Esteban, M.; Gasparatos, A. Climate change effects on mitigation measures: The case of extreme wind events and Philippines' biofuel plan. *Environ. Sci. Policy* **2011**, *14*, 1079–1090. [CrossRef]
39. Murdiyarso, D.; Brockhaus, M.; Sunderlin, W.D.; Verchot, L.V. Some lessons learned from the first generation of REDD+ activities. *Curr. Opin. Environ. Sustain.* **2012**, *4*, 678–685. [CrossRef]
40. Korhonen-Kurki, K.; Brockhaus, M.; Bushley, B.; Babon, A.; Gebara, M.F.; Kengoum, F.; Pham, T.T.; Rantala, S.; Moeliono, M.; Dwisatrio, B.; et al. Coordination and cross-sectoral integration in REDD+: Experiences from seven countries. *Clim. Dev.* **2016**, *8*, 458–471. [CrossRef]
41. Atela, J.O.; Quinn, C.H.; Minang, P.A.; Duguma, L.A.; Houdet, J.A. Implementing REDD+ at the national level: Stakeholder engagement and policy coherences between REDD+ rules and Kenya's sectoral policies. *For. Policy Econ.* **2016**, *65*, 37–46. [CrossRef]
42. Pramova, E. *Integrating Adaptation into REDD+: Potential Impacts and Social Return on Investment in Setulang, Malinau District, Indonesia*; Center for International Forestry Research (CIFOR): Bogor, Indonesia, 2013.

43. Robledo, C.; Kanninen, M.; Pedroni, L. *Tropical Forests and Adaptation to Climate Change*; Center for International Forestry Research (CIFOR): Bogor, Indonesia, 2005.
44. Forsius, M.; Anttila, S.; Arvola, L.; Bergström, I.; Hakola, H.; Heikkinen, H.I.; Helenius, J.; Hyvärinen, M.; Jylhä, K.; Karjalainen, J.; et al. Impacts and adaptation options of climate change on ecosystem services in Finland: A model based study. *Curr. Opin. Environ. Sustain.* **2013**, *5*, 26–40. [CrossRef]
45. Schoene, D.H.F.; Bernier, P.Y. Adapting forestry and forests to climate change: A challenge to change the paradigm. *For. Policy Econ.* **2012**, *24*, 12–19. [CrossRef]
46. Kalame, F.B.; Nkem, J.; Idinoba, M.; Kanninen, M. Matching national forest policies and management practices for climate change adaptation in Burkina Faso and Ghana. *Mitig. Adapt. Strateg. Glob. Chang.* **2008**, *14*, 135–151. [CrossRef]
47. Munang, R.; Thiaw, I.; Alverson, K.; Mumba, M.; Liu, J.; Rivington, M. Climate change and ecosystem-based adaptation: A new pragmatic approach to buffering climate change impacts. *Curr. Opin. Environ. Sustain.* **2013**, *5*, 67–71. [CrossRef]
48. Munang, R.; Thiaw, I.; Alverson, K.; Liu, J.; Han, Z. The role of ecosystem services in climate change adaptation and disaster risk reduction. *Curr. Opin. Environ. Sustain.* **2013**, *5*, 47–52. [CrossRef]
49. Locatelli, T.; Binet, T.; Kairo, J.G.; King, L.; Madden, S.; Patenaude, G.; Upton, C.; Huxham, M. Turning the tide: How blue carbon and payments for ecosystem services (PES) might help save mangrove forests. *AMBIO J. Hum. Environ.* **2014**, *43*, 981–995. [CrossRef] [PubMed]
50. Pramova, E.; Locatelli, B.; Djoudi, H.; Somorin, O.A. Forests and trees for social adaptation to climate variability and change. *WIREs Clim. Chang.* **2012**, *3*, 581–596. [CrossRef]
51. Seppälä, R.; Buck, A.; Katila, P. *Adaptation of Forests and People to Climate Change—A Global Assessment Report*; International Union of Forest Research Organizations (IUFRO) World Series vol. 22; IUFRO: Vienna, Austria, 2009.
52. Vignola, R.; Locatelli, B.; Martinez, C.; Imbach, P. Ecosystem-based adaptation to climate change: What role for policy-makers, society and scientists? *Mitig. Adapt. Strateg. Glob. Chang.* **2009**, *14*, 691–696. [CrossRef]
53. Sikor, T. (Ed.) *The Justices and Injustices of Ecosystem Services*; Routledge: London, UK, 2013.
54. Chong, J. Ecosystem-based approaches to climate change adaptation: Progress and challenges. *Int. Environ. Agreem.* **2014**, *14*, 391–405. [CrossRef]
55. Hoa, N.T.; Hasegawa, T.; Matsuoka, Y. Climate change mitigation strategies in agriculture, forestry and other land use sectors in Vietnam. *Mitig. Adapt. Strateg. Glob. Chang.* **2012**, *19*, 15–32. [CrossRef]
56. Dilley, M.; Chen, R.S.; Deichmann, U. *Natural Disaster Hotspots: A Global Risk Analysis Synthesis Report*; International Bank for Reconstruction and Development/The World Bank and Columbia University: Washington, DC, USA, 2005.
57. McElwee, P.D.; Nghiem, P.T.; Van Hue, L.T.; Huong, V.; Be, N.V.; Tri, L.Q.; Trung, N.H.; Tuan, L.A.; Dung, L.C.; Duat, L.Q.; et al. *Social Dimensions of Climate Change in Vietnam*; World Bank Discussion Paper No 17; World Bank: Washington, DC, USA, 2010.
58. Dang, H.; Michaelowa, A.; Tuan, D. Synergy of adaptation and mitigation strategies in the context of sustainable development: The case of Vietnam. *Clim. Policy* **2003**, *3*, S81–S96. [CrossRef]
59. Beckman, M. Converging and conflicting interests in adaptation to environmental change in central Vietnam. *Clim. Dev.* **2011**, *3*, 32–41. [CrossRef]
60. Bruun, O. Sending the right bill to the right people: Climate change, environmental degradation, and social vulnerabilities in Central Vietnam. *Weather Clim. Soc.* **2012**, *4*, 250–262. [CrossRef]
61. Adger, W.N.; Kelly, P.M.; Nguyen, H.N. (Eds.) *Living with Environmental Change*; Routledge: London, UK, 2001.
62. Garschagen, M. Resilience and organisational institutionalism from a cross-cultural perspective: An exploration based on urban climate change adaptation in Vietnam. *Nat. Hazards* **2013**, *67*, 25–46. [CrossRef]
63. Tran, T.; Neefjes, K.; Ta, T.T.H.; Le, N.T. *Vietnam Special Report on Managing the Risks of Extreme Events and Disasters to Advance Climate Change Adaptation*; Vietnam Publishing House of Natural Resources, Environment and Cartography: Institute for Hydrology, Environment and Meteorology (IMHEM) and UN Development Programme (UNDP): Hanoi, Vietnam, 2015; pp. 1–456.
64. Huynh, P.; Resurreccion, B.P. Women's differentiated vulnerability and adaptations to climate-related agricultural water scarcity in rural Central Vietnam. *Clim. Dev.* **2014**, *6*, 226–237. [CrossRef]

65. Delisle, S.; Turner, S. "The weather is like the game we play": Coping and adaptation strategies for extreme weather events among ethnic minority groups in upland northern Vietnam. *Asia Pac. Viewpo.* **2016**. [CrossRef]
66. Socialist Republic of Vietnam. *National Strategy on Climate Change*; Socialist Republic of Vietnam, Prime Minister's Office: Hanoi, Vietnam, 2011.
67. Socialist Republic of Vietnam. *Vietnam's Submission on Reference Levels for REDD+ Results-Based Payments under the UNFCCC*; Submission to the UN Framework Convention on Climate Change by the Socialist Republic of Vietnam: Hanoi, Vietnam, 2016.
68. Socialist Republic of Vietnam. *Intended Nationally Determined Contribution of Viet Nam*; Submission to the UN Framework Convention on Climate Change by the Socialist Republic of Vietnam: Hanoi, Vietnam, 2015.
69. Pham, T.T.; Di Gregorio, M.; Carmenta, R.; Brockhaus, M.; Le, D.N. The REDD+ policy arena in Vietnam: Participation of policy actors. *Ecol. Soc.* **2014**, *19*, 22. [CrossRef]
70. Pham, T.T. *REDD+ Politics in the Media: A Case Study from Vietnam*; Center for International Forestry Research (CIFOR): Bogor, Indonesia, 2011; pp. 1–48.
71. Pham, T.T.; Moeliono, M.; Hien, N.T.; Tho, N.H.; Hien, V.T. *The Context of REDD+ in Vietnam: Drivers, Agents and Institutions*; Center for International Forestry Research (CIFOR): Bogor, Indonesia, 2012; pp. 1–98.
72. Socialist Republic of Vietnam. *National REDD+ Action Plan Vietnam*; Socialist Republic of Vietnam, Prime Minister's Office: Hanoi, Vietnam, 2012; pp. 1–18.
73. Ministry of Agriculture and Rural Development. *Action Plan Framework for Adaptation and Mitigation of Climate Change of the Agricultural and Rural Development Sector Period 2008–2020*; Ministry of Agriculture and Rural Development: Hanoi, Vietnam, 2009.
74. Dung, N.V.; Van, N.H. *Payment for Forest Environmental Services Policy in Vietnam and Its Implications on Sub-National Forest Management Institutions*; Policy Brief; People and Nature Reconciliation (PanNature): Hanoi, Vietnam, 2015.
75. Dung, N.V.; Van, N.H. *Payment for Forest Ecosystem Services Impact Assessment at Sub-National Level. Presentation at National Workshop on PFES and Multi-Stakeholder Participation in Vietnam*; People and Nature Reconciliation (PanNature) and Vietnam Fund for Forests: Hanoi, Vietnam, 2015.
76. Pham, T.T.; Dung, L.N.; Vũ, T.P.; Nguyen, H.T.; Nguyen, V.T. *Forest Land Allocation and Payments for Forest Environmental Services in Four Northwestern Provinces of Vietnam*; Center for International Forestry Research (CIFOR) Occasional Paper 155; CIFOR: Bogor, Indonesia, 2016.
77. Pham, T.T.; Moeliono, M.; Brockhaus, M.; Le, D.N.; Wong, G.Y.; Le, T.M. Local preferences and strategies for effective, efficient, and equitable distribution of PES revenues in Vietnam: Lessons for REDD. *Hum. Ecol.* **2014**, *42*, 885–899. [CrossRef]
78. Dung, L.N.; Loft, L.; Tjajadi, J.S.; Pham, T.T.; Wong, G. *Being Equitable is not Always Fair: An Assessment of PFES Implementation in Dien Bien, Vietnam*; Center for International Forestry Research (CIFOR) Working Paper 205; CIFOR: Bogor, Indonesia, 2016.
79. Ministry of Agriculture and Natural Resources and Japanese International Cooperation Agency. *PRAP Preparation Handbook*; Ministry of Agriculture and Natural Resources and Japanese International Cooperation Agency: Dien Bien, Vietnam, 2014.
80. Adger, W.N. Social vulnerability to climate change and extremes in coastal Vietnam. *World Dev.* **1999**, *27*, 249–269. [CrossRef]
81. Bruun, O.; Casse, T. *On the Frontiers of Climate and Environmental Change: Vulnerabilities and Adaptations in Central Vietnam*; Springer: Dordrecht, The Netherlands, 2013.
82. Socialist Republic of Vietnam. *Decision 204 QD-TTG on the Designation of Poor, Remote and Border Communes for Programme 135*; Socialist Republic of Vietnam: Hanoi, Vietnam, 2016.
83. Van, N.T.H. Embedding Forest Carbon in Vietnam's Forestland Property Relations. Master's Thesis, Forest and Nature Conservation Policy Group, Wageningen University, Wageningen, The Netherlands, 2014.
84. Asian Development Bank. *Ecosystem-Based Approaches to Address Climate Change Challenges in the Greater Mekong Subregion*; Asian Development Bank—Greater Mekong Subregion Environment Operations Center: Manila, Philippines, 2015; pp. 1–8.
85. Institute for Strategy and Planning on Natural Resources and Environment. *Mainstreaming Ecosystem-Based Adaptation in Vietnam*; Institute for Strategy and Planning on Natural Resources and Environment, Ministry of Natural Resources and Environment: Hanoi, Vietnam, 2013.

86. Kengoum, F. *Adaptation Policies and Synergies with REDD+ in Democratic Republic of Congo*; Center for International Forestry Research (CIFOR) Occasional Paper 135; CIFOR: Bogor, Indonesia, 2015.
87. Nkem, J.; Kalame, F.B.; Idinoba, M.; Somorin, O.A.; Ndoye, O.; Awono, A. Shaping forest safety nets with markets: Adaptation to climate change under changing roles of tropical forests in Congo Basin. *Environ. Sci. Policy* **2010**, *13*, 498–508. [CrossRef]
88. Innes, J.L.; Hickey, G.M. The importance of climate change when considering the role of forests in the alleviation of poverty. *Int. For. Rev.* **2006**, *8*, 406–416. [CrossRef]
89. Wang, G.; Wang, T.; Kang, H.; Mang, S.; Riehl, B.; Seely, B.; Liu, S.; Guo, F.; Li, Q.; Innes, J.L. Adaptation of Asia-Pacific forests to climate change. *J. For. Res.* **2016**, *27*, 469–488. [CrossRef]
90. Wang, G.; Mang, S.; Kryzanowski, J.; Guo, F.; Wang, T.; Riehl, B.; Kang, H.; Li, Q.; Innes, J.L. Climate change and forest adaptation in the Asia-Pacific. *J. Geogr. Res.* **2015**, *63*, 1–36. [CrossRef]
91. McElwee, P.D. Forest environmental income in Vietnam: Household socioeconomic factors influencing forest use. *Environ. Conserv.* **2008**, *35*, 147–159. [CrossRef]
92. Red Cross. *Mangrove Plantation in Viet Nam: Measuring Impact and Cost Benefit*; International Federation of Red Cross and Red Crescent Societies: Hanoi, Vietnam, 2012.
93. Schmitt, K. *Protection and Sustainable Use of Coastal Wetlands through Co-Management and Mangrove Rehabilitation with Emphasis on Resilience to Climate Change*; Gesellschaft fur Internationale Zusammenarbeit (GIZ): Soc Trang, Vietnam, 2015.
94. Nam, V.N.; Sasmito, S.D.; Murdiyarso, D.; Purbopuspito, J.; MacKenzie, R.A. Carbon stocks in artificially and naturally regenerated mangrove ecosystems in the Mekong Delta. *Wetl. Ecol. Manag.* **2016**, *24*, 231–244. [CrossRef]
95. McElwee, P.D.; Thanh, N.C. *Review of Three Years Policy of Payments for Forest Environmental Services in Vietnam*; Forest Protection and Development Fund, Ministry of Agriculture and Rural Development and Winrock International: Hanoi, Vietnam, 2015.

13

Evolving Protected-Area Impacts in Mexico: Political Shifts as Suggested by Impact Evaluations

Alexander Pfaff [1,*,†], Francisco Santiago-Ávila [2,†] and Lucas Joppa [3]

1 Duke University, 302 Towerview Drive, Durham NC 27708, USA
2 University of Wisconsin, 1220 Linden Drive, Madison, WI 53706, USA; santiagoavil@wisc.edu
3 Microsoft Research, One Microsoft Way, Redmond, WA 98052, USA; lujoppa@microsoft.com
* Correspondence: alex.pfaff@duke.edu
† Co-lead authors.

Academic Editors: Esteve Corbera and Heike Schroeder

Abstract: For protected areas (PAs), variation in forest impacts over space—including types of PA—are increasingly well documented, while shifts in impacts over time receive less attention. For Mexico, in the 1990s, PAs effectively were 'paper parks'. Thus, achieving impacts on the forest would require shifts over time in the politics of PA siting and PA implementation. We rigorously analyze the impacts of Mexican PAs on 2000–2005 loss of natural land cover, using matching to reduce location bias caused by typical land-use economics and politics. We find a 3.2% lower loss, on average, due to PAs. Since politics often vary by type of PA, we also show that in Mexico stricter PAs are closer to cities and have greater impact than mixed-use PAs. These shifts in impacts suggest some potential for PAs to conserve forests.

Keywords: Mexico; deforestation; conservation; protected areas; impact evaluation; matching

1. Introduction

In recent decades, national and international efforts to reduce forest loss have had some impact but have not substantially slowed tropical forest loss. Adding climate change to concern about species, such forest losses now account for about one-sixth of anthropogenic greenhouse gas emissions. Climate-related incentives could enhance various programs and policies that affect deforestation, most likely by emphasizing performance with monitoring, reporting, and verification of outcomes. Such emphasis, with incentives, could increase policies' deforestation impacts [1].

Understanding of past policy impact also supports greater future impacts, through better design. It is crucial for countries to know what has (not) worked to date in reducing forest loss, and why. Here, we follow a recent explosion of literature that has provided improved impact evaluations for conservation policies, such as protected areas (PAs—impacts reviewed in [2]). Improved methods often reduce estimates of average PA impacts, while also clearly demonstrating variation in impacts across space and PA types [3,4].

First, to highlight the actors and incentives relevant for variation in impacts of PAs upon forests, we present conceptual models of individual land-use choices and the political economy of PA siting. They suggest key influences on PAs' impacts. Our results shed empirical light on those influences, suggesting that the politics of PA implementation in Mexico clearly have shifted over time.

We start with a classic land-use model, taking as given both the PAs' locations and their enforcement. The clear prediction that typical landscapes vary in their baseline deforestation rates is the motivation for our efforts to control for characteristics of sites that affect individual decisions that yield deforestation. Only by removing the influence of such characteristics, which might well be expected to differ for protected versus unprotected locations [5], can we infer the impact of the PAs.

As to why PA locations might differ from unprotected sites in relevant characteristics, we highlight political and economic tradeoffs. Those tradeoffs predict variations in both locations and enforcement across PA types and, thus, also varied impacts [6]. In sum, these models suggest that, depending on the political economic context, either strict or mixed-use PAs can have more impact.

For Mexico, an important basis for comparison is recent work showing no PA impacts in the 1990s ([7] on 'paper parks'). Thus, we consider PA impacts on more recent changes in land cover (2000–2005) after a shift in the politics of conservation in Mexico, with a change in administration in 2000 to President Fox, under whom Ernesto Enkerlin shifted management of PAs. In addition, GEF provided stable funds for some PAs (https://www.thegef.org/country/mexico). The time period we study, coming just after the 1990s, offers a useful window on a possible sharp shift in politics. We estimate the average PA impact and then differences in impacts between strict and mixed-use PAs.

Our application of matching to improve estimates of PA impact—following both of our models—reduces estimated impacts, as hypothesized, relative to when ignoring key location characteristics. Nonetheless, improved estimates show PAs in Mexico did lower land-cover change. Specifically, PAs across Mexico reduced the 2000–2005 loss of land cover by 3.2%, on average.

In comparing PA types, following the political economy model, we find that strict PAs avoid more land-cover change (5.2%) than mixed-use PAs (2.7%). That is intuitive, if enforcement is tougher. However, we show that greater impact from strict PAs is influenced by the relative locations of the PA types. That may seem counterintuitive but it is consistent with the PAs' sites having been chosen at a time when PA enforcement was low.

The paper proceeds as follows. Section 2 provides background for Mexico and prior literature evaluating policy impacts. Section 3 presents both of the conceptual models just discussed above. Section 4 then describes our data and methods, Section 5 conveys results, and Section 6 concludes.

2. Background and Related Literature

Mexican forests cover 67 million hectares, about one-third of the country (198 million hectares). Along with agriculture and fishing, forestry accounted for about 5% of the country's GDP in 2006. Agriculture and forestry are areas where greenhouse gas (GHG) emission can be reduced, having generated about 135 $MtCO_2e$ or 21% of Mexican emissions in 2002 [8], with two-thirds from the forest sector. Proximate causes of both deforestation and degradation are conversion to grassland, slash-and-burn agriculture, illegal logging, and fire. Underlying forces include a lack of investment in the forestry sector, low income from forest activities, multiple agricultural and livestock activities, uncertainty related to use rights, poverty, and a general lack of opportunities for forest owners [9]. The drivers are complex and they vary between regions.

It is widely acknowledged that successful forest interventions could not only reduce emissions but also generate ecosystem services, income, and employment—among other co-benefits [8]. Interventions such as reforestation and commercial plantations are what account for 85% of the state's proposed mitigation in agriculture and forestry. Success would depend on institutional changes, better public financing, and sustainable forest product markets [8].

This direction for policy appears to have some support in Mexico, despite a large illegal timber market, a lack of financial and human resources for operational capability, and (drug-related) insecurity. Mexico currently is emphasizing a national, multi-functional and multi-scale mechanism for monitoring, reporting and verification (MRV) based on remote sensing and ground-based forest inventory methodologies [9]. It could include early detection for changes in land cover and land use [10]. Such a mechanism could provide the relevant authorities with more precise measures to address concern about detection of small-scale land-cover changes.

2.1. Evaluating PA Impacts

Joppa and Pfaff [2] review the PA literature—as do Naughton-Treves [11], Nagendra [12], and Campbell et al. [13]—emphasizing obstacles to inferring PA impacts on forest. This follows global documentation [5] that, at least on average, the locations of PAs across national landscapes are significantly biased in ways relevant to deforestation rates. They stress that observing fully forested PAs (e.g., [14,15]) falls short of observing impacts, as impacts require a comparison to what would have happened without PAs. Given that it is literally impossible to observe such 'counterfactuals'—scenarios without PAs—one must estimate what would have happened on PA land without protection, using outcomes from unprotected sites. Siting bias means the average outcome for all unprotected lands (as in [16–18]) is a poor counterfactual estimate for PAs, since sites are different on average. Comparing to areas around PAs [15,19–23] can help. Yet, without explicitly comparing land characteristics, it is hard to know how similar they are. At least as troubling, those nearby lands could be affected by the establishment of the PA if that generates local spatial spillovers such as 'leakage'.

Matching methods explicitly compare land characteristics between sites, aiming to increase the similarity of controls to PA observations. Andam et al. [24]'s application of matching to Costa Rican PAs reduces the estimated impact from about 44%, over decades, to about 11% of the protected area. Joppa and Pfaff [3] provide analogous results for each of over 100 countries around the globe. Given those average impacts, Pfaff et al. [4] revisit Costa Rican PAs in subsets, using matching, to confirm predicted variation in PAs' impacts over the landscape. Impacts are higher near roads and cities and on flat land. Shah and Baylis [25] also show heterogeneous PA impacts on forest, for Indonesia, while Joppa and Pfaff [3] confirm such predictable variation in impacts in their global study of PAs.

2.2. Evaluating Conservation Impacts in Mexico

Early conservation evaluations for Mexico included payments by a federal agency to upstream land, which can affect water quality. Munoz [26] finds significant location bias in payments contracts, lowering impact. Alix-Garcia et al. [27] extend this with similar results while Alix-Garcia et al. [28], using additional data, consider both forest and poverty impacts of these hydro-services payments. Recently, Pfaff et al. [29] and Kaczan et al. [30] consider institutional adjustments in such payments.

Concerning PAs, Blackman et al. [7] is analogous to our study except for the time period, i.e., the 1990s, when it was widely asserted that PAs were under-resourced 'paper parks' (while Sims and Alix-Garcia 2016's analogous ongoing work extends further forward in time and compares outcomes of PAs to PES [31]). Blackman et al. [7] find no PA effect. Figueroa and Sánchez-Cordero [32] also find limited effect for the 1990s. Low intention or ability to enforce can affect not only PA outcomes given PA sites but also PAs' sites because, without enforcement, there is little reason for locals to push back. Mas [33] and Durán-Medina et al. [34] find some impact for particular PA sites, as do Honey-Roses et al. [35,36], for a site where PES and a PA are combined, while Miteva et al. [37] examine tenure.

3. Modeling PA Impact

3.1. PA Impact by Location

From von Thunen [38] to the 'monocentric model', landscape analysts assert that the pressure to clear forest falls as we move outward from the market center (in Figure 1, a city where the axes cross). Transport costs imply falling profits from agricultural products to be sold in the city, as we move out. If all land is originally forested and only transport matters, forests will remain farther from markets (in Figure 1, forest will remain to the right of where the 'Expected Profits' line crosses the '0'axis).

Of course, factors other than transport also affect relative profits from agriculture versus forests: e.g., high slopes near markets may stay forested; and good soils far from market may be cleared. From an analyst's point of view, some of these factors are observed while others are unobserved, as there are limits on all datasets. The empirical analyses we review include observable factors; however, Figure 1 does not explicitly depict the observables, focusing on representing unobserved factors in the form of a distribution, or varying density, of land-parcel profits around the Expected Profit line.

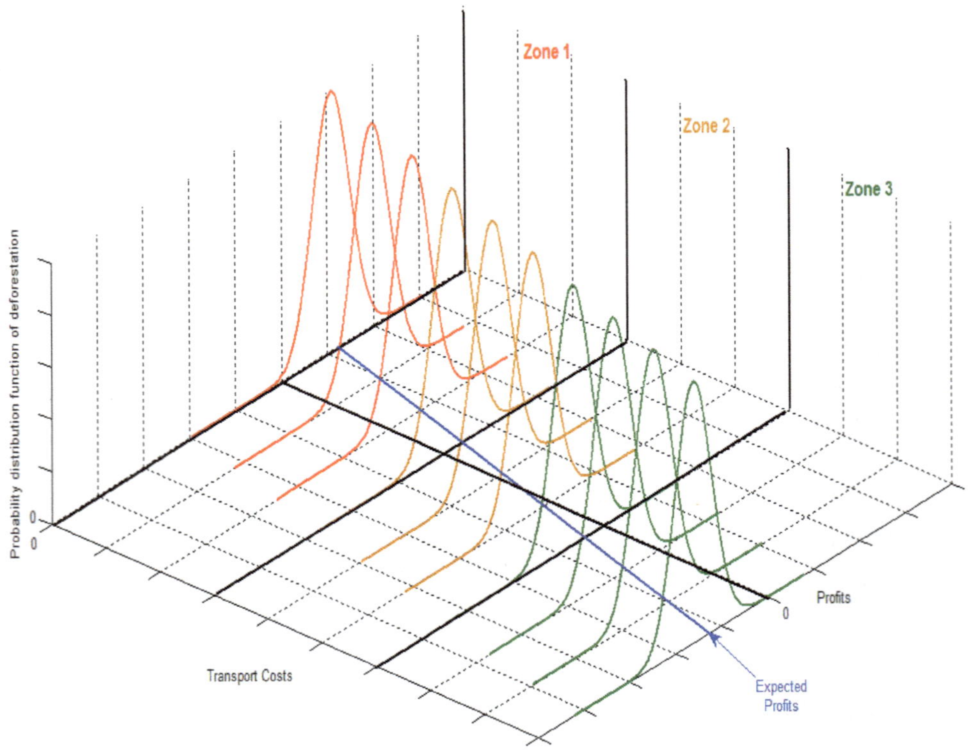

Figure 1. Expected Profits Across Landscape with Implications for PA Impact & Siting.

Conservation policies keep forests standing, using enforcement if there is pressure to clear forest. Without pressure, a fully forested PA does not imply impact, i.e., avoidance of baseline deforestation. Thus, if private pressure would not have led to clearing of the forest anyway, the policy did not have impact. More generally, a conservation policy's impact equals the private or 'baseline' deforestation rate that would have arisen without the policy minus the deforestation rate observed with the policy. Within Figure 1, if transport cost is a significant factor in the private (or 'baseline') rate of clearing, then a PA that is far to the right may not have had much impact on the forest, even if fully forested.

3.2. PA Impact by Type

Given the underlying logic above, we can see that PA sites determine average PA impacts. Those sites will be affected by the political economy of development: PAs block production; thus, lobbying by producers tends to push PAs to low-profit locations. Where agricultural profit is high, lands are expensive to buy for conservation, plus lobbying against allocating public lands to conservation will be significant. Formally, a simple theory of PA location would be that for a PA to be established its cost—including local profit in Figure 1 that is foregone—must be below benefit (perceptions of which vary by decision maker, as World Bank [39] shows for Brazil). For any given level of PA benefit, this theory predicts that PA siting may avoid costs by avoiding pressures (as confirmed by, e.g., [5]).

The political economy can vary with the PA type, as well as by country and across time periods. Development tradeoffs for 'strict' PAs differ from those for mixed-use PAs that permit some local extraction and, therefore, induce less lobbying against the establishment of the PA. Thus, PA types can end up with different locations. For instance, if differences in PA regulations by type are actually enforced, we might well expect to see stricter PAs further to the right in Figure 1 [40,41].

Yet where public actors locate new PAs depends on whether they intend to enforce them. For instance, if locals believed that PAs in Mexico would not be enforced, i.e., expected the 1990s 'paper parks' outcome in [7], then they would have had little reason to lobby against a PA. Even strict PAs might be permitted near cities. Once PAs are sited, enforcement may occur but may fall with isolation [42,43]. Thus, interactions with pressure could produce varied patterns across space of PAs' impacts.

4. Data and Matching Methods

4.1. Land Cover Data

Our data are from a global data set with land cover for 2000 [44] and land cover for 2005 (ESA) [45]. The 2000 data, GLC2000, have 23 classifications of land cover. They were reclassified to 'natural' or 'human modified', the latter including categories 16 (cultivated and managed areas), 17 (mosaic of cropland with tree cover or other natural vegetation), 18 (mosaics of cropland, with shrubs or grass cover), 19 (bare areas), and 22 (artificial surfaces and associated areas). The same was done for 2005 GLOBCOVER300 data. Again, multiple categories are placed into "natural" and "modified", the latter including categories 11 (irrigated croplands), 14 (rainfed croplands), 20 (mosaic cropland (50%–70%)), 30 (mosaic cropland (20%–50%)), and 190 (urban areas >50%). These datasets were not constructed for precise intertemporal comparison, yet this transformation allows for a reasonable comparison of years. In each dataset, the spatial scale of land observations is 1-km^2 polygons.

The change between the datasets—which we call 'vegchange'—was calculated after that process. With the transformation, the data in principle track change from a 'natural' to a 'human modified' landscape and this 2000–2005 change is the dependent variable upon which we focus. We want to study recent changes that can be affected by the presence of PAs (though they are net changes, i.e., from 'natural' to 'human modified' and vice versa, which matters for indicators of species habitat). Given any issues with comparing datasets, we also analyze spatial patterns in the 2005 land cover.

4.2. Land Characteristics Data

Table 1 provides descriptive statistics for land characteristics in our analysis. Elevation (m) is from the Shuttle Radar Topography Mission [46], with slope in degrees from horizontal. The roads and urban areas used to compute distances (km) are from VMAP0 Roads of the World [47] and the Global Rural Urban Extent data [48]. While not the best quality, these VMAP0 data are all that is freely accessible to define the global road network.

Table 1. Land characteristics and land cover (including PA type).

	Unprotected	Protected	Protected–Stricter	Protected–Mixed-Use
Protected, Stricter Subset (1/0)	0	0.21	1	0
Protected, Mixed-use Subset (1/0)	0	0.79	0	1
IUCN Category (1–6)	0	5.0	1.3	6.0
Distance Outside PA Edge (km)	55.8	−14.0	−10.5	−15.0
Urban Distance (km)	35.6	95.7	58.4	105.3
Road Distance (km)	8.5	12.6	14.3	12.2
Elevation (m)	1081	622	885	554
Slope (degrees)	2.81	2.85	3.42	2.70
Agricultural Suitability (0–9)	6.17	6.81	6.57	6.88
Relatively Low Agric. Suitability (1/0)	0.39	0.56	0.38	0.60
Relatively High Agric. Suitability (1/0)	0.37	0.21	0.22	0.21
Fires (number detected in 2000–2006)	0.005	0.003	0.004	0.003
Dummy for Whether Any Fires (1/0)	0.005	0.003	0.004	0.003
Pine Oak Forest Dummy (1/0)	0.115	0.003	0.005	0.003
Chihuahua Desert Dummy (1/0)	0.163	0.064	0.012	0.078
GLC 2000 Natural Land Cover (1/0)	0.840	0.920	0.912	0.921
Globcover 2005 Natural Land Cover (1/0)	0.862	0.957	0.970	0.954
2000–2005 Loss of Natural Cover (1/0)	0.092	0.029	0.022	0.031
# observations	*1,801,935*	*121,847*	*25,096*	*96,751*

PAs (protected areas) are from World Database on Protected Areas [49]. In these data, Categories I–II allow less human intervention, while Categories III–VI tend to be less protected, allowing for multiple uses. For any overlaps of categories, the polygon was categorized into the stricter category. We created three dummy variables: 'protected', for any protection (regardless of IUCN category); 'mixed-use PA' for IUCN categories III–VI; and 'strict PA', if the IUCN category is I or II. In order to check robustness, we also compared categories I–IV versus V–VI (following Joppa and Pfaff [3], as well as Nelson and Chomitz [40]). Also, for our time period, we used only the PAs created by 2000.

The World Wildlife Fund classified ecoregions [50]. Unclassified ecoregions were dropped ($n = 1747$) and we created dummy variables for the two ecoregions with the highest frequencies: 'pineoakdum' (10.7% of total) if the ecoregion is Sierra Madre Occidental Pine Oak Forest; and 'chidesertdum' (15.6% of total) if the ecoregion is Chihuahuan desert. The agricultural suitability is from International Institute for Applied Systems Analysis' Global Agro-Ecological Zones data [51]. It uses climate, soil type, land cover, and slope to assign a value to each polygon, ranging from 0 (meaning no constraints) to 9 (severe constraints). We created two dummy variables: 'low' suitability for more agricultural constraints (i.e., agricultural suitability score of 8 or higher); and 'high' for situations with fewer agricultural constraints, (i.e., agricultural suitability score of 5 or lower). Additionally, as a robustness check, we use the full variable (which does not affect any of the results).

The fire variable, as a continuous metric, simply captures the number of fires from 2001 to 2006. The 'firedum' dummy we created takes a value of '1' if that polygon experienced any fires ($\geq$1). Finally, the distance-to-edge variable measures the distance to the edge of a protected area (km) from the polygon in question. A negative distance value indicates that the observation is inside of the PA.

These variables are not expected to fully explain land-cover change or the location of protection. Yet as all influence profit, they predict deforestation as well as the local resistance to protected areas. It is a combination of relevance to protection and deforestation that makes them useful for matching. The data contain approximately 1,935,301 observations (1-km^2 polygons of land) and 13 variables. We have run many analyses with random 10% samples to confirm our results are robust to sampling. As noted, Table 1 presents summary statistics for all the aforementioned variables. For our results, we explore their impacts upon land-cover change as well as their significance for locations of PAs.

4.3. Matching Methods

If PAs were sited randomly, PA impacts would be easy to estimate using the differences between deforestation inside and outside of PAs. Deforestation outside would be an unbiased estimate of what would have been occurred inside PAs had there been no protection, as other factors would cancel out. Yet neither PAs in general, nor any PA type, seems to be distributed as if at random. Further, that non-randomness often appears to be along dimensions that can affect deforestation. To isolate PA impact by removing the influences of these differences, we use 'matching' methods to improve controls. One 'matches' each protected polygon with the most similar unprotected polygon(s) to get as close as possible to 'apples-to-apples' comparisons. Thus, PAs are compared not to all unprotected land but, instead, to the most similar unprotected land. Here, we apply propensity-score matching in our many initial analyses and then, for our final results, confirm robustness to covariate matching.

Propensity-score matching assesses the 'similarity' of sites using the predicted probability of a polygon being in a PA. PA polygons are then compared to unprotected polygons with similar enough characteristics to yield a similar probability of protection. Probabilities are generated by a probit model using factors in protection and deforestation to explain where protection occurred [52]. More weight is given, then, to the variables that are important determinants of protection. In covariate matching, similarity is assessed using the distance between polygons in the covariate space. For either method, we matched each treated polygon with the single most similar unprotected polygon.

However, selecting the most similar polygon does not guarantee that controls are, in fact, similar. Thus, we also check explicitly whether the selected unprotected polygons are similar to the protected.

We examine balance, i.e., if the characteristics' values are distinguishable between the protected and matched unprotected observations. Ideally, they should not be. Assuming large differences to begin with, we would expect at least a significant reduction in differences between groups, due to matching.

Given balanced characteristics, the deforestation in matched unprotected sites is an improved estimate of the deforestation that would have occurred at PA locations without the protection. PA impact is calculated as that counterfactual rate minus the observed deforestation (given protection). However, still there will be differences between these groups, in terms of characteristics relevant for deforestation. Thus, our preferred matching estimates involve first matching and then regressions. We refer to the latter as 'bias adjustment', as this addresses remaining differences in characteristics.

If the unobservable or omitted factors are correlated with "treatment", i.e., with where PAs are, that could bias estimates of impact. Matching can control only for the included observable factors. For instance, we do not know anything about the populations in the PAs versus unprotected sites. We suspect that factors we do observe, however, such as the road and city distances, correlate with the unobservables. Thus, given our observables, we cannot be sure of the sign of any residual biases. As a robustness check, we compute 'Rosenbaum bounds' to estimate how sensitive are these results.

5. Results

5.1. Descriptive Statistics

Table 1 provides means for our outcomes variables as well as our metrics for drivers, including Mexico's protection interventions—both in general and by PA type. Note that 79% of protection is a 'mixed-use' type, with an average IUCN category of 6, while 21% of the PA observations are stricter protection, having an average IUCN category of 1.3.

In terms of drivers of deforestation, Table 1 leads with a very important driver, urban distance. It is the driver that differs most in these data between PAs and unprotected polygons, and between the two types of PAs. Protected polygons are roughly three times as far as unprotected. Mixed-use PAs are roughly twice as far from cities as strict PAs—key for different PA impacts across types. We note that these relative locations of PA types in Mexico differ from the Brazilian Amazon [41] and from the averages across the globe (see Nelson and Chomitz [40] concerning PAs' impacts on fires).

The relative location of PAs on average fits our political economy model in which, for a given societal benefit, opportunity costs drive PAs to relatively unprofitable sites. Table 1 shows this for urban distance and, to a lesser extent, for road distance, as well as for the fractions of high and low quality soils. Without considering such differences in location, Table 1's last row provides a form of estimate of PAs' average impact: there is a 2.9% loss of natural land cover in PAs, lower than the 9.2% loss for unprotected observations. Further, Table 1 suggests that there was actually some enforcement difference across stricter and mixed-use PAs during our time period (after alleged political shifts). The land-cover loss of 2.2% is lower for strict PAs, despite the fact that the only major difference in characteristics is that strict are *closer* to market (recall, siting occurred before our time period). However, actually estimating PA impact requires revisiting with control for all land characteristics.

5.2. Drivers of Protection

Table 2 uses regression to consider which locations are protected. Considering urban distance, PAs are further from urban areas even controlling for other influences. As expected from Table 1, the coefficient on urban distance for mixed-use PAs is larger than for strict, i.e., mixed-use are farther. For road distance, the smaller coefficients also support Table 1, i.e., the mixed-use PAs are closer (PAs may be near rural *ejidos* with road access, as 70% of land with forest is owned by *ejidos*, as Mexico's 1992 Agrarian Law, Article 1 gave these communities legal status and land ownership and they can participate in conservation programs such as Mexico's PES (The REDD Desk, 2012).) The other findings in Table 2, e.g., for agricultural suitability, also reflect the differences seen in Table 1.

Table 2. Drivers of protection (including by type).

	Protected	Protected–Stricter Subset	Protected–Mixed-Use Subset
Urban Distance	0.00251 ***	0.00032 ***	0.00248 ***
	(0.000)	(0.000)	(0.000)
Road Distance	0.00031 *	0.00066 ***	−0.00035 ***
	(0.000)	(0.000)	(0.000)
Elevation	−0.00001 ***	0.00000 ***	−0.00001 ***
	(0.000)	(0.000)	(0.000)
Slope	0.00120 ***	0.00048 ***	−0.00044 ***
	(0.000)	(0.000)	(0.000)
High Agricultural Suitability	−0.01533 ***	−0.01053 ***	−0.00564 ***
	(0.000)	(0.000)	(0.000)
Pine Oak Forest	−0.10585 ***	−0.03090 ***	−0.08272 ***
	(0.000)	(0.000)	(0.000)
Chihuahuan Desert	−0.10554 ***	−0.03271 ***	−0.08405 ***
	(0.000)	(0.000)	(0.000)
constant	−0.00390 ***	−0.00566 ***	0.00461 ***
	(0.000)	(0.000)	(0.000)
# obs	1,923,782	1,827,031	1,898,686
R^2	0.182	0.021	0.196
Adjusted R^2	0.182	0.021	0.196

Robust standard errors in parentheses: *** $p < 0.001$; ** $p < 0.01$; * $p < 0.05$.

5.3. *Drivers of Land-Cover Change*

Before estimating impacts of PAs on land cover, with controls for effects of other determinants, we would like to empirically confirm for our data what was driving the loss of natural land cover in Mexico in this time period. To do this, we examine rates of loss in just the unprotected locations. Table 3 presents OLS regressions of both land-cover loss and 2005 land cover on characteristics (note that we would expect the coefficients in these two outcomes columns to have the opposite signs).

Table 3. Drivers of natural land cover, explaining both loss & final amount (outside PAs).

	Natural Land Cover **Loss 2000–2005**	***Natural Land Cover*** **Amount 2005**
Urban Distance	−0.00101 ***	0.00137 ***
	(0.000)	(0.000)
Road Distance	−0.00074 ***	0.00064 ***
	(0.000)	(0.000)
Elevation	−0.00004 ***	0.00006 ***
	(0.000)	(0.000)
Slope	−0.00445 ***	0.00572 ***
	(0.000)	(0.000)
High Agricultural Suitability	0.03979 ***	−0.06338 ***
	(0.001)	(0.001)
Pine Oak Forest	0.06457 ***	−0.06964 ***
	(0.001)	(0.001)
Chihuahuan Desert	−0.02074 ***	0.01979 ***
	(0.001)	(0.001)
Constant	0.19813 ***	0.75283 ***
	(0.001)	(0.001)
# obs	1,514,249	1,801,935
R^2	0.03	0.06
Adjusted R^2	0.03	0.06

Robust standard errors in parentheses: *** $p < 0.001$, ** $p < 0.01$, * $p < 0.05$.

In Table 3, as expected, urban distance has a significant effect on the profitability of land-cover loss. Also expected, road distance matters, even controlling for urban distance. Transport is critical. Higher elevation and slope also discourage whatever land uses involve the loss of natural cover, again as expected. Further, also expected, better conditions for agriculture increase the loss of cover. Both the Sierra Madre pine oak forest and the Chihuahuan desert are statistically significant controls.

5.4. Matching on Drivers of Land-Cover Change

Table 4 communicates whether our matching approach has actually made controls more similar. Again we start with the urban distance. For all PA polygons, or the strict and mixed-use subsets, Table 4 shows over 95% reductions in the differences between the PAs and unprotected controls. As the other differences noted in Table 1 were for road distance and high agricultural suitability, Table 4 shows similar reductions for them, as is the case for the significant and large ecoregions. These reductions in group differences are, we feel, a solid basis to claim improved estimates of PA impact.

Table 4. Matching balance.

Propensity Score Matching	Protected–Unprotected Characteristics Differences	Protected–Stricter Subset	Protected–Mixed-Use Subset
Urban Distance			
Pre-match difference	60.064 ***	19.273 ***	69.409 ***
Post-match difference	−2.580 ***	−0.817	−2.759 ***
% residual difference	−4.3%	−4.2%	−4.0%
Road Distance			
Pre-match difference	4.128 ***	5.584 ***	3.622 ***
Post-match difference	0.223 ***	0.174	−0.029
% residual difference	5.4%	3.1%	−0.8%
Elevation			
Pre-match difference	−459.298 ***	−169.814 ***	−524.714 ***
Post-match difference	−126.659 ***	26.358 **	−61.156 ***
% residual difference	27.6%	−15.5%	11.7%
Slope			
Pre-match difference	0.039 ***	0.617 ***	−0.118 ***
Post-match difference	−0.684 ***	−0.133 **	−0.003
% residual difference	†	−21.6%	2.5%
High Agric. Suitability			
Pre-match difference	−0.163 ***	−0.139 ***	−0.165 ***
Post-match difference	−0.049 ***	0.013 **	−0.002
% residual difference	30%	−9.4%	1.2%
Pine Oak Forest			
Pre-match difference	−0.111 ***	−0.105 ***	−0.110 ***
Post-match difference	−0.007 ***	−0.002 **	−0.005 ***
% residual difference	6.3%	1.9%	4.5%
Chihuahuan Desert			
Pre-match difference	−0.099 ***	−0.147 ***	−0.084 ***
Post-match difference	−0.021 ***	−0.001	−0.025 ***
% residual difference	21.2%	0.7%	29.8%

† increased small difference; *** $p < 0.001$; ** $p < 0.01$; * $p < 0.05$.

That said, matching always leaves residual characteristics differences between treated and control, noting that we also tested matching 'calipers' of 0.01 and 0.001, yet that did not improve balances. Within Table 4, many differences are statistically significant, in part due to the very large sample (a reason we checked for robustness to random 10% samples (Supplemental Materials: Tables S1 and S2). Thus, the post-matching regressions that we have employed are well motivated in Table 4—but also we simply flag these residual differences (another reason we test robustness to covariate matching). Finally, we note that unobservable differences could also exist—always a key issue for matching.

5.5. Average PA Impact

For loss of natural land cover and the 2005 amount of land cover, Table 5 presents PAs' impacts. The table's first column presents average impact, i.e., the interventions are any kind of protection. We discuss here the 2000–2005 loss of natural land cover (as impacts upon 2005 cover are analogous).

Table 5. PA impact estimates.

	Protected	Protected–Strict Subset	Protected–Mixed-Use Subset
2000–2005 Natural Land Cover Loss			
Pre-match difference in means	−6.3 % ***	−7.0 % ***	−6.1%***
# observations	*1,923,782*	*1,827,031*	*1,898,686*
Pre-match regression (all the data)	−3.5 % ***	−5.8 % ***	−2.6 % ***
SE	*(0.001)*	*(0.001)*	*(0.001)*
# observations	*1,626,295*	*1,537,139*	*1,603,405*
Post-match regression (most similar)	−3.2 % ***	−5.2 % ***	−2.7 % ***
SE	*(0.002)*	*(0.003)*	*(0.001)*
# observations	*181,844*	*38,633*	*143,318*
2005 Natural Land Cover			
Pre-match difference in means	9.5 % ***	10.8 % ***	9.2 % ***
# observations	*1,923,782*	*1,827,031*	*1,898,686*
Pre-match regression (all the data)	3.7 % ***	6.8 % ***	2.5 % ***
SE	*(0.001)*	*(0.001)*	*(0.001)*
# observations	*1,923,782*	*1,827,031*	*1,898,686*
Post-match regression (most similar)	3.5 % ***	6.3 % ***	2.9 % ***
SE	*(0.002)*	*(0.003)*	*(0.001)*
# observations	*201,569*	*43,655*	*157,826*

*** $p < 0.001$; ** $p < 0.01$; * $p < 0.05$.

Table 5's first row shows means differences, analogous to the impact estimate implied by Table 1 (a little different because, as seen in Table 3's first column, this outcome has fewer observations). Its second row introduces controls for the effects of other determinants of natural land-cover loss using regression analysis with all observations. This lower the estimated impact by almost half (as in Andam et al. [24] for Costa Rica and globally, for 141 countries, within Joppa and Pfaff [3]). The matching estimate in Table 5's third row confirms robustness to matching's use of observables, finding a 3.2% reduction in the 2000–2005 loss of land cover, using just the matched observations.

5.6. Impacts by PA Type

Table 5's second and third columns separate the PA treatment into the strict versus mixed-use PAs. We note that enforcement choices are illuminated by simple differences in means, in the top row, where the loss rate in each PA type is compared to the loss rate within all unprotected polygons. Despite the fact that mixed-use PAs are further from cities, which should lower clearing pressures, more loss of natural land cover is occurring within mixed-use PAs. This simplest 'impact estimate' is lower for mixed-use PAs (6.1%) than for strict PAs (7.0%). Enforcement seems to occur on average and, it seems here, enforcement choices differed by PA type in that clearing differs across PA types.

The differences between Table 5's first row and its second and third rows, which employ controls, in turn illuminate the different location choices for the two types of PAs. The more biased are PAs' locations towards low pressure—as we predict given apparent enforcement on average for this time period (again, per Blackman et al. [7], such enforcement did not always occur)—the greater the reduction in estimated PA impact we should see when controlling for all the land characteristics.

When we control using observables, impact estimates for strict and mixed-use PAs both fall, consistent with the effect in the column for all of the PAs. The estimate for the mixed-use PAs falls by more, however, consistent with greater urban distance for mixed-use reducing the true impact, given the apparent impact. Given enforcement and locations, stricter PAs have greater impact (5.2% versus 2.7%, noting that this difference in estimated impacts across types is statistically significant). These results may be explained by siting in the 'paper park' era, before increases in enforcement.

These results are supported by our robustness checks making use of 10% samples, which also addresses potential issues with spatial autocorrelation. Further, calculation of Rosenbaum bounds, given that there could be important unobserved factors for which matching clearly cannot control, shows that to eliminate the significance of strict impacts requires unobserved factors to make non-PA sites almost three times as likely to be cleared–versus only twice as likely for impacts of all PAs.

6. Conclusions

With two goals in mind, we estimated a suite of PA impacts upon land-cover changes in Mexico. One motivation was to study a period after the 1990s, as conservation politics were alleged to have shifted and prior work had demonstrated that, during the 1990s, PAs functioned as 'paper parks'. Another was to demonstrate the need to address how land-use decisions imply the possibility of bias in PA impact estimates and how the political economy of public PA choices implies its likelihood.

PAs did reduce losses of natural land cover within their boundaries during 2000–2005. Thus, it would appear that conservation politics shifted, at least as revealed by our impact assessment. Further, it was important in estimating the impacts to check for, and then control for, differences in characteristics of protected (versus unprotected) sites, as they directly influence deforestation. Controlling for them lowers the estimated average impact of PAs by about half. Further, controls generate different adjustments to the estimated impacts for the strict versus the mixed-use PAs.

Across PA types, a combination of siting and enforcement differences yielded different impacts, with strict PAs generating greater reductions in loss within their boundaries than mixed-use PAs. These results may reflect shifts over time in conservation politics, as PA locations seem consistent with a lack of intention to enforce during the 1990s that, as our results suggest, later was reversed.

Extensions could improve on these analyses. Our data for land cover are not ideal and Mexico's new MRV mechanism may offer options. For moving forward to future analyses, e.g., INEGI has invested in data on roads and vegetation, for example, and the Nature Conservancy has invested in tracking sites and types of protection. Finally, more study of differences in dynamics across space could further contribute to an understanding of the conditions that drive the impacts of protection.

Acknowledgments: We are grateful for helpful suggestions from Maria Carnovale and Elizabeth Shapiro. This paper derives from Francisco J. Santiago-Ávila's master's project (or MP) at Duke University, concluded in 2013.

Author Contributions: Alexander Pfaff conceived of the paper; all the authors designed analyses; Lucas Joppa organized most of the data; Francisco J. Santiago-Ávila performed the analyses with guidance from Alexander Pfaff and Lucas Joppa; and all of the authors wrote the paper.

Conflicts of Interest: The authors declare no conflict of interest.

References

1. Pfaff, A.; Amacher, G.S.; Sills, E.O. Realistic REDD: Improving the forest impacts of domestic policies in different settings. *Rev. Envir. Eco. Policy* **2013**, *7*, 114–135. [CrossRef]
2. Joppa, L.; Pfaff, A. Re-assessing the forest impacts of protection: The challenge of non-random protection & a corrective method. *Annal. NY Acad. Sci* **2010**, *1185*, 135–149.
3. Joppa, L.; Pfaff, A. Global Park Impacts. *Proc. Royal Society B.* **2010**. [CrossRef]
4. Pfaff, A.; Robalino, J.A.; Sanchez-Azofeifa, G.A.; Andam, K.; Ferraro, P. Park Location Affects Forest Protection: Land Characteristics Cause Differences in Park Impacts across Costa Rica. *BE J. Econ. Anal. Poli.* 2009. Available online: http://www.bepress.com/bejeap/vol9/iss2/art5 (accessed on 24 December 2016).

5. Joppa, L.; Pfaff, A. High & Far: Biases in the location of protected areas. *PLoS ONE* **2009**, *4*, e8273. [CrossRef]
6. Pfaff, A.; Robalino, J. Protecting Forests, Biodiversity and the Climate: Predicting policy impact to improve policy choice. *Oxford Rev. Econ. Policy* **2012**, *28*, 164–179. [CrossRef]
7. Blackman, A.; Pfaff, A.; Robalino, J. Paper Park Performance: Mexico's natural protected areas in the 1990s. *Glob. Environ. Chang.* **2015**, *31*, 50–61. [CrossRef]
8. Johnson, T.; Alatorre, C.; Romo, Z.; Feng, L. *Low-Carbon Development for Mexico*; The World Bank: Washington, DC, USA, 2010.
9. The REDD Desk. REDD in Mexico. 2011. Available online: www.theredddesk.org/countries/mexico/readiness_overview (accessed on 3 February 2013).
10. The Forest Carbon Partnership. REDD Readiness Progress Fact Sheet. 2012. Available online: www.forestcarbonpartnership.org/ (accessed on 3 February 2013).
11. Naughton-Treves, L.; Holland, M.B.; Brandon, K. The role of protected areas in conserving biodiversity and sustaining local livelihoods. *Annual Rev. Environ. Resour.* **2005**, *30*, 219–252. [CrossRef]
12. Nagendra, H. Do Parks Work? Impact of Protected Areas on Land Cover Clearing. *Ambio* **2008**, *37*, 330–337. [CrossRef] [PubMed]
13. Campbell, A.; Clark, S.; Coad, L.; Miles, L.; Bolt, K.; Roe, D. Protecting the future: Carbon, forests, protected areas and local livelihoods. *Biodiversity* **2008**, *9*, 117–122. [CrossRef]
14. Fuller, D.; Jessup, T.; Salim, A. Loss of forest cover in Kalimantan, Indonesia, since the 1997–1998 El Nino. *Conserv. Biol.* **2004**, *18*, 249–254. [CrossRef]
15. Sanchez-Azofeifa, G.A.; Quesada-Mateo, C.; Gonzalez-Quesada, P.; Dayanandan, S.; Bawa, K.S. Protected areas and conservation of biodiversity in the tropics. *Conserv. Biol.* **1999**, *13*, 407–411. [CrossRef]
16. DeFries, R.; Hansen, A.; Newton, A.C.; Hansen, M.C. Increasing isolation of protected areas in tropical forests over the past twenty years. *Ecol. Appl.* **2005**, *15*, 19–26. [CrossRef]
17. Gaveau, D.L.A.; Wandono, H.; Setiabudi, F. Three decades of deforestation in southwest Sumatra: Have protected areas halted forest loss and logging, and promoted re-growth? *Biol. Conserv.* **2007**, *134*, 495–504. [CrossRef]
18. Messina, J.P.; Walsh, S.J.; Mena, C.F.; Delamater, P.L. Land tenure and deforestation patterns in the Ecuadorian Amazon: Conflicts in land conservation in frontier settings. *Appl. Geogra.* **2006**, *26*, 113–128. [CrossRef]
19. Bruner, A.; Gullison, R.E.; Rice, R.E.; Da Fonseca, G.A. Effectiveness of parks in protecting tropical biodiversity. *Science* **2001**, *291*, 125–128. [CrossRef] [PubMed]
20. Curran, L.; Trigg, S.N.; McDonald, A.K.; Astiani, D.; Hardiono, Y.M.; Siregar, P.; Caniago, I.; Kasischke, E. Lowland forest loss in protected areas of Indonesian Borneo. *Science* **2004**, *303*, 1000–1003. [CrossRef] [PubMed]
21. Kinnaird, M.F.; Sanderson, E.W.; O'Brien, T.G.; Wibisono, H.T.; Woolmer, G. Deforestation trends in a tropical landscape and implications for endangered large mammals. *Conserv. Biol.* **2003**, *17*, 245–257. [CrossRef]
22. Liu, J.G.; Linderman, M.; Ouyang, Z.; An, L.; Yang, J.; Zhang, H. Ecological degradation in protected areas: The case of Wolong Nature Reserve for giant pandas. *Science* **2001**, *292*, 98–101. [CrossRef] [PubMed]
23. Sader, S.; Hayes, D.J.; Hepinstall, J.A.; Coan, M.; Soza, C. Forest change monitoring of a remote biosphere reserve. *Int. J. Remote Sens.* **2001**, *22*, 1937–1950. [CrossRef]
24. Andam, K.; Ferraro, P.J.; Pfaff, A.; Sanchez-Azofeifa, G.A.; Robalino, J.A. Measuring the effectiveness of protected area networks in reducing deforestation. *Proc. Natl. Acad. Sci. USA* **2008**, *105*, 16089–16094. [CrossRef] [PubMed]
25. Shah, P.; Baylis, K. Evaluating Heterogeneous Conservation Effects of Forest Protection in Indonesia. *PLoS ONE* **2015**, *10*, e0124872. [CrossRef] [PubMed]
26. Munoz-Pina, C.; Guevara, A.; Torres, J.M.; Brana, J. Paying for the hydrological services of Mexico's forests: Analysis, negotiations and results. *Ecol. Econ.* **2008**, *65*, 725–736. [CrossRef]
27. Alix-Garcia, J.M.; Shapiro, E.N.; Sims, K.R.E. Forest Conservation and Slippage: Evidence from Mexico's National Payments for Ecosystem Services Program. *Land Econ.* **2012**, *88*, 613–638. [CrossRef]
28. Alix-Garcia, J.M.; Sims, K.R.E.; Yanez-Pagans, P. Only One Tree from Each Seed? Environmental effectiveness and poverty alleviation in Mexico's Payments for Ecosystem Services Program. *Am. Econ. J. Econ. Policy* **2015**, *7*, 1–40. [CrossRef]

29. Rodriguez, L.A. On the Regulation of Small Actors: Three Experimental Essays about Policies based on Voluntary Compliance and Decentralized Monitoring. Ph.D. Thesis, University Program in Environmental Policy. Duke University, Durham, NC, USA, 2016.
30. Kaczan, D.; Pfaff, A.; Rodriguez, L.; Shapiro, E.N. Increasing the Impact of Collective Incentives: Conditionality on additionality within PES in Mexico. Unpublished work, 2016.
31. Sims, K.R.E.; Alix-Garcia, J.M. Parks versus PES: Evaluating direct and incentive-based land conservation in Mexico. *J. Environ. Econ. Manag.* **2016**. [CrossRef]
32. Figueroa, F.; Sánchez-Cordero, V. Effectiveness of Natural Protected Areas to Prevent Land Use and Land Cover Change in Mexico. *Biodivers. Cons.* **2008**, *17*, 3223–3240. [CrossRef]
33. Mas, J.F. Assessing Protected Areas Effectiveness Using Surrounding (Buffer) Areas Environmentally Similar to the Target Area. *Envir. Monit. Assess.* **2005**, *105*, 69–80. [CrossRef]
34. Durán, E.; Mas, J.F.; Velázquez, A. Land Use/Cover Change in Community-Based Forest Management Regions and Protected Areas in Mexico. In *The Community Forests of Mexico*; Bray, D.B., Merino-Pérez, L., Barry, D., Eds.; University of Texas Press: Austin, TX, USA, 2005; pp. 215–238.
35. Honey-Roses, J.; Lopez-Garcia, J.; Rendon-Salinas, E.; Peralta-Higuera, A.; Galindo-Leal, C. To pay or not to pay? Monitoring performance and enforcing conditionality when paying for forest conservation in Mexico. *Environ. Conserv.* **2009**, *36*, 120–128. [CrossRef]
36. Honey-Roses, J.; Baylis, K.; Ramirez, M.I. A Spatially Explicit Estimate of Avoided Forest Loss. *Conserv. Biol.* **2011**, *25*, 1032–1043. [CrossRef] [PubMed]
37. Miteva, D.; Ellis, E.; Ellis, P.; Griscom, B. The role of property rights in resisting forest loss in the Yucatan Peninsula. Unpublished work. 2016.
38. Von Thünen, J.H. Der Isolierte Staat in Beziehung der Landwirtschaft und Nationalökonomie (1996, trans.). In *von Thünen's The Isolated State*; Hall, P., Ed.; Pergamon Press: Oxford, United Kingdom, 1826.
39. World Bank. *Policies and Deforestation in the Brazilian Amazon: Roads, Protected Areas, Their Interactions, and Their Impacts*; Technical Paper; World Bank: Washington DC, USA, 2013.
40. Nelson, A.; Chomitz, K. Effectiveness of Strict vs. Multiple Use Protected Areas in Reducing Tropical Forest Fires: A Global Analysis Using Matching Methods. *PLoS ONE* **2011**, *6*, 322722. [CrossRef] [PubMed]
41. Pfaff, A.; Robalino, J.; Lima, E.; Sandoval, C.; Herrera, L.D. Governance, Location and Avoided Deforestation from Protected Areas: Greater restrictions can have lower impact, due to differences in location. *World Develop.* 2014. Available online: http://dx.doi.org/10.1016/j.worlddev.2013.01.011 (accessed on 24 December 2016).
42. Albers, H.J. Spatial modeling of extraction and enforcement in developing country protected areas. *Resour. Energy Econ.* **2010**, *32*, 165–179. [CrossRef]
43. Borner, J.; Kis-Katos, K.; Hargrave, J.; Konig, K. Post-Crackdown Effectiveness of Field-Based Forest Law Enforcement in the Brazilian Amazon. *PLoS ONE* **2015**, *10*, e0121544. [CrossRef] [PubMed]
44. Bartholome, E.; Belward, A. GLC2000: A new approach to global land cover mapping from Earth observation data. *Int. J. Remote Sens.* **2005**, *26*, 1959–1977. [CrossRef]
45. European Space Agency (ESA); ESA GlobeCover Project lead by MEDIAS-France. Ionia GlobCover. Available online: http://due.esrin.esa.int/page_globcover.php (accessed on 24 December 2009).
46. United States Geological Survey. Shuttle Radar Topography Mission, 30 Arc Second scene SRTM_GTOPO_u30, Mosaic. 2006. Available online: http://www2.jpl.nasa.gov/srtm/ (accessed on 24 December 2008).
47. National Imagery and Mapping Agency (NIMA). Vector Map Level. Available online: http://egsc.usgs.gov/nimamaps/ (accessed on 24 December 2010).
48. United Nations Environment Program-Center for International Earth Science Information Network (UNEP-CIESEN). Global Rural-Urban Mapping Project (GRUMP), Alpha Version: Urban Extent. 2006. Available online: http://sedac.ciesin.columbia.edu/gpw/ancillaryfigures.jsp (accessed on 24 December 2010).
49. World Conservation Monitoring Center. World Database on Protected Areas (WDPA). World Conservation Union (IUCN) and UNEP-World Conservation Monitoring Center Cambridge, UK. 2007. Available online: http://www.wdpa.org/ (accessed on 24 December 2010).

50. Olson, D.; Dinerstein, E.; Wikramanayake, E.D.; Burgess, N.D.; Powell, G.V.; Underwood, E.C.; D'Amico, J.A.; Itoua, I.; Strand, H.E.; Morrison, J.C.; et al. Terrestrial ecoregions of the world: A new map of life on Earth. *BioScience* **2001**, *51*, 933–938. [CrossRef]
51. Fischer, G.; van Velthuizen, H.; Nachtergaele, F.; Medow, S. Global Agro-Ecological Zones. 2002. Available online: http://www.fao.org/nr/gaez/en/ (accessed on 24 December 2010).
52. Rosenbaum, P.R.; Rubin, D.B. The central role of the propensity score in observational studies for causal effects. *Biometrika* **1983**, *70*, 41–55. [CrossRef]

PERMISSIONS

All chapters in this book were first published in FORESTS, by MDPI; hereby published with permission under the Creative Commons Attribution License or equivalent. Every chapter published in this book has been scrutinized by our experts. Their significance has been extensively debated. The topics covered herein carry significant findings which will fuel the growth of the discipline. They may even be implemented as practical applications or may be referred to as a beginning point for another development.

The contributors of this book come from diverse backgrounds, making this book a truly international effort. This book will bring forth new frontiers with its revolutionizing research information and detailed analysis of the nascent developments around the world.

We would like to thank all the contributing authors for lending their expertise to make the book truly unique. They have played a crucial role in the development of this book. Without their invaluable contributions this book wouldn't have been possible. They have made vital efforts to compile up to date information on the varied aspects of this subject to make this book a valuable addition to the collection of many professionals and students.

This book was conceptualized with the vision of imparting up-to-date information and advanced data in this field. To ensure the same, a matchless editorial board was set up. Every individual on the board went through rigorous rounds of assessment to prove their worth. After which they invested a large part of their time researching and compiling the most relevant data for our readers.

The editorial board has been involved in producing this book since its inception. They have spent rigorous hours researching and exploring the diverse topics which have resulted in the successful publishing of this book. They have passed on their knowledge of decades through this book. To expedite this challenging task, the publisher supported the team at every step. A small team of assistant editors was also appointed to further simplify the editing procedure and attain best results for the readers.

Apart from the editorial board, the designing team has also invested a significant amount of their time in understanding the subject and creating the most relevant covers. They scrutinized every image to scout for the most suitable representation of the subject and create an appropriate cover for the book.

The publishing team has been an ardent support to the editorial, designing and production team. Their endless efforts to recruit the best for this project, has resulted in the accomplishment of this book. They are a veteran in the field of academics and their pool of knowledge is as vast as their experience in printing. Their expertise and guidance has proved useful at every step. Their uncompromising quality standards have made this book an exceptional effort. Their encouragement from time to time has been an inspiration for everyone.

The publisher and the editorial board hope that this book will prove to be a valuable piece of knowledge for researchers, students, practitioners and scholars across the globe.

LIST OF CONTRIBUTORS

Emanuele Sibona, Fabio Meloni and Renzo Motta
DISAFA: Dipartimento di Scienze Agrarie, Forestali e Alimentari, University of Torino, Largo Paolo Braccini 2, 10095 Grugliasco (TO), Italy

Alessandro Vitali
D3A: Department of Agricultural, Food and Environmental Sciences, Marche Polytechnic University, Via Brecce Bianche 10, 60131 Ancona (AN), Italy

Lucia Caffo and Alberto Dotta
CFAVS: Consorzio Forestale Alta Val di Susa, Via Pellousiere 6, 10056 Oulx (TO), Italy

Emanuele Lingua
TESAF: Department of Land, Environment, Agriculture and Forestry, University of Padova, Viale dell'Università 16, 35020 Legnaro (PD), Italy

Matteo Garbarino
DISAFA: Dipartimento di Scienze Agrarie, Forestali e Alimentari, University of Torino, Largo Paolo Braccini 2, 10095 Grugliasco (TO), Italy
D3A: Department of Agricultural, Food and Environmental Sciences, Marche Polytechnic University, Via Brecce Bianche 10, 60131 Ancona (AN), Italy

Weili Kou, Changxian Liang, Lili Wei and Xuejing Yang
School of Computer and Information, Southwest Forestry University, Kunming, Yunnan 650224 China

Alexander J. Hernandez
Remote Sensing and GIS Laboratories, Quinney College of Natural Resources, Utah State University, Logan, UT 84322, USA

Jessica S. Veysey Powell and Kimberly J. Babbitt
Department of Natural Resources and the Environment, University of New Hampshire, 56 College Road, Durham, NH 03824, USA

Lars Drössler, Nils Fahlvik and Natalia K. Wysocka
Southern Swedish Forest Research Centre, Swedish University of Agricultural Sciences, Box 49, 230 53 Alnarp, Sweden

Karin Hjelm
Skogforsk; Ekebo 2250, 268 90 Svalöv, Sweden

Christian Kuehne
School of Forest Resources, University of Maine; 5755 Nutting Hall, Orono, ME 04469, USA

Miho Morimoto and Glenn Juday
School of Natural Resources and Extension, University of Alaska Fairbanks, Fairbanks, AK 99775, USA

Brian Young
Natural Science, Landmark College, Putney, VT 05346, USA

Aviva J. Gottesman and William S. Keeton
Rubenstein School of Environment and Natural Resources, University of Vermont, Burlington, VT 05405, USA

Irmeli Mustalahti, Mathias Cramm and Sabaheta Ramcilovic-Suominen
Department of Geographical and Historical Studies, University of Eastern Finland, Yliopistokatu 7, 80100 Joensuu, Finland

Yitagesu T. Tegegne
European Forest Institute, Yliopistokatu 6, 80100 Joensuu, Finland

Courtney Work
International Institute of Social Studies, Erasmus University, 2502 LT The Hague, Netherlands

Leif Tore Trædal and Pål Olav Vedeld
Department of International Environment and Development Studies (Noragric), Faculty of Land and Society, Norwegian University of Life Sciences, Ås, N-1432, Norway

Lukas R. Jarron, Txomin Hermosilla and Nicholas C. Coops
Department of Forest Resources Management, University of British Columbia, 2424 Main Mall, Vancouver, BC V6T 1Z4, Canada

Michael A. Wulder, Joanne C. White, GeordieW. Hobart and Donald G. Leckie
Canadian Forest Service (Pacific Forestry Centre), Natural Resources Canada, 506 West Burnside Road, Victoria, BC V8Z 1M5, Canada

Dorit Julich, Stefan Julich and Karl-Heinz Feger
Institute of Soil Science and Site Ecology, TU Dresden, Pienner Str. 19, 01737 Tharandt, Germany

Pamela McElwee
Department of Human Ecology, School of Environmental and Biological Sciences, Rutgers, The State University of New Jersey, New Brunswick, NJ 08901, USA

Van Hai Thi Nguyen and Dung Viet Nguyen
People and Nature Reconciliation (PanNature), Hanoi 10000, Vietnam

Nghi Huu Tran
Tropenbos International Vietnam (TBI), Hue 530000, Vietnam; nghi@tropenbos.vn

Hue Van Thi Le, Tuyen Phuong Nghiem and Huong Dieu Thi Vu
Centre for Natural Resources and Environmental Studies (CRES), Vietnam National University, Hanoi 10000, Vietnam

Alexander Pfaff
Duke University, 302 Towerview Drive, Durham NC 27708, USA

Francisco Santiago-Ávila
University of Wisconsin, 1220 Linden Drive, Madison, WI 53706, USA; santiagoavil@wisc.edu

Lucas Joppa
Microsoft Research, One Microsoft Way, Redmond, WA 98052, USA

Index